ABHANDLUNGEN
DER AKADEMIE DER WISSENSCHAFTEN DER DDR
Abteilung Mathematik — Naturwissenschaften — Technik

Jahrgang 1979 · Nr. 4N

Convergence Structures and Applications to Analysis

Proceedings of the International Summer School
held at Frankfurt/Oder, GDR
from May 8 to 12, 1978

Herausgegeben von

S. Gähler, W. Gähler und G. Kneis

AKADEMIE-VERLAG · BERLIN · 1980

Herausgegeben im Auftrag des Präsidenten
der Akademie der Wissenschaften der DDR
von Vizepräsident Prof. Dr. Heinrich Scheel

Erschienen im Akademie-Verlag, DDR-108 Berlin, Leipziger Straße 3—4

Lizenznummer: 202 · 100/543/80
Gesamtherstellung: VEB Druckhaus „Maxim Gorki", 74 Altenburg
Bestellnummer: 762 803 0 (2001/79/4N) · LSV 1035
Printed in GDR
DDR 28,— M

Preface

The Mathematical Society of the GDR and the Central Institute of Mathematics and Mechanics of the Academy of Sciences of the GDR organized an International Summer School

"Convergence structures and applications to analysis"

from May 8 to 12, 1978, at Frankfurt/Oder.

Participating in the summer school were: V. I. AVERBUH (Moscow), E. BINZ (Mannheim), H. BOSECK (Greifswald), Á. CSÁSZÁR (Budapest), K. CSÁSZÁR (Budapest), J. FLACHSMEYER (Greifswald), R. FRIČ (Žilina), S. GÄHLER (Berlin), W. GÄHLER (Berlin), M. P. KAZ (Leningrad), G. KNEIS (Berlin), V. KOUTNÍK (Prague), A. LECHICKI (Szczecin), B. MÜLLER (Mannheim), H. POPPE (Wustrow), W. PREUSS (Wismar), W. PRIESS (Berlin), S. PRÖSSDORF (Berlin), K.-P. RUDOLPH (Greifswald), R. SCHMID (Zürich), A. SPAKOWSKI (Szczecin), A. STÄRK (Wustrow).

This volume consists of the proceedings of the summer school; the starred title in the table of contents corresponds to a paper which was submitted in addition to those read at the conference.

Seventeen one-hour talks were presented; these and the additionally submitted paper may be divided according to the following topics:

1. Categorical aspects in topology, general theory of convergence structures (Á. CSÁSZÁR, W. GÄHLER, G. KNEIS)
2. Sequential structures (J. FLACHSMEYER, R. FRIČ/D. C. KENT, R. FRIČ/K. MCKENNON/G. D. RICHARDSON, R. FRIČ/V. KOUTNÍK)
3. Pseudo-topological vector spaces, convergence spaces of mappings (V. I. AVERBUH, M. P. KAZ, B. MÜLLER)
4. General differential calculus (V. I. AVERBUH, H. BOSECK, W. GÄHLER, K.-P. RUDOLPH)
5. Manifolds of mappings (E. BINZ, R. SCHMID)
6. Further applications of convergence structures to analysis, other applications (S. GÄHLER, W. PREUSS).

Berlin, 1979 The editors

Contents

Выпуклая функция, дифференцируемая по Гато, дифференцируема по Майклу-Бастиани

Владимир И. Авербух, Москва

Gâteaux differentiable convex functions are Michal-Bastiani differentiable

Editor's summary

In this note, the result of Lloyd [1], stating the Hadamard differentiability of a continuous Gâteaux differentiable polynorm on a topological linear space, is sharpened and generalized. The author proves the Michal-Bastiani differentiability of an arbitrary continuous convex real-valued function f on a pseudo-topological linear space in every point in which f is Gâteaux differentiable. The proof is based on a generalization of Dini's theorem on monotonous convergence.

Вьедение

Лойд [1] обнаружил, что для непрерывных полунорм в топологических линейных пространствах (т. л. п.) из дифференцируемости по Гато в какой-нибудь точке вытекает дифференцируемость по Адамару (т. е. компактная дифференцируемость) [2] в этой точке. Оказывается, это верно для *произвольных* непрерывных выпуклых функций. Более того, из дифференцируемости таких функций по Гато вытекает даже более сильная дифференцируемость, а именно дифференцируемость по Майклу-Бастиани [2].

В доказательстве Лойда существенно используются свойства полунорм. Приводимое ниже (в п. 3) доказательство основано на обобщении теоремы Дини о монотонной сходимости (п. 2). В п. 1 даются необходимые определения.

1. Определения

Пусть X — псевдотопологическое пространство (п. п.), Y — псевдотопологическое линейное пространство (п. л. п.) [3] (т. (л.) п. — частный случай п. (л.) п.). Через $F_{\mathrm{MB}}(X, Y)$ мы обозначаем пространство $F(X, Y)$ всех отображений из X в Y, наделенное псевдотопологией Майкла-Бастиани [4]. Обычно в теории псевдотопологических пространств используется язык фильтров. Здесь нам будет удобнее работать с сетями (обобщенными последовательностями). Сеть $(f_\alpha)_{\alpha\in A}$ в $F_{\mathrm{MB}}(X, Y)$ сходится к нулю тогда и только тогда, когда двухиндексная сеть $\big(f_\alpha(x_\beta)\big)_{\alpha\in A,\beta\in B}$ сходится к нулю в Y для каждой сети $(x_\beta)_{\beta\in B}$, сходящейся (к какой-либо точке) в X.

В случае когда п. л. п. X является топологическим (или, более общим образом, правильным [5]) пространством, псевдотопология Майкла-Бастиани сильнее псевдотопологии равномерной сходимости на системе всех компактных множеств в X; для конечномерных т. л. п. X эти две псевдотопологии совпадают.

Отображение f п. л. п. X в п. л. п. Y *дифференцируемо в точке* $x \in X$ *по Гато*, соответственно *по Майклу-Бастиани*, если

$$f(x+h) = f(x) + f'(x) \cdot h + r(h) \qquad (h \in X),$$

где $f'(x)$ — непрерывное линейное отображение из X в Y (производная отображения f в точке x), а отображение $r\colon X \to Y$ удовлетворяет условию

$$r_t(h) \underset{t\downarrow 0}{\to} 0 \text{ в } Y \text{ для всякого } h \in X,$$

соответственно условию

$$r_t \underset{t\downarrow 0}{\to} 0 \text{ в } F_{\mathrm{MB}}(X, Y).$$

Здесь r_t для всякого положительного числа t — это отображение из X в Y, определяемое равенством

$$r_t(h) = \frac{r(th)}{t} \quad (h \in X).$$

Дифференцируемость по Майклу-Бастиани сильнее дифференцируемости по Адамару, а для секвенциальных [2] (в частности для метризуемых) X совпадает с ней; в случае нормированных X и Y дифференцируемость по Майклу-Бастиани слабее дифференцируемости по Фреше, в случае конечномерных X и Y совпадает с ней [3].

2. Обобщение теоремы Дини

Классическая теорема Дини утверждает, что монотонная поточечная сходимость к нулю последовательности непрерывных вещественных функций на топологическом пространстве X равномерна на каждом компакте. Поскольку топология равномерной сходимости на системе всех компактных множесть слабее псевдотопологии Майкла-Бастиани в $F(X, \boldsymbol{R})$ (см. п. 1), следующая теорема представляет собой обобщение теоремы Дини.

Теорема 1. *Пусть* X *— произвольное псевдотопологическое пространство*, $(f_\alpha)_{\alpha \in A}$ *— сеть непрерывных вещественных функций на* X. *Если* $f_\alpha(x) \downarrow 0$ *для всякой точки* $x \in X$, *то* $f_\alpha \to 0$ *в* $F_{\mathrm{MB}}(X, \boldsymbol{R})$.

Доказательство. Надо доказать, что двухиндексная сеть $f_\alpha(x_\beta)$ сходится к нулю для всякой сети $(x_\beta)_{\beta \in B}$, сходящейся в X. Пусть $x_\beta \to x$, и пусть $\varepsilon > 0$. Так как $f_\alpha(x) \downarrow 0$, то найдется такое $\alpha_0 \in A$, что $0 \leqq f_\alpha(x) \leqq \varepsilon/2$ для всех $\alpha \geqq \alpha_0$. Так как f_{α_0} — непрерывная функция, то $f_{\alpha_0}(x_\beta) \to f_{\alpha_0}(x)$, следовательно, найдется такое $\beta_0 \in B$, что

$$|f_{\alpha_0}(x_\beta) - f_{\alpha_0}(x)| \leqq \varepsilon/2$$

для всех $\beta \geqq \beta_0$. Поэтому для всех $\alpha \geqq \alpha_0$, $\beta \geqq \beta_0$

$$f_\alpha(x_\beta) \leqq f_{\alpha_0}(x_\beta) \leqq f_{\alpha_0}(x) + |f_{\alpha_0}(x_\beta) - f_{\alpha_0}(x)| \leqq \frac{\varepsilon}{2} + \frac{\varepsilon}{2} = \varepsilon.$$

3. Основной результат

Теорема 2. *Пусть X — произвольное п. л. п. и f — непрерывная выпуклая вещественная функция на X. Тогда если функция f дифференцируема по Гато в некоторой точке x_0, то она дифференцируема в этой точке по Майклу-Бастиани.*

Доказательство. Не уменьшая общности, можно считать, что $x_0 = 0$ и $f(0) = 0$, $f'(0) = 0$. Надо доказать, что $f_t \underset{t\downarrow 0}{\to} 0$ в $F_{\mathrm{MB}}(X, \boldsymbol{R})$. Так как f дифференцируема в нуле по Гато с нулевой производной, то $f_t(x) \underset{t\downarrow 0}{\to} 0$ в $\boldsymbol{R}$ для всякого $x \in X$. Так как f выпукла, то $f_t(x)$ для всякого $x \in X$ монотонно убывает при $t \downarrow 0$: если $0 < t_1 < t_2$, то

$$f_{t_1}(x) = \frac{f(t_1 x)}{t_1} = \frac{f\left(\frac{t_1}{t_2} t_2 x\right)}{t_1} \leqq \frac{\frac{t_1}{t_2} f(t_2 x)}{t_1} = \frac{f(t_2 x)}{t_2} = f_{t_2}(x).$$

Остается применить теорему 1.

Замечание 1. В доказательстве использовалась лишь выпуклость функции f на лучах, выходящих из рассматриваемой точки.

Замечание 2. Во всех результатах Лойда [1] гладкости т. л. п. дифференцируемость на Адамару можно заменить на дифференцируемость по Майклу-Бастиани.

Замечание 3. Было бы интересно выяснить, какие еще из определений дифференцируемости (см. [2]) „слипаются" для выпуклых функций.

Литература

[1] J. W. Lloyd, Inductive and projective limits of smooth topological vector spaces, Bull. Austral. Math. Soc. **6** (1972), No. 2, 227—240.

[2] В. И. Авербух и О. Г. Смолянов, Различные определения производной в линейных топологических пространствах, УМН **23** (1968), № 4, 67—116.

[3] А. Фрёлихер и В. Бухер, Дифференциальное исчисление в векторных пространствах без нормы, Мир, Москва 1970.

[4] В. И. Авербух и О. Г. Смолянов, Дополнения к книге А. Фрёлихера и В. Бухера [3].

[5] В. И. Авербух и О. Г. Смолянов, Дифференцирование и псевдотопологии, Вестник МГУ **1** (1972), 3—7.

Anschrift des Verfassers:

121099 Москва
пер. Воеводина 9 кв. 14
СССР

Analytic Mappings of Topological Groups — Sheaf Theoretic Approach —

HELMUT BOSECK, Greifswald

0. In this paper we define a notion of analyticity on topological groups, which is based on considerations of one parameter subgroups and is the analytic counterpart to the notion of differentiability given in [2]. As in [2] we are especially interested in locally compact topological groups and sheaf theory seems to be the best way in dealing with questions of analyticity on such groups. In a forthcoming paper we shall show how the notion of analytic vectors enables us to overcome in analogy to the well-known case of LIE groups the difficulties arising in the connections between a representation of a locally compact group and the corresponding infinitesimal representation of its topological LIE algebra (see also [1]).

1. Assume G to be a HAUSDORFF topological group. By $\Lambda = \Lambda(G) = \mathrm{Hom}\,(\boldsymbol{R}^+, G)$ we denote the set of continuous one parameter subgroups of G. Equipped with the compact open topology Λ is a HAUSDORFF topological space. Let us choose a fixed system of open neighborhoods $\mathfrak{U}$ of the identity in G. If $U \in \mathfrak{U}$ and $(\lambda_1, \ldots, \lambda_n) \in \Lambda^n$ is an n-tuple of one parameter subgroups, we denote by $U^{(\lambda_1,\ldots,\lambda_n)}$ the following open neighborhood of the origin in $\boldsymbol{R}^n$

$$U^{(\lambda_1,\ldots,\lambda_n)} = \{(t_1, \ldots, t_n) \mid \lambda_1(t_1) \cdots \lambda_n(t_n) \in U\}. \tag{1}$$

Let F denote a locally convex linear space. For every open set $O \subseteqq G$ we denote by $A(O; F)$ the set of continuous F-valued functions f on O satisfying the following property

(a) for every $x \in O$ there exists a neighborhood $U \in \mathfrak{U}$, such that $xU \subseteqq O$ and for all $n = 1, 2, \ldots$ and all $(\lambda_1, \ldots, \lambda_n) \in \Lambda^n$ the function

$$\tilde{f}_x^{(\lambda_1,\ldots,\lambda_n)}(t_1, \ldots, t_n) = f(x\lambda_1(t_1) \cdots \lambda_n(t_n)) \tag{2}$$

possesses a TAYLOR series with respect to the origin in $\boldsymbol{R}^n$, which converges for all $(t_1, \ldots, t_n) \in U^{(\lambda_1,\ldots,\lambda_n)}$ with coefficients jointly continuous on $O \times \Lambda^n$

$$\tilde{f}_x^{(\lambda_1,\ldots,\lambda_n)}(t_1, \ldots, t_n) = \sum_{j_1,\ldots,j_n=0,\ldots,0}^{\infty,\ldots,\infty} a_{j_1\cdots j_n}(x; \lambda_1, \ldots, \lambda_n)\, t_1^{j_1} \cdots t_n^{j_n}, \tag{3}$$

$$(t_1, \ldots, t_n) \in U^{(\lambda_1,\ldots,\lambda_n)},\ a_{j_1\cdots j_n} \in C(O \times \Lambda^n).$$

Obviously $A(O; F)$ is a linear space. If F is a locally convex algebra, then $A(O; F)$ is an algebra by pointwise multiplication.

Remark 1⁰. If differentiability of continuous F-valued functions on G is defined by means of one parameter subgroups as in [2] and even if it is defined in a more restrictive way assuming joint continuity of the derivatives $D^n f(x; \lambda_1, \ldots, \lambda_n)$ on $O \times \Lambda^n$, a definition which is more suitable in dealing with higher derivatives, than the original definition in [2], the inclusion $A(O; F) \subseteq C^\infty(O; F)$ is valid. If the group G is locally compact, the assumed joint continuity of the coefficients $a_{j_1 \cdots j_n}(x; \lambda_1, \ldots, \lambda_n)$ or — what is the same — of the derivatives $D^n f(x; \lambda_1, \ldots, \lambda_n)$ on $O \times \Lambda^n$ is implied by continuity on O. In this case in condition (a) it is enough to assume continuity of the coefficients of the TAYLOR series with respect to $x \in O$ and the original notion of differentiability from [2] can be used.

Remark 2⁰. Replacing the function $\tilde{f}_x^{(\lambda_1, \ldots, \lambda_n)}$ in condition (a) by

$$\tilde{\tilde{f}}_x^{(\lambda_1, \ldots, \lambda_n)}(t_1, \ldots, t_n) = f\big(\lambda_1(t_1) \cdots \lambda_n(t_n)\, x\big)$$

and the inclusion $xU \subseteq O$ by $Ux \subseteq O$ it is easy to prove, that the space $A(O; F)$ is independent from the fact, that the product $\lambda_1(t_1) \cdots \lambda_n(t_n)$ is multiplied by x from the left in (a). (See the proof of this fact for the concept of differentiability given in [2].)

If O_1 and O_2 are open subsets of G and $O_1 \subseteq O_2$, we denote by $\varrho^{O_2}_{O_1}$ the map which maps every function on O_2 onto its restriction to O_1. Let $\alpha^{O_2}_{O_1}$ be the restriction of $\varrho^{O_2}_{O_1}$ to the function space $A(O_2; F)$, then $\alpha^{O_2}_{O_1}$ is a linear mapping of $A(O_2; F)$ into $A(O_1; F)$.

By $\tilde{\mathfrak{A}}(G; F) = \{A(O; F); \alpha^{O_2}_{O_1}\}$ we denote the pre-sheaf of linear spaces over the base space G defined by the spaces $A(O; F)$ and the mappings $\alpha^{O_2}_{O_1}$.

The corresponding sheaf $\mathfrak{A}(G; F) = \{A(F), \alpha, G\}$ is called the *sheaf of germs of locally analytic F-valued functions* on G.

The pre-sheaf $\tilde{\mathfrak{A}}(G; F)$ is canonical by definition, i.e. $A(O; F)$ equals the space $\Gamma(O, \mathfrak{A})$ of continuous sections from O into the sheaf $\mathfrak{A} = \mathfrak{A}(G; F)$.

The elements of $A(G; F)$ are defined to be the *analytic F-valued functions* on the group G.

Remark 3⁰. Using the notion of differentiability mentioned in Remark 1⁰ and denoting by $\delta^{O_2}_{O_1}$ the restriction of $\varrho^{O_2}_{O_1}$ to the function space $C^\infty(O_2; F)$ of smooth F-valued functions on O_2, an analogous procedure yields a *sheaf $\mathfrak{L}^\infty(G; F)$ of germs of locally smooth F-valued functions* on G. The sheaf $\mathfrak{A}(G; F)$ is a subsheaf of $\mathfrak{L}^\infty(G; F)$.

If G is a connected LIE group, then using canonical coordinates of the second kind, the following proposition can easily be shown.

Proposition 1.1. *For a connected Lie group G the sheaf $\mathfrak{A}(G; F)$ equals the sheaf of germs of F-valued functions which are analytic in the usual sense with respect to the analytic structure of the manifold G.*

Remark 4⁰. It is evident, that the definition of analyticity given above is of some interest only for such topological groups having in some sense "enough" one parameter subgroups. We shall not discuss this important question here, since in this paper we are primarily interested in connected locally compact groups. These groups are representable as projective limits of connected LIE groups and by lifting processes we are sure to have enough one parameter subgroups. In a locally compact groups there exists a neighborhood of the identity with the property, that the images of one parameter subgroups are dense in this neighborhood. A connected locally compact group is locally connected if and only if there is a neighborhood of the identity which is covered by the images of one parameter subgroups (see [2] and [3]).

Let N denote a closed normal subgroup of G and π denote the canonical epimorphism mapping G onto the quotient group $\bar{G} = G/N$. There is an obvious π-comonomorphism $\bar{p}$ of the sheaf $\mathfrak{A}(\bar{G}; F)$ into the sheaf $\mathfrak{A}(G; F)$. If $\bar{O}$ is open in G the monomorphism $\bar{p}_{\bar{O}}$ from $A(\bar{O}; F)$ into $A(\pi^{-1}\bar{O}; F)$ maps the function $\bar{f} \in A(\bar{O}; F)$ onto $f = \bar{f} \circ \pi \in A(\pi^{-1}\bar{O}; F)$, which is constant on the cosets modulo N. Since every π-comonomorphism admits a unique factorization with respect to the inverse image $\check{\pi}\mathfrak{A}(\bar{G}; F)$ of the sheaf $\mathfrak{A}(\bar{G}; F)$: $\bar{p} = p \circ \check{\pi}$ and p is a monomorphism of the sheaf $\check{\pi}\mathfrak{A}(\bar{G}; F)$, which is a sheaf with base space G, into the sheaf $\mathfrak{A}(G; F)$.

Proposition 1.2. *The inverse image $\check{\pi}\mathfrak{A}(G/N; F)$ is canonically isomorphic to a subsheaf $\mathfrak{A}_N(G; F)$ of the sheaf $\mathfrak{A}(G; F)$. The continuous sections of $\mathfrak{A}_N(G; F)$ are constant on the cosets modulo N.*

2. Let G and G' denote HAUSDORFF topological groups, and $\mathfrak{F}$ a fixed class of locally convex linear spaces. For every open subset O of G by $A_{\mathfrak{F}}(O; G')$ we denote the set of continuous mappings φ from O into G' satisfying the following condition

(ag) for every open subset $O' \subseteqq G'$ such that $\varphi(O) \subseteqq O'$ and every $f' \in \mathfrak{A}(O'; F)$ where $F \in \mathfrak{F}$, $f = f' \circ \varphi \in \mathfrak{A}(O; F)$.

$A_{\mathfrak{F}}(O; G')$ is a group by pointwise multiplication. Let $\alpha_{O_1}^{O_2}$, $O_1 \subseteqq O_2$, denote the homomorphism from $A_{\mathfrak{F}}(O_2; G')$ into $A_{\mathfrak{F}}(O_1; G')$ defined by restricting the mappings from O_2 to O_1. We get a pre-sheaf $\tilde{\mathfrak{A}}_{\mathfrak{F}}(G; G') = \{A_{\mathfrak{F}}(O; G'); \alpha_{O_1}^{O_2}\}$ and the corresponding sheaf $\mathfrak{A}_{\mathfrak{F}}(G; G') = \{A_{\mathfrak{F}}(G'), \alpha, G\}$ with base space G is called the *sheaf of germs of locally $\mathfrak{F}$-analytic mappings* from G into G'.

The elements of the group $A_{\mathfrak{F}}(G; G')$ are the *$\mathfrak{F}$-analytic mappings* from G into G'.

Analogous considerations as in the proof of proposition 1.2 yield the $\mathfrak{F}$-analyticity of the canonical projection π from G to the quotient group $G/N = G'$. But it is easy to prove

Proposition 2.1. *Every continuous homomorphism φ mapping the topological group G into the topological group G' is an $\mathfrak{F}$-analytic mapping: for every class $\mathfrak{F}$ of locally convex linear spaces we have*

$$\mathrm{Hom}\,(G, G') \subseteqq A_{\mathfrak{F}}(G; G').$$

To prove Proposition 2.1 we mention first, that the definition of analyticity implies $\mathrm{Hom}\,(R^+, G) \subseteqq A_{\mathfrak{F}}(R^+; G)$. The function $f \circ \lambda(t) = f(\lambda(t)) = f(\lambda(t_0)\,\lambda(t')) = \bar{f}^{(\lambda)}_{\lambda(t_0)}(t')$ is analytic if $f \in A(O; F)$, $F \in \mathfrak{F}$, and $\lambda(R) \subseteqq O$. Every continuous homomorphism $\varphi \in \mathrm{Hom}\,(G, G')$ defines a mapping of $\Lambda(G)$ into $\Lambda(G')$ which is continuous in the compact open topologies: $\lambda' = \varphi \circ \lambda$. Assume O' to be an open subset of G' such that $\varphi(G) \subseteqq O'$. If $f' \in A(O'; F)$, $F \in \mathfrak{F}$, then the function $f = f' \circ \varphi$ satifies the following equation

$$\begin{aligned} \bar{f}_x^{(\lambda_1,\ldots,\lambda_n)}(t_1, \ldots, t_n) &= f\big(x\lambda_1(t_1) \cdots \lambda_n(t_n)\big) \\ &= f'\big(\varphi(x)\,(\varphi \circ \lambda_1)\,(t_1) \cdots (\varphi \circ \lambda_n)\,(t_n)\big) \\ &= \bar{f}'^{(\lambda_1',\ldots,\lambda_n')}_{\varphi(x)}(t_1, \ldots, t_n). \end{aligned} \tag{4}$$

Possibly some of the $\lambda_i' = \varphi \circ \lambda_i$ are trivial and the function $\bar{f}_x^{(\lambda_1,\ldots,\lambda_n)}$ is independent of some t_i. Nevertheless the function $\bar{f}^{(\lambda_1',\ldots,\lambda_n')}_{\varphi(x)}$ satisfies (3) by assumption for $(t_1, \ldots, t_n)$

$\in U'^{(\lambda_1',\ldots,\lambda_n')}$, U' denoting a neighborhood of the identity in G'. Take $U = \varphi^{-1}(U')$, then U is a neighborhood of the identity in G and $U^{(\lambda_1,\ldots,\lambda_n)} = U'^{(\lambda_1',\ldots,\lambda_n')}$, hence $\tilde{f}_x^{(\lambda_1,\ldots,\lambda_n)}$ satisfies (3) for $(t_1, \ldots, t_n) \in U^{(\lambda_1,\ldots,\lambda_n)}$ and $f \in A(G; F)$, $F \in \mathfrak{F}$. This completes the proof.

By definition of the space $A(G; F)$ of F-valued analytic functions on G it is evident, that $A(G; F)$ is invariant under left translations. We define a mapping l_a, $a \in G$, of G onto G by $l_a(x) = a^{-1}x$. The equation

$$L(a)\, f = f \circ l_a \tag{5}$$

defines a linear operator $L(a)$ on the space $A(G; F)$ and the mapping $L: a \to L(a)$ is a representation of the group G with representation space $A(G; F)$, which is called the *left regular representation of G on $A(G; F)$*. The invariance of the space $A(G; F)$ under left translations is equivalent to $l_a \in A_{\mathfrak{F}}(G; G)$ for every $a \in G$ and every class $\mathfrak{F}$ of locally convex linear spaces. If r_a is defined by $r_a(x) = xa$ and i_a denotes the inner automorphism $i_a(x) = a^{-1}xa$ then $r_a = l_{a^{-1}} \circ i_a$. By proposition 2.1, $i_a \in A_{\mathfrak{F}}(G; G)$ holds and the following proposition yields $r_a \in A_{\mathfrak{F}}(G; G)$.

Proposition 2.2. *If $\varphi \in A_{\mathfrak{F}}(O; G')$, $\varphi' \in A_{\mathfrak{F}}(O'; G'')$ and $\varphi(O) \subseteq O'$, then $\varphi' \circ \varphi \in A_{\mathfrak{F}}(O; G'')$.*

To prove this let us take an open subset $O'' \subseteq G''$ such that $\varphi' \circ \varphi(O) \subseteq O''$ and a function $f'' \in A(O''; F)$, $F \in \mathfrak{F}$. By continuity of φ' the set $(\varphi')^{-1}(O'') \cap O' = O_1'$ is open in G' and $\varphi(O) \subseteq O_1'$. By $\varphi_1' = \alpha_{O_1'}^{O'}\varphi'$ we denote the restriction of φ' to O_1', then $\varphi_1' \circ \varphi = \varphi' \circ \varphi$. Since $\varphi_1' \in A_{\mathfrak{F}}(O_1'; G'')$ and $\varphi_1'(O_1') \subseteq O''$ it follows that $f'' \circ \varphi_1' \in A(O_1'; F)$, $F \in \mathfrak{F}$, and hence $(f'' \circ \varphi_1') \circ \varphi = f'' \circ (\varphi' \circ \varphi) \in A(O; F)$, $F \in \mathfrak{F}$.

Remark 5⁰. The invariance of the space $A(G; F)$ under right translations, which is equivalent to $r_a \in A_{\mathfrak{F}}(G; G)$ for every $a \in G$ and every class $\mathfrak{F}$ of locally convex linear spaces can be proved directly by considerations analogous to those in [2] for the space of differentiable functions — or using Remark 2⁰.

Proposition 2.3. *Left and right translations of the group G are analytic mappings of G onto G: $l_a, r_a \in A_{\mathfrak{F}}(G; G)$ for every $a \in G$ and every class $\mathfrak{F}$ of locally convex linear spaces.*

Proposition 2.4. a) *The inversion $\iota: x \to x^{-1}$ is an analytic mapping of G onto G: $\iota \in A_{\mathfrak{F}}(G; G)$ for every class $\mathfrak{F}$ of locally convex linear spaces;*

b) *The multiplication $\mu: (x, y) \to xy$ is an analytic mapping of $G \times G$ onto G: $\mu \in A_{\mathfrak{F}}(G \times G; G)$ for every class $\mathfrak{F}$ of locally convex linear spaces.*

For every $f \in A(G; F)$, $F \in \mathfrak{F}$ we have

$$f \circ \iota\big(x\lambda_1(t_1) \cdots \lambda_n(t_n)\big) = \tilde{f}_x^{(\lambda_n,\ldots,\lambda_1)}(-t_n, \ldots, -t_1). \tag{6}$$

Statement a) follows immediately using Remark 2⁰. Since $\Lambda(G \times G) = \Lambda(G) \times \Lambda(G)$, the following equation holds

$$\begin{aligned} &f \circ \mu\big((x, y)\,(\lambda_{11}(t_1), \lambda_{12}(t_1)) \cdots (\lambda_{n1}(t_n), \lambda_{n2}(t_n))\big) \\ &= f \circ \mu\big((x\lambda_{11}(t_1) \cdots \lambda_{n1}(t_n), y\lambda_{12}(t_1) \cdots \lambda_{n2}(t_n))\big) \\ &= f\big(x\lambda_{11}(t_1) \cdots \lambda_{n1}(t_n)\, y\lambda_{12}(t_1) \cdots \lambda_{n2}(t_n)\big) \\ &= f\big(xy\lambda_{11}^y(t_1) \cdots \lambda_{n1}^y(t_n)\, \lambda_{12}(t_1) \cdots \lambda_{n2}(t_n)\big). \end{aligned} \tag{7}$$

In this equation, $\lambda^{y} = i_{y} \circ \lambda$ denotes the "conjugate" one parameter subgroup. If $f \in A(G; F), F \in \mathfrak{F}$, equation (7) implies $f \circ \mu \in A(G \times G; F), F \in \mathfrak{F}$, hence statement b) is proved.

3. In the sequel G denotes a connected locally compact Hausdorff group. By $\mathfrak{N}$ we denote the lattice of compact normal subgroups of finite codimension, i.e. dim $G/N < \infty$ for $N \in \mathfrak{N}$, and by $\mathfrak{K}$ we denote the sublattice of co-Lie subgroups, i.e. G/K is a Lie group for $K \in \mathfrak{K} \subseteqq \mathfrak{N}$. Let us choose a system $\mathfrak{U}$ of symmetric neighborhoods of the identity having compact closure and satisfying the property

(k) for every $U \in \mathfrak{U}$ there exists a compact normal co-Lie subgroup $K \in \mathfrak{K}$ such that $UK \subseteqq U$.

By the theorem of Yamabe the connected locally compact group G is the limit of the inverse system $\{G_K, \pi^{K_1}_{K_2}\}_{\mathfrak{K}}$ of Lie groups $G_K = G/K$. $\pi^{K_1}_{K_2}$, $K_1 \subseteqq K_2$, denotes the canonical projection from G_{K_1} onto G_{K_2}. By π_K we denote the canonical projection from G onto G_K. By proposition 1.2 the subsheaf $\mathfrak{A}_K(G; F)$ of $\mathfrak{A}(G; F)$ is canonically isomorphic to the inverse image $\overset{*}{\pi}_K \mathfrak{A}(G_K; F)$.

If $f \in A(O; F)$ and if there exists a co-Lie subgroup $K \in \mathfrak{K}$, such that f is constant on cosets modulo K, we call f a *cylindrical* analytic F-valued function on O.

The continuous sections of the sheaf $\mathfrak{A}_K(G; F)$ are the K-cylindrical analytic functions, i.e. $\mathfrak{A}_K(G; F)$ is the *sheaf of germs of locally analytic K-cylindrical F-valued functions.*

The equation $\pi_{K_2} = \pi^{K_1}_{K_2} \circ \pi_{K_1}$, $K_1 \subseteqq K_2$, implies the equation $\overline{p}^{K} = \overline{p}^{K_1} \circ \overline{p}^{K_2}_{K_1}$ of the corresponding π-comonomorphisms and the factorization with respect to the corresponding coimages yields

$$p^{K_2} \circ \overset{*}{\pi}_{K_2} = p^{K_1} \circ \overset{*}{\pi}_{K_1} \circ p^{K_2}_{K_1} \circ \overset{*}{\pi}^{K_1}_{K_2}. \tag{8}$$

The sheaf $\mathfrak{A}_{K_2}(G; F) = p^{K_2}\big(\overset{*}{\pi}_{K_2}\mathfrak{A}(G_{K_2}; F)\big)$ is a subsheaf of $p^{K_1}\big(\overset{*}{\pi}_{K_1}\mathfrak{A}(G_{K_1}; F)\big) = \mathfrak{A}_{K_1}(G; F)$. The following diagram is commutative

$$\begin{array}{ccccc}
\mathfrak{A}(G_{K_2}; F) & \longrightarrow & \overset{*}{\pi}^{K_1}_{K_2}\mathfrak{A}(G_{K_2}; F) & \xrightarrow{p^{K_2}_{K_1}} & \mathfrak{A}(G_{K_1}; F) \\
\downarrow & & & & \downarrow \\
\overset{*}{\pi}_{K_2}\mathfrak{A}(G_{K_2}; F) & & & & \overset{*}{\pi}_{K_1}\mathfrak{A}(G_{K_1}; F) \\
\downarrow{\scriptstyle p^{K_2}} & & & & \downarrow{\scriptstyle p^{K_1}} \\
\mathfrak{A}_{K_2}(G; F) & & \xrightarrow{\quad \varepsilon^{K_2}_{K_1} \quad} & & \mathfrak{A}_{K_1}(G; F)
\end{array} \tag{9}$$

and $\varepsilon^{K_2}_{K_1}$, $K_1 \subseteqq K_2$, denotes the canonical embedding of $\mathfrak{A}_{K_2}(G; F)$ into $\mathfrak{A}_{K_1}(G; F)$.

If $K_1 \subseteqq K_2 \subseteqq K_3$, then the equation $\pi^{K_1}_{K_3} = \pi^{K_1}_{K_2} \circ \pi^{K_2}_{K_3}$ implies $\varepsilon^{K_3}_{K_1} = \varepsilon^{K_3}_{K_2} \circ \varepsilon^{K_2}_{K_1}$, which is easily verified by commutativity of the following diagram implied by (9).

$$\begin{array}{ccccc}
\mathfrak{A}(G_{K_3}; F) & \xrightarrow{\overline{p}^{K_3}_{K_2}} & \mathfrak{A}(G_{K_2}; F) & \xrightarrow{\overline{p}^{K_2}_{K_1}} & \mathfrak{A}(G_{K_1}; F) \\
\downarrow & & \downarrow & & \downarrow \\
\overset{*}{\pi}_{K_3}\mathfrak{A}(G_{K_3}; F) & & \overset{*}{\pi}_{K_2}\mathfrak{A}(G_{K_2}; F) & & \overset{*}{\pi}_{K_1}\mathfrak{A}(G_{K_1}; F) \\
\downarrow & & \downarrow & & \downarrow \\
\mathfrak{A}_{K_3}(G; F) & \xrightarrow{\varepsilon^{K_3}_{K_2}} & \mathfrak{A}_{K_2}(G; F) & \xrightarrow{\varepsilon^{K_2}_{K_1}} & \mathfrak{A}_{K_1}(G; F).
\end{array} \tag{10}$$

The direct limit of the direct spectrum of subsheaves $\{\mathfrak{A}_K(G;F); \varepsilon_{K_1}^{K_2}\}_{\mathfrak{K}}$ is canonically isomorphic to a subsheaf $\mathfrak{A}_1(G;F)$ of the sheaf $\mathfrak{A}(G;F)$

$$\mathfrak{A}_1(G;F) = \varinjlim \{\mathfrak{A}_K(G;F); \varepsilon_{K_1}^{K_2}\}_{\mathfrak{K}}. \tag{11}$$

$\mathfrak{A}_1(G;F)$ is the *sheaf of germs of locally analytic locally cylindrical F-valued functions* on the locally compact group G. The continuous sections of this sheaf are analytic functions, which are locally cylindrical i.e. locally constant on cosets modulo some co-LIE subgroup $K \in \mathfrak{K}$.

The vector space $A_1(G;F) = \Gamma(G;\mathfrak{A}_1)$ is the *space of locally cyclindrical analytic F-valued functions* on G and it holds $A_1(G;F) \subseteqq A(G;F)$.

Remark 6⁰. Applying the analogous procedure to the sheaf of germs of locally smooth F-valued functions $\mathfrak{L}^\infty(G;F)$ instead of $\mathfrak{A}(G;F)$ (see also Remark 3⁰) we get a subsheaf $\mathfrak{L}_1^\infty(G;F)$ of *germs of locally smooth locally cyclindrical F-valued functions* on G. It is easy to verify that the space $C_1^\infty(G;F) = \Gamma(G;\mathfrak{L}_1^\infty)$ is exactly the space of smooth functions on G defined by F. BRUHAT [6], M. G. KAZ [8], and K. MAURIN [10]. Once more using Remark 3⁰ we get the inclusions

$$\mathfrak{A}_1(G;F) \subseteqq \mathfrak{L}_1^\infty(G;F) \quad \text{and} \quad A_1(G;F) \subseteqq C_1^\infty(G;F)$$

in the sense of subsheaves and subspaces respectively. Moreover $A_1(G;F) = A(G;F) \cap C_1^\infty(G;F)$.

Let us restrict further considerations to functions with values in a BANACH space B. If O is an open subset in G, $f \in A(O;B)$ and $x \in O$ then there exists a neighborhood $U \in \mathfrak{U}$ and a smooth realvalued function g_U with compact support such that $xU \subseteqq O$ and $g_U(y) = 1$ for every $y \in U$. Hence the smooth B-valued function $(g_U \circ l_x)\, f$ coincides with f on the subset xU and is a function with compact support: $(g_U \circ l_x)\, f \in C_0^\infty(G;B)$. From this it follows ([1] and [2]) that the function $(g_U \circ l_x)\, f$ is constant on cosets modulo some $N \in \mathfrak{N}$.

Proposition 3.1. *Every analytic function $f \in A(O;B)$ with values in a Banach space B is locally constant on the cosets modulo some compact normal subgroup N of finite codimension, N depending on the neighborhood U.*

It was shown in [3] that every compact normal subgroup of finite codimension of a connected, locally connected, locally compact group G is a co-LIE subgroup, hence

Proposition 3.2. *If G is a connected, locally connected, locally compact group, then every analytic function $f \in A(O;B)$ with values in a Banach space is locally cylindrical. The following equation holds*

$$\mathfrak{A}_1(G;B) = \mathfrak{A}(G;B).$$

We used the theorem of YAMABE to represent the connected locally compact group G as a projective limit of LIE groups $G_K = G/K$, $K \in \mathfrak{K}$. It is evident, that the connected group G is also the limit of the inverse spectrum $\{G_N; \pi_{N_2}^{N_1}\}_{\mathfrak{N}}$, $G_N = G/N$, $N \in \mathfrak{N}$. Considerations analogous to those given above yield to a subsheaf of the sheaf $\mathfrak{A}(G;B)$ which is isomorphic to the direct limit $\varinjlim \{\mathfrak{A}_N(G;F); \varepsilon_{N_1}^{N_2}\}_{\mathfrak{N}}$ and is the sheaf of germs of locally analytic B-valued functions locally constant on the cosets modulo compact normal subgroups of finite codimension. Proposition 3.1 implies the equality with the whole sheaf $\mathfrak{A}(G; B)$.

Theorem 3.3. *If G is a connected locally compact group and B is a Banach space, then the sheaf of germs of locally analytic B-valued functions on G is isomorphic to the direct limit of the subsheaves of functions constant on the cosets modulo the compact normal subgroups of finite codimension*

$$\mathfrak{A}(G; B) \cong \varinjlim \{\mathfrak{A}_N(G; B); \varepsilon_{N_1}^{N_2}\}_{\mathfrak{N}}.$$

If the group G is locally connected, then $\mathfrak{A}(G; B)$ is the sheaf of germs of locally analytic locally cylindrical B-valued functions

$$\mathfrak{A}(G; B) \cong \varinjlim \{\mathfrak{A}_K(G; B); \varepsilon_{K_2}^{K_1}\}_{\mathfrak{K}}.$$

We conclude the considerations of this paper proving the following version of the *principle of Monodromie.*

Proposition 3.4. *Let G denote a connected separable locally compact group. If two functions $f, g \in A(G; B)$ coincide on some neighborhood of a point $x_0 \in G$, then $f = g$.*

Let $\{O_i\}_{i=1}^{\infty}$ denote an increasing sequence of connected open subsets with compact closure such that $x_0U = O_1$, $\bigcup_{i=1}^{\infty} O_i = G$ and $f(x) = g(x)$ for every $x \in x_0U$. The restrictions of the functions f and g to the sets O_i are constant on the cosets modulo some N_i' and N_i'' from $\mathfrak{N}$ by compactness of the closure $\bar{O}_i$. Take $N_i \subseteqq N_i' \cap N_i''$, then f and g are constant on the cosets modulo N_i if they are restricted to O_i. We can choose the compact normal subgroups N_i in such a way that the sequence $\{N_i\}_{i=1}^{\infty}$ is decreasing. Let $f_i, g_i \in A(\pi_i O_i; B)$ be the functions corresponding to the restrictions of f and g: $f|_{O_i} = f_i \circ \pi_i$ and $g|_{O_i} = g_i \circ \pi_i$. By assumption f_i and g_i coincide on $\pi_i(x_0U)$ and hence $f_i = g_i$. This is evident if $N_i = K_i \in \mathfrak{K}$ is a co-Lie subgroup, but in the general case G/N_i is a finite-dimensional connected locally compact group and there exists a Lie group G_i which is a dense algebraic subgroup of G/N_i such that the functions f_i and g_i, when restricted to G_i, belong to $A(G_i \cap \pi_i O_i; B)$ ([4]). Using the principle of monodromie for analytic functions on a connected open subset of a Lie group the functions f_i and g_i coincide on a dense subset and hence everywhere. This proves the proposition.

References

[1] H. Boseck, Über Darstellungen lokalkompakter topologischer Gruppen, Math. Nachr. 74 (1976), 233–251.

[2] H. Boseck und G. Czichowski, Grundfunktionen und verallgemeinerte Funktionen auf topologischen Gruppen, I; II, Math. Nachr. 58 (1973), 215–240; 66 (1975), 319–331.

[3] H. Boseck und G. Czichowski, Connectedness properties of locally compact groups, Math. Nachr. 89 (1979), 17–24.

[4] G. Czichowski, The structure of connected LP-groups with finite-dimensional Lie algebra, Math. Nachr. 62 (1974), 77–81.

[5] G. E. Bredon, Sheaf theory, Mc Graw Hill, New York 1967.

[6] F. Bruhat, Distributions sur un group localement compact et applications à l'étude des représentations des groups p-adiques, Bull. Soc. Math. France 89 (1961), 43–75.

[7] В. М. Глушков, Алгебры Ли локально бикомпактных групп, Усп. мат. наук **XII** (1957), 137—143.

[8] Г. И. Кац, Обобщенные функции на локально компактной группе и разложений унитарных представлений. Труды моск.-ого мат.-ого общ-ва **10** (1961), 3—40.

[9] K. Lashof, Lie-algebras of locally compact groups, Pacific J. Math. **VII** (1957), 1145—1162.

[10] K. Maurin, Distributionen auf Yamabe-Gruppen. Harmonische Analyse auf einer lokalkompakten Gruppe, Bull. Acad. Polon. **9** (1961), 845—850.

[11] K. Maurin, General eigenfunction expansions and unitary representations of topological groups, PWN Polish scientific publishers, Warszawa 1968.

[12] J. Riss, Elements de calcul differentielle et théorie des distributions sur les groupes abéliens localement compacts, Acta Math. **89** (1953), 45—105.

Anschrift des Verfassers:

Ernst-Moritz-Arndt-Universität
Sektion Mathematik
22 Greifswald
Friedrich-Ludwig-Jahn-Str. 15a
DDR

Topogenous Spaces and Zero-Set Spaces

ÁKOS CSÁSZÁR, Budapest

0. The concept of a topogenous space has been introduced — as a special case of the more general concept of a syntopogenous space — in the monograph [1] in order to give a common generalization of the concepts of topological space and proximity space. The purpose of the present paper is to show that this concept embrasses also that one of zero-set spaces introduced in [2] and to discuss some problems connected with the embedding of the category of zero-set spaces into that of topogenous spaces.

Detailed proofs will be contained — in a somewhat more general setting — in another paper.

1. Concerning topogenous spaces, let us recall some definitions and results contained in [1].

A *topogenous order* on a set X is a binary relation defined for the subsets of X (i.e. a subset of $2^X \times 2^X$), denoted by $<$, and satisfying the following conditions:

$$\emptyset < \emptyset, \quad X < X; \tag{1.1}$$

$$A < B \quad \text{implies} \quad A \subset B; \tag{1.2}$$

$$A' \subset A, \quad A < B, \quad B \subset B' \quad \text{implies} \quad A' < B'; \tag{1.3}$$

$$A_i < B_i \quad (i = 1, 2) \quad \text{implies} \quad A_1 \cap A_2 < B_1 \cap B_2 \quad \text{and} \quad A_1 \cup A_2 < B_1 \cup B_2. \tag{1.4}$$

A *topogenous structure* on X is a topogenous order on X that is *idempotent*, i.e. such that $< = < \circ <$ or equivalently $< \subset < \circ <$, in other words satisfying

$$A < B \text{ implies the existence of a } C \text{ such that } A < C < B. \tag{1.5}$$

Here $\circ$ denotes the composition and $\subset$ the inclusion for subsets of $2^X \times 2^X$.

E.g. if τ is a topology on X and we define

$$A < (\tau)\, B \quad \text{iff} \quad A \subset \operatorname{int} B \tag{1.6}$$

then $< (\tau)$ is a topogenous structure on X. If δ is a proximity relation on X and we put

$$A < (\delta)\, B \quad \text{iff} \quad A \bar{\delta} X - B \tag{1.7}$$

then $<(\delta)$ is again a topogenous structure. In both cases $<(\tau)$ and $<(\delta)$ determine τ and δ respectively in a unique manner since $G \subset X$ is τ-open iff $G < (\tau)\, G$ and $A\bar{\delta}B$ iff $A < (\delta)\, X - B$.

A topogenous structure $<$ on X is said to be *coarser* than another such structure $<'$ (or $<'$ is said to be *finer* than $<$) iff $< \subset <'$.

A *topogenous space* is a pair $(X, <)$ where X is a set and $<$ is a topogenous structure on X.

If $(X, <)$ and $(X', <')$ are topogenous spaces and $f\colon X \to X'$, then f is said to be $(<, <')$-*continuous* iff $A' <' B'$ implies $f^{-1}(A') < f^{-1}(B')$. In the case when τ and τ' are topologies on X and X' respectively, $\big(<(\tau), <(\tau')\big)$-continuity coincides with ordinary continuity with respect to τ and τ'. Similarly, if δ and δ' are proximity relations, then $\big(<(\delta), <(\delta')\big)$-continuity coincides with proximal continuity with respect to δ and δ'.

This can be summarized in the following way: If **Tpg** denotes the category with the topogenous spaces as objects and $(<, <')$-continuous maps as morphisms, then the category **Top** of topological spaces and continuous maps can be isomorphically embedded by means of (1.6) into **Tpg**, and the same holds with the help of (1.7) for the category **Prox** of proximity spaces and proximally continuous maps. If we denote the corresponding subcategories of **Tpg** by **Top** and **Prox** as well, then both of them are full subcategories of **Tpg**.

The category **Tpg** is a properly fibred topological category in the sense of [3]. The initial structure corresponding to the maps $f_i\colon X \to X_i$ and topogenous spaces $(X_i, <_i)$ is given by

$$< \;=\; \bigvee_{i \in I} f_i^{-1}(<_i) \tag{1.8}$$

where

$$A f_i^{-1}(<_i)\, B \quad \text{iff} \quad f_i(A) <_i X_i - f_i(X - B) \tag{1.9}$$

and $< \;=\; \bigvee\limits_{i \in I} <_i'$ is defined by

$$A < B \quad \text{iff} \quad A = \bigcup_{1}^{m} A_j, \quad B = \bigcap_{1}^{m} B_k, \quad \text{and} \quad A_j <'_{i(j,k)} B_k \tag{1.10}$$

with suitable $i(j, k) \in I$.

The subcategories **Top** and **Prox** of **Tpg** can be easily characterized. In fact, $(X, <)$ is an object of **Top** iff the topogenous order $<$ is *perfect*, i.e. iff

$$A_i < B_i \quad (i \in I) \quad \text{implies} \quad \bigcup_{i \in I} A_i < \bigcup_{i \in I} B_i, \tag{1.11}$$

and $(X, <)$ is an object of **Prox** iff $<$ is *symmetric*, i.e. iff

$$A < B \quad \text{implies} \quad X - B < X - A\,. \tag{1.12}$$

Both **Top** and **Prox** are bicoreflective subcategories of **Tpg**; the corresponding coreflections of $(X, <)$ are $(X, <^p)$ and $(X, <^s)$ respectively where

$$A <^p B \quad \text{iff} \quad A = \bigcup_{i \in I} A_i, \quad B = \bigcup_{i \in I} B_i \quad \text{and} \quad A_i < B_i \quad \text{for each } i, \tag{1.13}$$

and

$$<^s \;=\; < \vee <^c \tag{1.14}$$

where

$$A <^c B \quad \text{iff} \quad X - B < X - A. \tag{1.15}$$

2. According to [2], a zero-set structure on X is a system $\mathfrak{Z}$ of subsets of X such that

$$\emptyset, X \in \mathfrak{Z}; \tag{2.1}$$

$$Z_i \in \mathfrak{Z} \quad (i = 1, 2) \quad \text{implies} \quad Z_1 \cup Z_2 \in \mathfrak{Z}; \tag{2.2}$$

$$Z_i \in \mathfrak{Z} \quad (i \in \mathbf{N}) \quad \text{implies} \quad \bigcap_1^\infty Z_i \in \mathfrak{Z}; \tag{2.3}$$

$Z \in \mathfrak{Z}$ implies the existence of $Z_i \in \mathfrak{Z}$ such that

$$X - Z = \bigcup_1^\infty Z_i; \tag{2.4}$$

$Z_1, Z_2 \in \mathfrak{Z}$, $Z_1 \cap Z_2 = \emptyset$ implies the existence of

$$C_1, C_2 \text{ such that} \quad X - C_i \in \mathfrak{Z}, \quad Z_i \subset C_i \quad \text{and} \quad C_1 \cap C_2 = \emptyset. \tag{2.5}$$

A *zero-set space* is a pair $(X, \mathfrak{Z})$ of a set X and a zero-set structure $\mathfrak{Z}$ on X. If $(X, \mathfrak{Z})$ and $(X', \mathfrak{Z}')$ are zero-set spaces then $f: X \to X'$ is said to be a *zero-set map* iff $Z' \in \mathfrak{Z}'$ implies $f^{-1}(Z') \in \mathfrak{Z}$. Let us denote by **Zer** the category with zero-set spaces as objects and zero-set maps as morphisms.

If X is a topological space and $\mathfrak{Z}$ denotes the system of the usual zero-sets of X (i.e. the sets having the form $\{x \in X : f(x) = 0\}$ where $f: X \to \mathbf{R}$ is a continuous function) then $(X, \mathfrak{Z})$ is a zero-set space (whence the terminology).

3. It is easy to show that the category **Zer** is isomorphic with a full subcategory of **Tpg**. In fact, we have the following facts:

Theorem 1. *If* $(X, \mathfrak{Z})$ *is a zero-set space and we define*

$$A < (\mathfrak{Z})\, B \quad \textit{iff there is a } C \textit{ such that} \quad X - C \in \mathfrak{Z} \quad \textit{and} \quad A \subset C \subset B, \tag{3.1}$$

then $< (\mathfrak{Z})$ *is a topogenous structure on* X. *The topogenous order* $< (\mathfrak{Z})$ *determines* $\mathfrak{Z}$ *in a unique manner since*

$$Z \in \mathfrak{Z} \quad \textit{iff} \quad X - Z < (\mathfrak{Z})\, X - Z. \tag{3.2}$$

Theorem 2. *If* $(X, \mathfrak{Z})$ *and* $(X', \mathfrak{Z}')$ *are zero-set spaces then a map* $f: X \to X'$ *is a zero-set map iff it is* $\big(< (\mathfrak{Z}), < (\mathfrak{Z}')\big)$*-continuous.*

Let us keep the notation **Zer** for the full subcategory of **Tpg** the objects of which are of the form $\big(X, < (\mathfrak{Z})\big)$. In order to characterize this subcategory we need some definitions.

Definition 1. A topogenous order $<$ is said to be *countably perfect* iff (1.11) holds for any countable index set I. A topogenous space $(X, <)$ is said to be countably perfect iff $<$ is countably perfect.

Theorem 3. *The countably perfect topogenous spaces constitute the objects of a bicoreflective full subcategory of* **Tpg**. *The corresponding coreflection of* $(X, <)$ *is given by* $(X, <^{\pi})$ *where*

$$A <^{\pi} B \quad \textit{iff} \quad A = \bigcup_{1}^{\infty} A_i, \quad B = \bigcup_{1}^{\infty} B_i, \quad A_i < B_i \quad \textit{for} \quad i \in \mathbf{N}. \tag{3.3}$$

Definition 2. A topogenous order $<$ is said to be *semi-symmetric* iff $< \subset <^c$.

Definition 3. A topogenous order $<$ is said to be *normal* iff $<^c \circ < \subset < \circ <^c$.

Definition 4. A topogenous order $<$ is said to be *strongly idempotent* iff $A < B$ implies the existence of a C such that $A \subset C < C \subset B$.

For the terminology we observe that a symmetric order is semi-symmetric, a strongly idempotent order is idempotent, and an order $< (\tau)$ (for a topology τ) is normal iff the topology τ is normal. A symmetric order is obviously normal.

Theorem 4. *A topogenous space* $(X, <)$ *is an object of* **Zer** *iff* $<$ *is countably perfect, semi-symmetric, normal, and strongly idempotent.*

4. We shall investigate the properties of **Zer** as a subcategory of **Tpg**. For this purpose, let us denote by $<^*$ the topogenous structure on $\mathbf{R}$ associated with the euclidean proximity relation by (1.7), i.e.

$$A <^* B \quad \text{iff there is} \quad \varepsilon > 0 \quad \text{such that} \quad S(A, \varepsilon) \subset B \tag{4.1}$$

where

$$S(A, \varepsilon) = \{x \in \mathbf{R}: |x - a| < \varepsilon \quad \text{for some} \quad a \in A\}. \tag{4.2}$$

Now we can define a functor from **Tpg** onto **Zer** as follows:

Definition 5. For a topogenous space $(X, <)$, let $A <^z B$ hold iff there is a $(<, <^*)$-continuous $f: X \to \mathbf{R}$ such that $x \in A$ implies $f(x) > 0$ and $f(x) > 0$ implies $x \in B$.

Theorem 5. *If* $(X, <)$ *is a topogenous space, then* $<^z$ *is a topogenous structure,* $<^{zz} = <^z$, *and* $(X, <)$ *is an object of* **Zer** *iff* $< = <^z$. *If* $f: X \to X'$ *is* $(<, <')$*-continuous, then it is* $(<^z, <'^z)$*-continuous as well. Hence* $< \mapsto <^z$ *induces a covariant functor* F_z *from* **Tpg** *onto* **Zer**.

Theorem 6. *Let* $(X, <)$ *be an object of* **Zer**. $<$ *is perfect iff* $< = < (\tau)$ *where* τ *is a regular and perfectly normal topology.* $<$ *is symmetric iff* $< = < (\mathfrak{Z})$ *where* $\mathfrak{Z}$ *is a* σ*-algebra in* X.

Hence F_z is neither a coreflector (since if τ denotes a topology which is not perfectly normal then $< = < (\tau)$ implies $<^z \subset <$, $<^z \neq <$) nor a reflector (because if $<$ is symmetric then $< \subset <^z$ and in general $< \neq <^z$). It can occur that $<$ and $<^z$ are not comparable with respect to $\subset$.

However, the functor F_z is the product of a reflector and a coreflector. In order to see this, let us introduce

Definition 6. If $(X, <)$ is a topogenous space, let $A <^{\sigma} B$ hold iff there exists a $(<, <^*)$-continuous $f: X \to \mathbf{R}$ such that $0 \leq f(x) \leq 1$ for $x \in X$, $f(x) = 0$ for $x \in A$, $f(x) = 1$ for $x \in X - B$.

Theorem 7. *For a topogenous space* $(X, <)$, $<^\sigma$ *is a symmetric topogenous structure on* X, *namely the finest symmetric topogenous structure coarser than* $<$. *Hence* $<^{\sigma\sigma} = <^\sigma$, *and* $<$ *is symmetric iff* $< = <^\sigma$.

Theorem 8. *If* $f: X \to X'$ *is* $(<, <')$*-continuous then it is* $(<^\sigma, <'^\sigma)$*-continuous as well. Hence* $< \mapsto <^\sigma$ *induces a reflector* F_σ *from* **Tpg** *onto* **Prox** *so that* **Prox** *is a bireflective subcategory of* **Tpg**.

Theorem 9. *If* $<$ *is a normal topogenous structure then* $<^\sigma = <^c \circ <$.

Definition 7. Let us denote by **Z** the full subcategory of **Tpg** the objects of which are those $(X, <)$ that satisfy $< \subset <^z$.

Theorem 10. *We have the inclusions*

$$\mathbf{Zer} \subset \mathbf{Z}, \quad \mathbf{Prox} \subset \mathbf{Z}.$$

The functor $F_z|\mathbf{Z}$ *is a coreflector so that* **Zer** *is a bicoreflective subcategory of* **Z**.

As a generalization of the first part of Theorem 10, we can state:

Theorem 11. *If* $<$ *is a semi-symmetric and normal topogenous structure then* $(X, <)$ *is an object of* **Z**.

Problem 1. If $(X, <)$ is an object of **Z**, is it true that it is semi-symmetric and normal?

Theorem 12. *The equality* $<^z = <^{\sigma z}$ *holds for any topogenous structure* $<$. *Hence the functor* F_z *is the product of a reflector from* **Tpg** *onto* **Prox** *by a coreflector from* $\mathbf{Z} \supset \mathbf{Prox}$ *onto* **Zer**.

5. **Definition 8.** For a topogenous structure $<$, let us write $<^\nu = <^{\sigma z}$.

Theorem 13. *For any topogenous structure* $<$, $<^\nu$ *is the finest symmetric topogenous structure* $<'$ *satisfying* $<'^z = <^z$.

In general, there are several symmetric topogenous structures $<'$ fulfilling the condition $<'^z = <^z$.

Theorem 14. *We have* $<^{\nu\nu} = <^\nu$ *and* $<^{\nu p} = <^{zp}$ *for any topogenous structure* $<$.

Definition 9. Let us denote by **Y** and $\mathbf{Y}_0$ the full subcategories of **Tpg** the objects of which are those $(X, <)$ that satisfy $< \subset <^\nu$ and $< = <^\nu$, respectively.

Theorem 15. *We have the inclusions*

$$\mathbf{Y}_0 \subset \mathbf{Prox} \subset \mathbf{Y}.$$

$\mathbf{Y}_0$ *is a bicoreflective subcategory of* **Y** *having as coreflector the functor induced by* $< \mapsto <^\nu$, *restricted to* **Y**.

Theorem 16. *For a topogenous structure* $<$, *the equalities* $< = <^{\nu p}$ *and* $< = <^{zp}$ *both hold iff* $< = <(\tau)$ *with a completely regular topology* τ.

Problem 2. Is the inclusion $\mathbf{Prox} \subset \mathbf{Y}$ a proper one? If the answer is yes, give a suitable characterization of **Y**.

References

[1] Á. Császár, Foundations of general topology, Budapest—Oxford—London—New York—Paris 1963.
[2] H. Gordon, Rings of functions determined by zero-sets, Pacific J. Math. **36** (1971), 133—157.
[3] H. Herrlich, Topological structures, Math. Centre Tracts No. 52, Amsterdam 1974.

Anschrift des Verfassers:

MTA Mat. Kut. Int.
Budapest V
Reáltanoda u. 13/15
Hungary

On Sequential Topologies

JÜRGEN FLACHSMEYER, Greifswald

0. Introduction

For general topological spaces sequences are not enough in studying convergence, continuity, adherence and so on, but more general nets or filters must be used. However the cases for which sequences suffice are of special interest especially in functional analysis. A number of fundamental theorems in functional analysis is known where only sequences occur and which may fail for general nets.

Convergence structures based on sequences have been studied in great detail by many authors. A well-known working group in this area is the Czech group around J. NOVÁK and V. KOUTNÍK (compare also the lecture of R. FRIČ and V. KOUTNÍK given on this conference). R. ENGELKING's book [4] contains several questions on spaces which topologies are described by sequences. A. WILANSKI's lecture notes [18] concern with some sequential properties in linear topology. Further investigations must be made and seem to be fruitfull.

1. Preliminaries. Basic definitions

We recall the known definitions of spaces in which convergence of sequences suffices to describe the topology (vid. also ENGELKING's book [4]).

1.1. Fréchet-Urysohn spaces. A topological space $(X, \mathcal{O})$ is called *FU-space* (FRÉCHET-URYSOHN) iff every adherence point x of a set A can be riched by a convergent sequence (x_n) from A:

$$x \in \bar{A} \Leftrightarrow (x_n) \subset A \quad \text{and} \quad x_n \underset{\mathcal{O}}{\rightarrow} x.$$

ENGELKING used the notation FRÉCHET space but for example in Russian literature the notation FRÉCHET-URYSOHN space is used. Because of the different notions of a FRÉCHET space in the area of topological vector spaces (complete metrizable vector space topology (Polish terminology) or complete metrizable locally convex vector space topology (BOURBAKI terminology)) we prefer the notation FU-space.

Of course, FU-spaces are such spaces for which the sequential closure of each set is closed in the topology (or the same: the sequential closure is the closure). But it is different from the condition that the sequential closure is idempotent (is a topological closure operation). Each FU-space has idempotent sequential closure. Example 1.2.c) below shows a space which is not FU but has idempotent sequential closure. Furthermore, every T_1-FU-space has tightness $t(X) \leqq \aleph_0$. But there are spaces of tightness $t(X) = \aleph_0$ which are not FU-spaces. Namely, there are countable spaces X which are not FU (see example below). Spaces of countable tightness are considered by R. C. MOORE and S. MROWKA [12].

Every first countable space (A_1-space) is FU but not conversely.

1.2. Example. Take infinite many straight lines of the plane R^2 through one point p. The finest topology in this bundle which induces on every line the usual Euclidean topology gives an FU-space. The point p has no countable base. If we take only countable many lines of such bundle and on every line only the "rational" points, then we get in this way a countable space without being a first countable space, but it is FU.

1.3. Sequential spaces. A subset F of a topological space $(X, \mathcal{O})$ is called *sequentially closed* iff F contains with every convergent sequence the limit points:

$$(x_n) \subset F \quad \text{and} \quad x_n \xrightarrow[\mathcal{O}]{} x \quad \Rightarrow \quad x \in F.$$

Of course, every closed subset is sequentially closed. The converse does not always hold.

1.4. Examples. a) The following fact is well-known from analysis. In the space R^R of all real valued functions on R with the simple topology (topology of pointwise convergence) the subset $\boldsymbol{Bo}(\mathsf{R})$ of all BOREL functions is sequentially closed but not closed.

$f \in \boldsymbol{Bo}(\mathsf{R}) \Leftrightarrow$ for all $r \in \mathsf{R}$ the set $[f \leqq r]$ is BOREL in R. Let be $f_n \xrightarrow[\text{simple}]{} f$, $f_n \in \boldsymbol{Bo}(\mathsf{R})$, then

$$[f \leqq r] = \bigcap_k \bigcup_m \bigcap_{n \geqq m} \left[f_n \leqq r + \frac{1}{k}\right].$$

Thus $f \in \boldsymbol{Bo}(\mathsf{R})$.

On the other hand: For every indicator function χ_F of a closed set $F \subset \mathsf{R}$ we have $\chi_F = \inf\{g : g \in \mathscr{C}(\mathsf{R}), g \geqq \chi_F\}$, where the infimum is taken pointwise and $\mathscr{C}(\mathsf{R})$ is the set of all continuous functions. Thus χ_F is a simple limit of a net of continuous functions. Now for every subset $A \subset \mathsf{R}$, $\chi_A = \sup\{\chi_F : F \subset A, F \text{ closed}\}$, where the supremum is taken pointwise. Every bounded function is the pointwise limit of stepfunctions. Therefore every function on R is a pointwise limit of a net of continuous functions. This means:

$\mathscr{C}(\mathsf{R})$ and because of $\mathscr{C}(\mathsf{R}) \subset \boldsymbol{Bo}(\mathfrak{R})$ also $\boldsymbol{Bo}(\mathsf{R})$ is dense with respect to the simple topology in R^R.

b) Let X be a nonvoid set and $\mathfrak{B} \subset \mathfrak{P}(X)$ a σ-field of subsets with $\{x\} \in \mathfrak{B}$ for every $x \in X$. Take in $\mathfrak{P}(X)$ the simple topology of the corresponding indicator functions. Then $\mathfrak{B}$ is sequentially closed but not closed in $\mathfrak{P}(X)$.

c) Take in R the following refinement $\mathcal{O}$ of the Euclidean topology: The family of sets $V = U \setminus A$, where U is Euclidean open and card $A \leqq \aleph_0$, is stable with respect to finite intersections. Thus these V's form a base for a topology $\mathcal{O}$ on R.

This $\mathcal{O}$ is not discrete but every subset of R is sequentially closed. Namely $x_n \xrightarrow[\mathcal{O}]{} x$ means $x_n = x$ eventually.

A set is called *sequentially open* iff the complement is sequentially closed. The sequentially open sets of a given topology $\mathcal{O}$ on X form another topology $\mathcal{O}_s \supset \mathcal{O}$ on X. $(X, \mathcal{O})$ is called *sequential* iff $\mathcal{O}_s = \mathcal{O}$, or it is the same that every sequentially closed set is closed. For a space $(X, \mathcal{O})$ the space $(X, \mathcal{O}_s)$ is called its *sequential refinement.*

Every FU-space is sequential.

1.5. Proposition. *Let $(X, \mathcal{O})$ be a topological space and $(X, \mathcal{O}_s)$ its sequential refinement. Then holds:*

(1) *$\mathcal{O}_s$ is the finest topology $\mathcal{P}$ on X with $x_n \underset{\mathcal{O}}{\to} x \Rightarrow x_n \underset{\mathcal{P}}{\to} x$. With other words: $\mathcal{O}_s$ is the underlying topology (the topological modification) of the convergence class: $x_n \underset{\mathcal{O}}{\to} x$.*

(2) *$\mathcal{O}_s$ is the finest topology $\mathcal{P}$ on X with the same convergent sequences as $\mathcal{O}$.*

(3) *$\mathcal{O}_s$ is the unique sequential topology on X with the same convergent sequences as $\mathcal{O}$.*

(4) *For a topology $\mathcal{P} \supset \mathcal{O}$ one has: $\mathcal{P} \subset \mathcal{O}_s$ iff $\mathcal{O}$ and $\mathcal{P}$ have the same convergent sequences.*

Proof. (1) Let be $x_n \underset{\mathcal{O}}{\to} x$ and $x \in U \in \mathcal{O}_s$. Then $x_n \in U$ eventually. Thus $x_n \underset{\mathcal{O}_s}{\to} x$. Let $\mathcal{P}$ be a topology on X for which every $\mathcal{O}$-convergent sequence is $\mathcal{P}$-convergent. If F is $\mathcal{P}$-closed, then F is sequentially closed with respect to $\mathcal{O}$, i.e. F is $\mathcal{O}_s$-closed.

(2) $\mathcal{O}_s \supset \mathcal{O}$, thus $x_n \underset{\mathcal{O}_s}{\to} x \Rightarrow x_n \underset{\mathcal{O}}{\to} x$. With (1) one has (2).

(3) $\mathcal{O}_s$ is sequential. Namely, let $F \subset X$ and F be sequentially closed with respect to $\mathcal{O}_s$. According to (2), F is sequentially closed with respect to $\mathcal{O}$, i.e. F is $\mathcal{O}_s$-closed. Let $\mathcal{P}$ be a sequential topology on X with the same convergent sequences as $\mathcal{O}$. (2) implies $\mathcal{O}_s \supset \mathcal{P}$. If F is closed with respect to $\mathcal{O}_s$, then F is sequentially closed with respect to $\mathcal{O}$. Thus F is sequentially closed with respect to $\mathcal{P}$ and therefore it is $\mathcal{P}$-closed.

(4) From (2) we have $\mathcal{P} \subset \mathcal{O}_s$. Let be conversely $\mathcal{P} \subset \mathcal{O}_s$. Then every $\mathcal{O}_s$-convergent sequence ($\equiv$ every $\mathcal{O}$-convergent sequence) is $\mathcal{P}$-convergent. Because of $\mathcal{O} \subset \mathcal{P}$ every $\mathcal{P}$-convergent sequence is $\mathcal{O}$-convergent. □

1.6. Proposition. *A continuous map from a topological space into another yields for every sequentially closed (sequentially open) set a sequentially closed (sequentially open) preimage*

Proof. Let $F \subset Y$ be sequentially closed, $f\colon X \to Y$. Then for $(x_n) \subset f^{-1}(F)$ and $x_n \to x$ one has $f(x_n) \to f(x)$ and $\big(f(x_n)\big) \subset F$. Thus $f(x) \in Z$, i.e. $x \in f^{-1}(F)$. □

Remarks. a) The last proposition gives that a continuous map from a sequential space into a space $(X, \mathcal{O})$ is continuous as a map from the space into the sequential refinement $(X, \mathcal{O}_s)$.

In the categorical language this means that $(X, \mathcal{O}_s)$ is the *coreflection* of the space $(X, \mathcal{O})$ with respect to the subcategory

SeqTop of **Top**.

b) The proposition 1.6 gives also that

$\mathcal{O}_s$ is the strongest sequential topology on X which is finer than $\mathcal{O}$.

1.7. Proposition. *A topological space $(X, \mathcal{O})$ is sequential iff for $(X, \mathcal{O})$ the Heine criterion of the continuity of maps holds, i.e. a map $f\colon (X, \mathcal{O}) \to (Y, \mathcal{P})$ into any topological space $(Y, \mathcal{P})$ is continuous iff $x_n \underset{\mathcal{O}}{\to} x$ implies $f(x_n) \underset{\mathcal{P}}{\to} f(x)$.*

Proof. For sequential spaces the continuity and the sequential continuity are the same. (This is remarked for example in ENGELKING's book). Let $F \subset Y$ be closed. Then $f^{-1}(F)$ is sequentially closed, i.e. $f^{-1}(F)$ is closed. If now conversely the HEINE-criterion holds then because of the sequential continuity of the identity id: $(X, \mathcal{O}) \to (X, \mathcal{O}_s)$ it must be continuous, i.e. $\mathcal{O}_s \subset \mathcal{O}$ and therefore $\mathcal{O} = \mathcal{O}_s$. □

2. The radiolar topology — an example of an sequential space which is not FU

There are known examples of sequential spaces which are not FU-spaces. S. P. FRANKLIN [5, 6] firstly examined sequential spaces more systematically. But it should be mentioned that the most earliest example is in a paper of J. NOVÁK [13, p. 24].

We should like to contribute the following example which may be of some interest. We start with a real vector space V of dimension $\geqq 2$. By a *radiolar* R_x at the point $x \in V$ we understand a set $R_x \subset V$ with the following property:

For every straight line g through x the trace $R_x \cap g$ is an interval containing x as an interior point. (In biology radiolaries are little animals of a similar form.) Of course, the intersection $R_x \cap R_x'$ of two radiolars of any point x is a radiolar. Thus the radiolars at a point yield a filter base at this point. By the *radiolar topology* $\mathcal{O}_{\mathrm{rad}}$ in V we understand the topology generated by the filter bases of radiolars.

$U \in \mathcal{O}_{\mathrm{rad}} \Leftrightarrow$ For every $x \in U$ there is a radiolar R_x at x with $R_x \subset U$. These sets are called sometimes in vector space considerations algebraic open sets.

We emphasize the fact that a radiolar R_x must not be a neighborhood at x in the topology $\mathcal{O}_{\mathrm{rad}}$ (vid. below the proof of (5) in the next statement).

2.1. Theorem. *In any real vector space V with* dim $V \geqq 2$ *with respect to the radiolar topology $\mathcal{O}_{\mathrm{rad}}$ the following holds:*

(1) *$\mathcal{O}_{\mathrm{rad}}$ is the finest topology in V which induces on each straight line the Euclidean topology.*

(2) *$\mathcal{O}_{\mathrm{rad}}$ is strictly finer than any Hausdorff vector topology on V.*

(3) *Every affine bijection on V is a topological map with respect to $\mathcal{O}_{\mathrm{rad}}$.*

(4) *If a sequence x_n converges radiolar to x, then on some straight line through x there is a subsequence of x_n which tends in Euclidean topology to x.*

(5) *$\mathcal{O}_{\mathrm{rad}}$ is sequential but not FU.*

(6) *The vector addition is not continuous in $\mathcal{O}_{\mathrm{rad}}$.*

Proof. (1) For the finest topology $\mathcal{O}$ on V which induces on every straight line $g \subset V$ the Euclidean topology a set U is $\mathcal{O}$-open iff $U \cap g$ is Euclidean open. Now $U \in \mathcal{O}_{\mathrm{rad}}$ implies $U \cap g$ Euclidean open, i.e. $\mathcal{O}_{\mathrm{rad}} \subset \mathcal{O}$. Now let be $U \cap g$ Euclidean open for every g. Then for each $x \in U$ there is a radiolar $R_x \subset U$, i.e. $U \in \mathcal{O}_{\mathrm{rad}}$. Thus $\mathcal{O} \subset \mathcal{O}_{\mathrm{rad}}$.

(2) By a well-known result of TICHONOV each finite-dimensional vector space admits exactly one HAUSDORFF vector topology. Thus the trace topology of a HAUSDORFF vector topology $\mathcal{O}$ on V gives on each straight line the Euclidean topology, i.e. $\mathcal{O}_{\mathrm{rad}} \supset \mathcal{O}$. From (6) it follows $\mathcal{O}_{\mathrm{rad}} \neq \mathcal{O}$.

(3) Let be $A: V \to V$ an affine bijection. Then the property to be a radiolar is invariant under A.

(4) Because of (3) the radiolar topology is a homogeneous one. Thus we have only to consider the situation at the point 0. Let be $x_n \underset{\text{rad}}{\longrightarrow} 0$ and $x_n \neq 0$. Then there must be a straight line g through 0 with infinite many $x_k \in g$. These x_k form a subsequence which tends to 0 in Euclidean topology on g.

Namely, if we assume that on each line g through 0 there are only finite many $x_n \in g$, then we have a radiolar R_0 at x with $x_n \notin R_0$ for all n. We take any HAUSDORFF vector topology $\mathcal{O}$ on V. Then $x_n \underset{\text{rad}}{\longrightarrow} 0$ implies $x_n \underset{\mathcal{O}}{\longrightarrow} 0$. Thus $V \setminus (\{x_n : n \in \mathsf{N}\} \cup \{0\})$ is $\mathcal{O}$-open and therefore radiolar open. But $V \setminus \{x_n : n \in \mathsf{N}\}$ is also radiolar open. It remains only the consideration at the point 0. We have the radiolar R_0 at 0 with $R_0 \subset V \setminus \{x_n : n \in \mathsf{N}\}$.

$$V \setminus \{x_n : n \in \mathsf{N}\} \in \mathcal{O}_{\text{rad}} \quad \text{contradicts} \quad x_n \underset{\text{rad}}{\longrightarrow} \mathcal{O}.$$

(5) Let U be sequentially open with respect to $\mathcal{O}_{\text{rad}}$. Then $U \cap g$ is sequentially open on each line g. Thus $U \cap g$ is Euclidean open, i.e. $U \in \mathcal{O}_{\text{rad}}$. But $\mathcal{O}_{\text{rad}}$ is not FU for $\dim V \geqq 2$.

We will construct a set $M \subset V$ which has 0 as an accumulation point in $\mathcal{O}_{\text{rad}}$, but there will be no sequence $(x_n) \subset M$ with $x_n \underset{\text{rad}}{\rightarrow} 0$. We take a plane R^2 through 0. Let g_n be the line in R^2 through 0 with arcus $\frac{\pi}{n}$. Take on each g_n the intervals $I_n := \left[\!\left[-1, \frac{1}{n}\right]\!\right]$. On each other g in V through 0 take the interval $I_g = [\![-1, 1]\!]$. Then $R_0 = \bigcup I_g \cup \bigcup I_n$ is a radiolar at 0. Put $M := V \setminus R_0$. There is no sequence $(x_n) \subset M$ such that $x_n \underset{\text{rad}}{\rightarrow} 0$, because on each line through 0 we have no sequence converging to 0 (compare (4)). Let U be radiolar open with $0 \in U$. On g_1 there is an interval $[\![0, \varepsilon]\!]$, $\varepsilon > 0$, such that $[\![0, \varepsilon]\!] \subset U$. Take the orthogonal line $h \subset \mathsf{R}^2$ which goes through ε. Then there is an interval $[\![-\delta, \delta]\!]$ on h such that $[\![-\delta, \delta]\!] \subset U$. All but finite many g_n must intersect this interval $[\![-\delta, \delta]\!]$ in the points y_n. If $\frac{1}{n} < \varepsilon$ then all this points y_n belong to M. The constructed radiolar R_0 fails to be a neighborhood at 0 with respect to $\mathcal{O}_{\text{rad}}$.

(6) We consider on two orthogonal axis g, h in $\mathsf{R}^2 \subset V$ sequences $(x_n) \subset g$, $(y_n) \subset h$ converging to 0 in Euclidean topology such that arcus $(x_n + y_n) = \frac{\pi}{n}$. Then $x_n \underset{\text{rad}}{\rightarrow} 0$, $y_n \underset{\text{rad}}{\rightarrow} 0$ but $x_n + y_n \underset{\text{rad}}{\not\rightarrow} 0$ because of (4). □

Remark. If we consider the radiolar topology on $\mathbb{Q}^2$ — the rational points of the plane — then we have a homogeneous sequential space of countable many points which is not FU.

3. The box topology and its sequential refinement

Let $(X_\alpha, \mathcal{O}_\alpha)_{\alpha \in A}$ be a family of topological spaces. In the Cartesian product $X = \prod X_\alpha$ the boxes $\prod U_\alpha$, $U_\alpha \in \mathcal{O}_\alpha$, form a base of a topology — the so-called *box topology*. $\mathcal{O}_{\text{box}}$ is finer than the product (= simple) topology.

Now we consider the case R^ω with the box topology. $(\mathsf{R}^\omega, \mathcal{O}_{\text{box}})$ is a topological group but not a topological vector space. The scalar multiplication is not continuous, namely

$$\frac{1}{n} \cdot x \to 0 \quad \text{for} \quad x = (1, 2, 3, \ldots, n, \ldots).$$

We take the subspace $\mathscr{C}_{00}(\mathsf{N})$ of R^ω. (For a locally compact space X the usual notation of the space of all real valued continuous functions on X with compact support supp $f = \overline{[f \neq 0]}$ is $\mathscr{C}_{00}(X)$. Thus in case $X = \mathsf{N}$ the space $\mathscr{C}_{00}(\mathsf{N})$ means those vectors $x \in \mathsf{R}^\omega$ which have only finite many coordinates $\xi_i \neq 0$.)

3.1. Theorem. *The vector space $\mathscr{C}_{00}(\mathsf{N})$ is a locally convex topological vector space with respect to the box topology $\mathcal{O}_{\text{box}}$. $\mathscr{C}_{00}(\mathsf{N})$ is closed in R^ω for the box topology. A sequence x_n converges in the box topology to 0 iff x_n simple converges to 0 and all x_n vanish outside a suitable compact set of N. The sequential refinement of the box topology is the finest vector topology on $\mathscr{C}_{00}(\mathsf{N})$. This is not an FU-space.*

Proof. The trace in $\mathscr{C}_{00}(\mathsf{N})$ of each box $U = \prod [\![-\varepsilon_n, \varepsilon_n]\!]$, $\varepsilon_n > 0$, is a convex, circled, absorbing neighborhood of the origin. Thus $(\mathscr{C}_{00}(\mathsf{N}), \mathcal{O}_{\text{box}})$ is a locally convex topological vector space. Let be $x = (\xi_1, \xi_2, \ldots, \xi_n, \ldots)$ and assume for every n there is an $m(n) > n$ with $\xi_m \neq 0$.

Choose the following box $U = \prod U_n$ with $U_n = \mathsf{R}$ for $\xi_n = 0$ and $U_n = \mathsf{R} \setminus \{0\}$ for $\xi_n \neq 0$. Then $x \in U$ but no $y \in \mathscr{C}_{00}(\mathsf{N})$ is in U. Let be $x_n \xrightarrow[\text{box}]{} 0$, $x_n = (\xi_1{}^n, \xi_2{}^n, \xi_3{}^n, \ldots, \xi_k{}^n, \ldots)$. Because of $\mathcal{O}_{\text{box}} \supset \mathcal{O}_{\text{simple}}$ we have $x_n \xrightarrow[\text{simple}]{} 0$. We assume that there is for every index i a $j = j(i) > i$ with $\xi_j{}^{n_j} \neq 0$ for some n_j. The set $\{n_j : j = j(i)\}$ is infinite because every x_n has only a finite number of coordinates $\neq 0$. We take $U = \prod\limits_j \left[\!\!\left[-\frac{1}{2} |\xi_j{}^{n_j}|, \frac{1}{2} |\xi_j{}^{n_j}| \right]\!\!\right] \times \prod\limits_{k \neq j} \mathsf{R}_k$. Then $x_{n_j} \notin U$. This contradicts $x_n \xrightarrow[\text{box}]{} 0$.

For the converse implication let $U = \prod\limits_i [\![-\varepsilon_i, \varepsilon_i]\!]$ be a box. Then there is a k_0 such that $\xi_k{}^n = 0$ for all $k > k_0$ and all n. Now $\xi_k{}^n \to 0$ for $n \to \infty$ and $k = 1, 2, \ldots, k_0$. Then $x_n \in U$ for all but finite many n.

Let $\mathcal{O}$ be the finest vector topology. By a theorem of KAKUTANI-KLEE [10] for a vector space V with dim $V \leqq \aleph_0$ the finest vector space topology on V is locally convex and is the finest topology on V which induces on every finite dimensional subspace the Euclidean topology. In other words: For $\mathsf{R}^1 \subset \mathsf{R}^2 \subset \cdots \subset \mathsf{R}^n \subset \cdots \subset \mathscr{C}_{00}(\mathsf{N})$ the inductive topology, the inductive vector topology and the inductive locally convex topology coincide. It is obvious that the finest topology on V which induces on every finite dimensional subspace the Euclidean topology is sequential.

Now we look for the convergence of sequences $x_n \xrightarrow[\mathcal{O}]{} 0$. Because of $\mathcal{O} \supset \mathcal{O}_{\text{box}}$, $x_n \xrightarrow[\mathcal{O}]{} 0$ implies $x_n \xrightarrow[\text{box}]{} 0$.

Now let be $x_n \xrightarrow[\text{box}]{} 0$. Then by the characterization of box convergence we have $x_n \in \mathsf{R}^k$ for a suitable k and all n. Thus for an open $U \in \mathcal{O}$ with $0 \in U$ the set $U \cap \mathsf{R}^k$ is Euclidean open and $x_n \in U \cap \mathsf{R}^k$ eventually. Thus $\mathcal{O}$ is sequential and finer than $\mathcal{O}_{\text{box}}$ with the same convergent sequences as $\mathcal{O}_{\text{box}}$. By Proposition 1.5. (3), $\mathcal{O}$ is the sequential refinement of $\mathcal{O}_{\text{box}}$. Now we show that $\mathcal{O}$ is not an FU-topology. We take the following set $S = \bigcup\limits_n \left]\!\!\left] -\frac{1}{n}, \frac{1}{n} \right[\!\!\right[^n$ in $\mathscr{C}_{00}(\mathsf{N})$. We assume that the complement $M = \mathscr{C}_{00}(\mathsf{N}) \setminus S$ contains a sequence x_k which converges in $\mathcal{O}$ to 0. Then there is an index m such that $x_k \in \mathsf{R}^m$ for all k. Now $x_k \notin S$ means that there is a coordinate $\xi_i{}^k$, $i \in \{1, 2, \ldots, m\}$, of x_k with $|\xi_i{}^k| \geqq \frac{1}{m}$. But this contradicts $x_k \xrightarrow[\text{simple}]{} 0$.

Thus no sequence from M converges in $\mathcal{O}$ to 0. But 0 is an adherence point of M. If not, then S would contain 0 as interior point. Now $\mathbf{R}^n \cap \mathrm{Int}_{\mathcal{O}} S \subset \left]-\frac{1}{n}, \frac{1}{n}\right[^n$ for all n, then no point $x \neq 0$ belongs to $\mathrm{Int}_{\mathcal{O}} S$, i.e. $\mathrm{Int}_{\mathcal{O}} S = \emptyset$.

We remark the following. U is open with respect to $\mathcal{O}$ iff for all $x \in U$ there is a sequence $\varepsilon_1 > \varepsilon_2 \cdots > \varepsilon_n > \cdots$, $\varepsilon_i \to 0$, such that $x + \bigcup_i \left]-\varepsilon_i, \varepsilon_i\right[^i \subset U$. But the sets of the form $x + \bigcup_i \left]-\varepsilon_i, \varepsilon_i\right[^i$ are not neighborhoods. □

4. Concluding remarks

In the given lecture at the conference some more results were explained respectively reviewed from the literature and some problems have been proposed. We should like to mention here only a little.

4.1. Diestel's problem. From a paper of A. Grothendieck it is known that in the case of an extremally disconnected compact Hausdorff space X in the space $\mathcal{M}(X)$ of Radon measures on X the two weak topologies $\sigma(\mathcal{M}(X), \mathcal{C}(X))$ and $\sigma(\mathcal{M}(X), \mathcal{M}'(X))$ have the same convergent sequences.

The question is how to characterize those compact Hausdorff spaces X with the same property. Particularly results are known. σ-extremal disconnectedness for X or X to be an F-space are sufficient (vid. Z. Semadeni [16], G. L. Seever [15]).

A necessary condition for X is that its topology $\mathcal{O}$ has a discrete sequential refinement $\mathcal{O}_s$. J. Creutziger [2] has given an example that this is not sufficient.

4.2. Let X be a compact Hausdorff space and $\mathcal{O}$ the weak topology $\sigma(\mathcal{C}(X), \mathcal{M}(X))$ in the space of continuous functions. What is $\mathcal{O}_s$? How depends $\mathcal{O}_s$ from X? Is $\mathcal{O}_s$ a vector topology?

4.3. Find classes of topological vector spaces $(E, \mathcal{O})$ such that $(E, \mathcal{O}_s)$ is not a topological vector space. An example was given by R. M. Dudley [3]. V. Koutník kindly informed the author that there is an example of a topological group $(G, \mathcal{O})$ such that $(G, \mathcal{O}_s)$ is not one, even $\mathcal{O}_s$ is not completely regular (vid. V. Koutník [11] and J. Novák [14]).

4.4. Is there a topological group $(G, \mathcal{O})$ with discrete $\mathcal{O}_s$ but $\mathcal{O}$ is not?

4.5. How to characterize topological spaces $(X, \mathcal{O})$ such that $\mathcal{O}_s$ is metrizable? For topological vector spaces it can be that $\mathcal{O}_s$ is normable but $\mathcal{O}$ is not. But the whole story is not clear.

References

[1] A. Arhangelskij, Some types of quotient mappings and the relations between classes of topological spaces, Engl. translation: Soviet Math. Dokl. 4 (1963), 1726–1729.
[2] J. Creutziger, Grothendieck-Räume, Diplomarbeit, Univ. Greifswald 1978.
[3] R. M. Dudley, On sequential convergence, Trans. Amer. Math. Soc. 112 (1964), 483–507.
[4] R. Engelking, General Topology, PWN, Warszawa 1977.
[5] S. P. Franklin, Spaces in which sequences suffice, Fund. Math. 57 (1965), 107–115.

[6] S. P. FRANKLIN, Spaces in which sequences suffice. II, Fund. Math. **61** (1967), 51—56.
[7] S. P. FRANKLIN and M. RAJAGOPALAN, Some examples in topology, Trans. Amer. Math. Soc. **155** (1971), 305—314.
[8] W. GÄHLER, Grundstrukturen der Analysis I, II, Akademie-Verlag Berlin/Birkhäuser Verlag, Basel und Stuttgart 1977; 1978.
[9] A. GROTHENDIECK, Sur les applications lineaires faiblement compactes d'espaces du type C(K), Cand. J. Math. **5** (1953), 129—173.
[10] S. KAKUTANI and V. KLEE, The finite topology of a linear space, Arch. Math. **14** (1963), 55—58.
[11] V. KOUTNÍK, On sequentially regular convergence spaces, Czech. Math. J. **17** (1967), 232—247.
[12] R. C. MOORE and S. MROWKA, Topologies determined by countable objects, Notices Amer. Math. Soc. **11** (1964), 554.
[13] J. NOVÁK, Sur les espaces (L) et sur les produits cartesiens (L), Publications Facult. des Sciences de l'Univers. Masaryk (1939), 1—28.
[14] J. NOVÁK, On convergence groups, Czech. Math. J. **20** (1970), 357—374.
[15] G. L. SEEVER, Measures on F-spaces, Trans. Amer. Math. Soc. **133** (1968), 267—280.
[16] Z. SEMADENI, On weak convergence of measures and σ-complete Boolean algebras, Colloquium Math. **12** (1964), 229—233.
[17] J. H. WEBB, Sequential convergence in locally convex spaces, Proc. Cambridge Philos. Soc. **64** (1968), 341—364.
[18] A. WILANSKY, Topics in Functional Analysis. Lecture Notes in Math. **45**, Springer-Verlag, Berlin—Heidelberg—New York 1967.

Anschrift des Verfassers:

Ernst-Moritz-Arndt-Universität
Sektion Mathematik
22 Greifswald
Friedrich-Ludwig-Jahn-Str. 15a
DDR

On c-Embedded Sequential Convergence Spaces

ROMAN FRIČ, Žilina (ČSSR), and DARRELL C. KENT, Pullman (U.S.A.)

A *convergence space* is said to be *sequential* (abbreviated to *S-space*) if every convergent filter majorizes some convergent sequential filter. We find necessary and sufficient conditions for a sequential convergence space to be *c-embedded* in the sense of E. BINZ.

Let X be a non-empty set and let X^N be the set of all sequences ranging in X. A relation $\mathfrak{L} \subseteq X^N \times X$ is said to be an *$\mathscr{L}$-structure* for X if the following conditions are satisfied:

($\mathscr{L}_1$) $(\langle x \rangle, x) \in \mathfrak{L}$ for each $x \in X$, where $\langle x \rangle$ is the constant sequence ranging in $\{x\}$.

($\mathscr{L}_2$) $(\langle x_n \rangle, x) \in \mathfrak{L}$ implies $(\langle x_n' \rangle, x) \in \mathfrak{L}$ for every subsequence $\langle x_n' \rangle$ of $\langle x_n \rangle$.

A pair $(X, \mathfrak{L})$, where $\mathfrak{L}$ is an $\mathscr{L}$-structure for X, is said to be an *$\mathscr{L}$-space.*

Let X be a non-empty set and let $\mathbf{F}(X)$ be the set of all proper filters on X. A relation $q \subseteq \mathbf{F}(X) \times X$ is said to be a *convergence structure* for X if the following conditions are satisfied:

(i) $(\dot{x}, x) \in q$ for each $x \in X$, where $\dot{x}$ denotes the fixed ultrafilter generated by $\{x\}$.

(ii) $(\mathscr{F}, x) \in q$ and $\mathscr{G} \supseteq \mathscr{F}$ imply $(\mathscr{G}, x) \in q$.

If, moreover,

(L) $(\mathscr{F}, x) \in q$ and $(\mathscr{G}, x) \in q$ imply $(\mathscr{F} \cap \mathscr{G}, x) \in q$ holds true,

then q is said to be a *Limitierung.* A pair (X, q), where q is a convergence structure for X, is said to be a (filter) *convergence space.*

Let (X, q) be a convergence space. The notations "$x \in q(\mathscr{F})$" and "$\mathscr{F}$ q-converge to x" are equivalent to $(\mathscr{F}, x) \in q$. For $\langle x_n \rangle, \langle y_n \rangle \in X^N$ denote by $\langle x_n \rangle \wedge \langle y_n \rangle$ the sequence $\langle x_1, y_1, x_2, y_2, x_3, y_3, \ldots \rangle$. For each $\langle x_n \rangle \in X^N$ let $\mathscr{F}(\langle x_n \rangle)$ denote the filter $\{F \subseteq X \mid x_n \in F$ for all but finitely many $n \in N\}$. Such filters are called *sequential.* Put $\mathfrak{L}(q) = \{(\langle x_n \rangle, x) \in X^N \times X \mid (\mathscr{F}(\langle x_n \rangle), x) \in q\}$. Let $\mathfrak{L}$ be an $\mathscr{L}$-structure for X. Put $q(\mathfrak{L}) = \{(\mathscr{F}, x) \in \mathbf{F}(X) \times X \mid$ there is $(\langle x_n \rangle, x) \in \mathfrak{L}$ such that $\mathscr{F} \supseteq \mathscr{F}(\langle x_n \rangle)\}$. In the sequel we shall use the following conditions:

($\mathscr{F}\mathscr{L}$) $(\langle x_n \rangle, x) \in \mathfrak{L}$ and $\mathscr{F}(\langle x_n \rangle) = \mathscr{F}(\langle y_n \rangle)$ imply $(\langle y_n \rangle, x) \in \mathfrak{L}$.

($\mathscr{M}\mathscr{L}$) $(\langle x_n \rangle, x) \in \mathfrak{L}$ and $(\langle y_n \rangle, x) \in \mathfrak{L}$ imply $(\langle x_n \rangle \wedge \langle y_n \rangle, x) \in \mathfrak{L}$.

Proposition 1. *Let (X, q) be a convergence space. Then:*

(i) *$\mathfrak{L}(q)$ satisfies $(\mathscr{F}\mathscr{L})$.*

(ii) *If q is a Limitierung, then $\mathfrak{L}(q)$ satisfies $(\mathscr{M}\mathscr{L})$.*

The easy proof is omitted.

Proposition 2. *Let $(X, \mathfrak{L})$ be an $\mathscr{L}$-space. Then:*

(i) *$q(\mathfrak{L})$ is a convergence structure for X.*

(ii) *$\mathfrak{L} = \mathfrak{L}(q(\mathfrak{L}))$ iff $\mathfrak{L}$ satisfies $(\mathscr{F}\mathscr{L})$.*

(iii) *If $\mathfrak{L}$ satisfies $(\mathscr{M}\mathscr{L})$, then $q(\mathfrak{L})$ is a Limitierung.*

The easy proof is omitted.

Proposition 3. *Let (X, q) be a convergence space. Then the following conditions are equivalent:*

(i) *There is an $\mathscr{L}$-structure $\mathfrak{L}$ for X such that $q = q(\mathfrak{L})$.*

(ii) *$(\mathscr{F}, x) \in q$ iff there is a sequential filter $\mathscr{F}(\langle x_n \rangle)$ such that $(\mathscr{F}(\langle x_n \rangle), x) \in q$ and $\mathscr{F} \supseteq \mathscr{F}(\langle x_n \rangle)$.*

The easy proof is omitted.

A convergence space (X, q) satisfying any of the above equivalent conditions is said to be an *S-space*. In this case q is said to be an *S-convergence structure.*

Note that if a convergence structure q for X is supposed to satisfy the additional condition "$(\mathscr{F}, x) \in q$ implies $(\mathscr{F} \cap \dot{x}, x) \in q$", then the notion of an S-space has to be modified as follows (cf. [2]): (X, q) is said to be an S-space if there is an $\mathscr{L}$-structure $\mathfrak{L}$ for X such that $q = \{(\mathscr{F}, x) \in \mathbf{F}(X) \times X \mid \text{there is } (\langle x_n \rangle, x) \in \mathfrak{L} \text{ such that } \mathscr{F} \supseteq (\mathscr{F}(\langle x_n \rangle \cap \dot{x})\}$. In this case condition (ii) in Proposition 3 reads as follows: $(\mathscr{F}, x) \in q$ iff there is a sequential filter $\mathscr{F}(\langle x_n \rangle)$ such that $(\mathscr{F}(\langle x_n \rangle), x) \in q$ and $\mathscr{F} \supseteq (\mathscr{F}(\langle x_n \rangle) \cap \dot{x})$.

A convergence space (X, q) is said to be *pseudo-topological*, resp. q is said to be a *pseudotopology*, if a filter q-converges to a point whenever every finer ultrafilter q-converges to the same point. A pseudotopology is obviously a Limitierung; the converse is generally false, but is shown in the next proposition to be true for S-convergence structures.

For the sake of the reader's convenience, we reproduce here Lemma 1.1 from [4] which will be used in the proof of Proposition 4.

Lemma. *Let $\mathscr{F}$ be a filter on the set X, and let $\{\mathscr{G}_\alpha \mid \alpha \in I\}$ denote the set of all ultrafilters finer than $\mathscr{F}$. For each $\alpha \in I$, choose $F_\alpha \in \mathscr{F}_\alpha$. Then there is a finite subset $\{\alpha_1, \ldots, \alpha_k\} \subseteq I$ such that $\bigcup \{F_{\alpha_i} \mid i = 1, \ldots, k\} \in \mathscr{F}$.*

Proposition 4. *An S-convergence structure for X is a pseudotopology iff q is a Limitierung.*

Proof. It suffices to prove that if q is an S-structure satisfying condition (L), then q is a pseudotopology. Let $\mathscr{F} \in \mathbf{F}(X)$, let $\{\mathscr{G}_\alpha \mid \alpha \in I\}$ denote the set of all ultrafilters finer than $\mathscr{F}$, and let x be a point in X such that each $\mathscr{G}_\alpha$, $\alpha \in I$, q-converges to x. We are to prove that $\mathscr{F}$ q-converges to x.

Since (X, q) is an S-space, for each $\alpha \in I$ there is a sequence $\langle x_n^\alpha \rangle \in X^N$ such that $(\mathscr{F}(\langle x_n^\alpha \rangle), x) \in q$ and $\mathscr{G}_\alpha \supseteq \mathscr{F}(\langle x_n^\alpha \rangle)$. Put $A_\alpha = \{x_n^\alpha \mid n \in N\}$. Then, for each $\alpha \in I$, $A_\alpha \in \mathscr{G}_\alpha$ and $\bigcup \{A_\alpha \mid \alpha \in I\} \in \mathscr{F}$. By Lemma 1.1 in [4], there are finitely many indices

$\alpha_1, \ldots, \alpha_k$, such that $B = \bigcup \{A_{\alpha_i} \mid i = 1, \ldots, k\} \in \mathscr{F}$. Let $\langle z_n \rangle$ denote the sequence $\langle x_1^{\alpha_1}, x_1^{\alpha_2}, \ldots, x_1^{\alpha_k}, x_2^{\alpha_1}, \ldots, x_2^{\alpha_k}, \ldots, x_n^{\alpha_1}, \ldots, x_n^{\alpha_k}, \ldots \rangle$. Then $\mathscr{F}(\langle z_n \rangle) = \bigcap \{\mathscr{F}(\langle x_n^{\alpha_i} \rangle) \mid i = 1, \ldots, k\}$. Since $(\mathscr{F}(\langle x_n^{\alpha_i} \rangle), x) \in q$, $i = 1, \ldots, k$, by (L), we have $(\mathscr{F}(\langle z_n \rangle), x) \in q$. Now, it suffices to prove that $\mathscr{F} \supseteq \mathscr{F}(\langle x_n \rangle)$. This will imply, by Proposition 3, that $\mathscr{F}$ q-converges to x.

Let $\mathscr{G}$ be an ultrafilter finer than $\mathscr{F}$. Then $\bigcup \{A_{\alpha_i} \mid i = 1, \ldots, k\} \in \mathscr{G}$ implies that for some j, $1 \leq j \leq k$, we have $\mathscr{G} \supseteq \mathscr{F}(\langle x_n^{\alpha_j} \rangle) \supseteq \mathscr{F}(\langle z_n \rangle)$. Consequently, $\mathscr{F} \supseteq \mathscr{F}(\langle z_n \rangle)$. This completes the proof.

Let (X, q) be a convergence space, let $C(X, q)$ denote the set of all real-valued continuous functions on (X, q). Let $C_c(X, q)$ denote the set $C(X, q)$ equipped with the *continuous convergence structure*, which is the coarsest convergence structure for $C(X, q)$ relative to which the mapping w of $C(X, q) \times (X, q)$ into the real line defined by $w(f, x) = f(x)$ is continuous. Notice that the continuous convergence structure is a Limitierung. We say that (X, q) is *c-embedded* if the evaluation mapping ev_X of (X, q) into $C_c(C_c(X, q))$ defined by $\mathrm{ev}_X(x) = \Phi_x$, where $\Phi_x(f) = f(x)$, is an embedding. For a more extensive discussion of $C_c(X, q)$ and c-embedded spaces, the reader if referred to [1] and [3].

Let (X, q) be a convergence space. Recall that the *completely regular modification* ωq of q is the weak topology for X induced by $C(X, q)$, (X, q) is HAUSDORFF if every filter q-converges to at most one point, (X, q) is ω-HAUSDORFF if $(X, \omega q)$ is HAUSDORFF (equivalently, $C(X, q)$ separates point of X), (X, q) is *ω-regular* if $\mathrm{cl}_{\omega q}\mathscr{F}$ q-converges to x whenever $\mathscr{F}$ q-converges to x ($\mathrm{cl}_{\omega q}$ denotes the closure operator relative to $(X, \omega q)$).

Proposition 5. *A convergence space is c-embedded iff it is Hausdorff, ω-regular, and pseudo-topological.*

Proof. The assertion is valid in the case when q is a Limitierung (cf. Theorem 2.4 in [3]). The general case is a trivial consequence.

Theorem. *An S-space (X, q) is c-embedded iff the following conditions are satisfied:*

(i) *$C(X, q)$ separates points of X.*

(ii) *q is a Limitierung.*

Proof. *The necessity.* The space (X, q) inherits properties (i) and (ii) from $C_c(C_c(X, q))$.

The sufficiency. The S-space (X, q) is, by (i), HAUSDORFF and, by Proposition 4, pseudo-topological. By Proposition 5, it remains to prove that (X, q) is ω-regular. Let $(\mathscr{F}, x) \in q$. Since (X, q) is an S-space, there is, by Proposition 3, a sequence $\langle x_n \rangle \in X^N$ such that $\mathscr{F}(\langle x_n \rangle)$ q-converges to x and $\mathscr{F} \supseteq \mathscr{F}(\langle x_n \rangle)$. By (ii), the filter $\mathscr{F}(\langle x_n \rangle) \cap \dot{x}$ q-converges to x. Since $\omega q \supseteq q$ and the completely regular space $(X, \omega q)$ is, by (i), HAUSDORFF, it is easy to see that $\mathrm{cl}_{\omega q}\mathscr{F} \supseteq \mathrm{cl}_{\omega q}(\mathscr{F}(\langle x_n \rangle) \cap \dot{x}) = \mathscr{F}(\langle x_n \rangle) \cap \dot{x}$. Consequently, $(\mathrm{cl}_{\omega q}\mathscr{F}, x) \in q$. This completes the proof.

Let $(X, \mathfrak{L})$ be an $\mathscr{L}$-space and let $C(X, \mathfrak{L})$ be the set of all real-valued $\mathscr{L}$-continuous functions on $(X, \mathfrak{L})$, i.e. $f \in C(X, \mathfrak{L})$ whenever $(\langle x_n \rangle, x) \in \mathfrak{L}$ implies $f(x) = \lim f(x_n)$. It is easy to show that $C(X, \mathfrak{L}) = C(X, q(\mathfrak{L})) = C(X, \mathfrak{L}(q(\mathfrak{L})))$.

Corollary 1. *Let $(X, \mathfrak{L})$ be an $\mathscr{L}$-space. Then $(X, q(\mathfrak{L}))$ is c-embedded iff the following conditions are satisfied:*

(i) *$C(X, \mathfrak{L})$ separates points of X.*

(ii) *If $(\langle x_n \rangle, x) \in \mathfrak{L}$ and $(\langle y_n \rangle, x) \in \mathfrak{L}$, then there is a sequence $\langle z_n \rangle \in X^N$ such that $(\langle z_n \rangle, x) \in \mathfrak{L}$ and $\mathscr{F}(\langle x_n \rangle) \cap \mathscr{F}(\langle y_n \rangle) \supseteq \mathscr{F}(\langle z_n \rangle)$.*

Corollary 2. *Let $(X, \mathfrak{L})$ be an $\mathscr{L}$-space satisfying $(\mathscr{F}\mathscr{L})$. Then $\big(X, q(\mathfrak{L})\big)$ is c-embedded iff the following conditions are satisfied:*

(i) *$C(X, \mathfrak{L})$ separates points of X.*

(ii) *$\mathfrak{L}$ satisfies $(\mathscr{M}\mathscr{L})$.*

References

[1] E. Binz, Continuous convergence in $C(X)$, Lecture Notes in Mathematics **469**, Springer-Verlag, Berlin—Heidelberg—New York 1975.

[2] R. Frič, On convergence spaces and groups, Math. Slovaca **27** (1977), 323—330.

[3] D. Kent, K. McKennon, G. Richardson, and M. Schroder, Continuous convergence in $C(X)$, Pacific. J. Math. **52** (1974), 271—279.

[4] D. C. Kent and G. D. Richardson, Locally compact convergence spaces, Michigan Math. J. **22** (1975), 353—360.

Anschriften der Verfasser:

Katedra matematiky
Fakulta strojnicka a elektrotechnická
Vysoká škola dopravná
Marxa-Engelsa 25
01088 Žilina
ČSSR

Department of Pure and Applied Mathematics
Washington State University
Pullmann, Washington 99163
U.S.A.

Sequential Structures

Roman Frič, Žilina, and Václav Koutník, Prague

Introduction

In this paper sequential structures are considered. By sequential structures we understand structures in the sense of N. Bourbaki generated by one auxiliary basic set N (the set of natural numbers). It seems reasonable, however, to focus on three of them: $\mathscr{L}$-structures (introduced by M. Fréchet), $\mathscr{UL}$-structures (introduced by A. Goetz), and $\mathscr{CL}$-structures (introduced by J. M. Irwin and D. C. Kent). These concepts arose as natural generalizations of metric spaces and are intuitively clear. They play a vital role in studying virtually any other sequential structure.

The paper is divided into four sections. Section one contains basic definitions and seeks a unifying approach to various sequential structures. It describes the mutual relations among the three basic structures mentioned above and their relations to other continuous structures. In Section two axiomatics for each of the three basic sequential structures is given and analysed from the viewpoint of these relations. Section three gives a survey of the basic categorical properties of these structures. In the last section sequential structures projectively generated by various families of continuous functions and their completions are considered. Each section contains references to pertinent papers; however, no attempt has been made to provide a complete list. Remarks of a historical nature are given and some problems and prospects for future research are mentioned.

Notation

N	denotes the set of natural numbers
R	denotes the set of real numbers
X, Y	with or without indices ... denote the underlying set of a structure
ξ, η	or $\langle x_n \rangle$, $\langle y_n \rangle$, ... denote sequences of points
$\xi(n) = x_n$	denotes the n-th term of the sequence $\xi = \langle x_n \rangle$
s, t	denote monotone mappings of N into N
$\xi \circ s = \langle x_n' \rangle$	denotes a subsequence of the sequence $\xi = \langle x_n \rangle$
$\xi \wedge \eta$	denotes the sequence whose $(2n-1)$-th term is $\xi(n)$ and $2n$-th term is $\eta(n)$
$\langle x \rangle$	denotes the constant sequence
$\mathsf{P}X$	denotes the power set $\{A \mid A \subseteq X\}$ of X

$\mathsf{F}X$	denotes the set of all filters in X
$\mathscr{F}(\xi)$	denotes the filter $\{F \subseteq X \mid \xi(n) \in F$ for all but finitely many $n \in N\}$
$\dot{x} = \mathscr{F}(\langle x \rangle)$	denotes the fixed ultrafilter generated by $\{x\}$
$\mathfrak{L}, \mathfrak{M}$	with or without indices ... denote $\mathscr{L}$-structures
$\mathfrak{n}, \mathfrak{m}$	with or without indices ... denote $\mathscr{U}\mathscr{L}$-structures
L, M	with or without indices ... denote $\mathscr{C}\mathscr{L}$-structures
$C(X)$	denotes the set of all real-valued continuous functions on (X, u)
$C^*(X)$	denotes the set of all bounded real-valued continuous functions on (X, u)
$C_s(X)$	denotes the set of all real-valued $\mathfrak{L}(u)$-continuous functions on (X, u)
ω_1	denotes the first uncountable ordinal

1. Review of some basic concepts

1.1. $\mathscr{L}$-structures

The notion of an $\mathscr{L}$-space was introduced by M. Fréchet in [13]. He assumes the axioms on uniqueness of limits, constant sequences, and subsequences. The important condition of maximality was defined by P. Urysohn in [1]. In his book [14] M. Fréchet showed that the maximal convergence is largest in the sense that no convergent sequences can be added without changing the induced closure operator. Various convergences in closure spaces were considered by F. Hausdorff in [32] (cf. [39]). Since then $\mathscr{L}$-structures have been investigated by many authors mainly as tools for generating topological and other continuous structures (see [41, 6, 7, 8, 42, 43, 44, 10, 4, 2, 3]). M. Dolcher [9] studied $\mathscr{L}$-structures from the categorical point of view. He considered only maximal convergences but did not assume the axiom on uniqueness of limits.

Definition. Let $X \neq \emptyset$ and let $\mathfrak{L} \subseteq X^N \times X$. Consider the following axioms:

($\mathscr{L}_0$) $(\xi, x) \in \mathfrak{L}$ and $(\xi, y) \in \mathfrak{L}$ imply $x = y$.

($\mathscr{L}_1$) $(\langle x \rangle, x) \in \mathfrak{L}$ for each $x \in X$.

($\mathscr{L}_2$) $(\xi, x) \in \mathfrak{L}$ implies $(\xi \circ s, x) \in \mathfrak{L}$ for every subsequence $\xi \circ s$ of ξ.

($\mathscr{L}_3$) Let $(\xi, x) \in X^N \times X$. If for each s there exists t such that $(\xi \circ s \circ t, x) \in \mathfrak{L}$, then $(\xi, x) \in \mathfrak{L}$.

We say that $\mathfrak{L}$ is an *$\mathscr{L}$-structure* if it satisfies axioms ($\mathscr{L}_1$) and ($\mathscr{L}_2$). An $\mathscr{L}$-structure is said to be *single-valued* if it satisfies ($\mathscr{L}_0$). An $\mathscr{L}$-structure $\mathfrak{L}$ is said to be *maximal* and denoted by $\mathfrak{L}^*$ if it satisfies ($\mathscr{L}_3$).

Remark. The following axioms are useful for the study of relations between $\mathscr{L}$-structures and other continuous structures:

($\mathscr{S}\mathscr{L}$) Let $\xi \in X^N$ and denote $\operatorname{Lim} \xi = \{x \in X \mid (\xi, x) \in \mathfrak{L}\}$. If $(\eta, y) \in \mathfrak{L}$ and $\eta(n) \in \operatorname{Lim} \xi$ for each $n \in N$, then $y \in \operatorname{Lim} \xi$.

($\mathscr{F}\mathscr{L}$) $(\xi, x) \in \mathfrak{L}$ and $\mathscr{F}(\xi) = \mathscr{F}(\eta)$ imply $(\eta, x) \in \mathfrak{L}$.

($\mathscr{M}\mathscr{L}$) $(\xi, x) \in \mathfrak{L}$ and $(\eta, x) \in \mathfrak{L}$ imply $(\xi \wedge \eta, x) \in \mathfrak{L}$.

Clearly ($\mathscr{L}_0$) implies ($\mathscr{S}\mathscr{L}$) and ($\mathscr{L}_3$) implies both ($\mathscr{F}\mathscr{L}$) and ($\mathscr{M}\mathscr{L}$). None of the converse implications holds.

M. DOLCHER [9] introduced the following weaker forms of ($\mathscr{L}_0$):

($\mathscr{L}_0'$) If $(\langle x\rangle, y) \in \mathfrak{L}$, then $x = y$.

We use this axiom when we consider relations between $\mathscr{L}$-structures and topological structures. It is also important for establishing isomorphisms between subcategories of $\mathscr{L}$-structurs and topological structures.

Remark. If $(\xi, x) \in \mathfrak{L}$, then ξ is said to $\mathfrak{L}$-converge to x, in symbols $\mathfrak{L}\text{-}\lim \xi = x$ or $\xi \xrightarrow{\mathfrak{L}} x$ (the letter $\mathfrak{L}$ being sometimes omitted). An $\mathscr{L}$-structure $\mathfrak{L}$ for X is sometimes called a convergence $\mathfrak{L}$ in X.

Examples. Important examples of $\mathscr{L}$-structures are: convergence of sequences in topological spaces, convergence of sequences of functions in function spaces, convergence of sequences of sets in systems of sets.

1.2. $\mathscr{U}\mathscr{L}$-structures

$\mathscr{U}\mathscr{L}$-structures were introduced by A. GOETZ ([29], cf. [30]). The relations between $\mathscr{U}\mathscr{L}$-structures and proximities were studied by V. Z. POLJAKOV [49] who developed the theory of completion for $\mathscr{U}\mathscr{L}$-structures. V. KOUTNÍK showed in [38] that the sequential envelope of a sequential convergence space can be considered as a completion for a particular $\mathscr{U}\mathscr{L}$-structure. A general theory of $\mathscr{U}\mathscr{L}$-structures and their relation to proximity and uniform structures was given by A. GOETZ in [31]. Some open problems mentioned there were solved in [17]. The relations between completions of $\mathscr{U}\mathscr{L}$-structures and completions of corresponding $\mathscr{L}$-structures resp. $\mathscr{C}\mathscr{L}$-structures are described in papers by R. FRIČ [18] resp. J. M. IRWIN and D. C. KENT [34, 35].

Definition. Let $X \neq \emptyset$ and let $\mathfrak{n} \subseteqq X^N \times X^N$. Consider the following axioms:

($\mathscr{U}\mathscr{L}_0$) $(\langle x\rangle, \langle y\rangle) \in \mathfrak{n}$ implies $x = y$.

($\mathscr{U}\mathscr{L}_1$) $\mathfrak{n}$ is an equivalence relation for X^N.

($\mathscr{U}\mathscr{L}_2$) $(\xi, \eta) \in \mathfrak{n}$ implies $(\xi \circ s, \eta \circ s) \in \mathfrak{n}$ for every s.

($\mathscr{U}\mathscr{L}_3$) Let $(\xi, \eta) \in X^N \times X^N$. If for every s there exists t such that $(\xi \circ s \circ t, \eta \circ s \circ t) \in \mathfrak{n}$, then $(\xi, \eta) \in \mathfrak{n}$.

We say that $\mathfrak{n}$ is a *$\mathscr{U}\mathscr{L}$-structure* if it satisfies axioms ($\mathscr{U}\mathscr{L}_1$) and ($\mathscr{U}\mathscr{L}_2$). A $\mathscr{U}\mathscr{L}$-structure is said to be *separated* if it satisfies ($\mathscr{U}\mathscr{L}_0$). A $\mathscr{U}\mathscr{L}$-structure $\mathfrak{n}$ is said to be *maximal* and denoted by $\mathfrak{n}^*$ if it satisfies ($\mathscr{U}\mathscr{L}_3$).

Remark. The following axiom is useful for the study of relations between $\mathscr{U}\mathscr{L}$-structures and proximity structures:

($\mathscr{U}\mathscr{L}^\square$) Let $(\xi, \eta) \in X^N \times X^N$. If for every s there exist t and u such that $(\xi \circ s \circ t, \eta \circ s \circ u) \in \mathfrak{n}$, then $(\xi, \eta) \in \mathfrak{n}$.

Clearly ($\mathscr{U}\mathscr{L}^\square$) implies ($\mathscr{U}\mathscr{L}_3$). An example of a $\mathscr{U}\mathscr{L}$-structure satisfying ($\mathscr{U}\mathscr{L}_3$) but not ($\mathscr{U}\mathscr{L}^\square$) is not known.

Remark. If $(\xi, \eta) \in \mathfrak{n}$, then the sequences ξ and η are said to be $\mathfrak{n}$-near, in symbols $\xi \mathfrak{n} \eta$.

Examples The set of all pairs of adjacent sequences (cf. [8]) in a uniform space and the set of all pairs of contiguous sequences of probability measures (cf. [51)] are $\mathscr{U}\mathscr{L}$-structures.

1.3. $\mathscr{C}\mathscr{L}$-structures

In completing $\mathscr{L}$- and $\mathscr{U}\mathscr{L}$-structures the ideal points are in fact limits of Cauchy sequences. A general theory of such sequences from the axiomatic point of view was developed by J. M. Irwin and D. C. Kent [34, 35]. Using a suitable modification of completing filter structures (cf. [28]) they developed a completion theory for $\mathscr{C}\mathscr{L}$- and $\mathscr{U}\mathscr{L}$-structures. The relations between completions of $\mathscr{C}\mathscr{L}$- and $\mathscr{L}$-structures were investigated by R. Frič and D. C. Kent [23].

Definition. Let $X \neq \emptyset$ and $L \subseteq X^N$. Consider the following axioms:

$(\mathscr{C}\mathscr{L}_0)$ $\langle x \rangle \wedge \langle y \rangle \in L$ implies $x = y$.

$(\mathscr{C}\mathscr{L}_1)$ $\langle x \rangle \in L$ for each $x \in X$.

$(\mathscr{C}\mathscr{L}_2)$ $\xi \in L$ implies $\xi \circ s \in L$ for every subsequence $\xi \circ s$ of ξ.

$(\mathscr{C}\mathscr{L}_3)$ $\xi, \eta \in L$ and $\xi \circ s = \eta \circ t$ for some s, t imply $\xi \wedge \eta \in L$.

$(\mathscr{C}\mathscr{L}_4)$ Let $\xi \in X^N$. If conditions

(a) For every s there exists t such that $\xi \circ s \circ t \in L$.
and
(b) $\xi \circ s \in L$ and $\xi \circ t \in L$ imply $\xi \circ s \wedge \xi \circ t \in L$.

are satisfied, then $\xi \in L$.
We say that L is a *$\mathscr{C}\mathscr{L}$-structure* if it satisfies axioms $(\mathscr{C}\mathscr{L}_1)$, $(\mathscr{C}\mathscr{L}_2)$, and $(\mathscr{C}\mathscr{L}_3)$. A $\mathscr{C}\mathscr{L}$-structure is said to be *separated* if it satisfies $(\mathscr{C}\mathscr{L}_0)$. A $\mathscr{C}\mathscr{L}$-structure L is said to be *maximal* and denoted by L^* if it satisfies $(\mathscr{C}\mathscr{L}_4)$.

Remark. The following axiom is useful for the study of relations between $\mathscr{C}\mathscr{L}$-structures and various filter structures:

$(\mathscr{F}\mathscr{C}\mathscr{L})$ $\xi \in L$ and $\mathscr{F}(\xi) = \mathscr{F}(\eta)$ imply $\eta \in L$.

Remark. If $\xi \in L$, then ξ is said to be L-Cauchy (the letter L being sometimes omitted).

Examples. The set of all Cauchy sequences in a metric space is a typical example of a $\mathscr{C}\mathscr{L}$-structure.

1.4. $\mathscr{N}$-structures

All sequential structures introduced so far are special cases of the notion of a structure generated by one auxiliary basic set N introduced by N. Bourbaki in [5]. For our purpose it is convenient to use the form given by Z. Semadeni [53] and called $\{A_1, A_2, \ldots, A_n\}$-structure. In our case $n = 1$ and $A_1 = N$. In conformity with our notation we shall denote $\{N\}$-structures by $\mathscr{N}$-structures.

Definition. Let $X \neq \emptyset$ and let $\mathfrak{S}(X, N)$ be the smallest class containing X and N and closed under the formation of power sets and finite products. We say that $\mathfrak{N}$ is an *$\mathcal{N}$-structure* if there is a set $S \in \mathfrak{S}(X, N)$ such that $\mathfrak{N} \in S$.

Remark. An $\mathcal{L}$-structure $\mathfrak{L}$ for X is an $\mathcal{N}$-structure since $\mathfrak{L} \in \mathsf{P}(\mathsf{P}(N \times X) \times X)$. A $\mathcal{UL}$-structure $\mathfrak{n}$ for X is an $\mathcal{N}$-structure since $\mathfrak{n} \in \mathsf{P}(\mathsf{P}(N \times X) \times \mathsf{P}(N \times X))$. A $\mathcal{CL}$-structure L for X is an $\mathcal{N}$-structure since $L \in \mathsf{P}(\mathsf{P}(N \times X))$.

1.5. $\mathcal{N}$-spaces

Definition. Let $X \neq \emptyset$ and let $\mathfrak{N}$ be an $\mathcal{N}$-structure for X. Then $(X, \mathfrak{N})$ is said to be an *$\mathcal{N}$-space.*

Remark. The important special cases of $\mathcal{N}$-spaces are $\mathcal{L}$-spaces $(X, \mathfrak{L})$, $\mathcal{UL}$-spaces $(X, \mathfrak{n})$, and $\mathcal{CL}$-spaces (X, L). The notions of an $\mathcal{N}$-continuous mapping, $\mathcal{N}$-subspace, $\mathcal{N}$-product, etc. can be introduced in categorical terms. First we define $\mathcal{N}$-objects and $\mathcal{N}$-morphisms and the other categorical concepts. However, to avoid excessive formalism we shall restrict ourselves to $\mathcal{L}$-spaces. The corresponding notions for $\mathcal{UL}$-spaces and $\mathcal{CL}$-spaces are defined analogously.

Definition. Let $(X_1, \mathfrak{L}_1)$ and $(X_2, \mathfrak{L}_2)$ be $\mathcal{L}$-spaces. We say that $f\colon X_1 \to X_2$ is an *$\mathcal{L}$-continuous mapping* if $(\langle x_n \rangle, x) \in \mathfrak{L}_1$ implies $(\langle f(x_n) \rangle, f(x)) \in \mathfrak{L}_2$. If $X_1 \subseteq X_2$ and $\mathfrak{L}_1 = \{(\langle x_n \rangle, x) \in \mathfrak{L}_2 \mid x_n, x \in X_1\}$, then $(X_1, \mathfrak{L}_1)$ is called an *$\mathcal{L}$-subspace* of $(X_2, \mathfrak{L}_2)$.

Let $\{(X_\iota, \mathfrak{L}_\iota) \mid \iota \in I\}$ be a set of $\mathcal{L}$-spaces. Then $(X, \mathfrak{L}) = \prod_{\iota \in I} (X_\iota, \mathfrak{L}_\iota)$ is an *$\mathcal{L}$-product* if $X = \prod_{\iota \in I} X_\iota$ and $\mathfrak{L} = \{(\langle (x_n^\iota) \rangle, (x^\iota)) \mid (\langle x_n^\iota \rangle, x^\iota) \in \mathfrak{L}_\iota \text{ for each } \iota \in I\}$.

1.6. Mutual relations between $\mathcal{L}$-, $\mathcal{UL}$-, and $\mathcal{CL}$-structures

Denote

$$\mathfrak{L}(\mathfrak{n}) = \{(\xi, x) \in X^N \times X \mid (\xi, \langle x \rangle) \in \mathfrak{n}\},$$

$$\mathfrak{L}(L) = \{(\xi, x) \in X^N \times X \mid \xi \wedge \langle x \rangle \in L\},$$

$$\mathfrak{n}(L) = \{(\xi, \eta) \in X^N \times X^N \mid \xi = \eta \quad \text{or} \quad \xi \wedge \eta \in L\},$$

$$\mathfrak{n}(\mathfrak{L}) = \{(\xi, \eta) \in X^N \times X^N \mid \xi = \eta \quad \text{or} \quad (\xi, x), (\eta, x) \in \mathfrak{L} \quad \text{for some } x \in X\},$$

$$L(\mathfrak{n}) = \{\xi \in X^N \mid (\xi, \xi \circ s) \in \mathfrak{n} \quad \text{for each subsequence} \quad \xi \circ s \text{ of } \xi\},$$

$$L(\mathfrak{L}) = \{\xi \in X^N \mid (\xi, x) \in \mathfrak{L} \text{ for some } x \in X\}.$$

Then $\mathfrak{L}(\mathfrak{n})$ and $\mathfrak{L}(L)$ are $\mathcal{L}$-structures, $\mathfrak{n}(L)$ and $\mathfrak{n}(\mathfrak{L})$ are $\mathcal{UL}$-structures and, under certain assumptions, $L(\mathfrak{n})$ and $L(\mathfrak{L})$ are $\mathcal{CL}$-structures.

If $\mathfrak{L}_1$ and $\mathfrak{L}_2$ are single-valued maximal $\mathcal{L}$-structures, then $L(\mathfrak{L}_1)$ and $L(\mathfrak{L}_2)$ are separated maximal $\mathcal{CL}$-structures and $L(\mathfrak{L}_1) \neq L(\mathfrak{L}_2)$ whenever $\mathfrak{L}_1 \neq \mathfrak{L}_2$. Similarly, if L_1 and L_2 are separated maximal $\mathcal{CL}$-structures, then $\mathfrak{n}^*(L_1)$ and $\mathfrak{n}^*(L_2)$ are separated maximal $\mathcal{UL}$-structures and $\mathfrak{n}^*(L_1) \neq \mathfrak{n}^*(L_2)$ whenever $L_1 \neq L_2$. From this point of view $\mathcal{UL}$-structures are more general than $\mathcal{CL}$-structures which in turn are more general than $\mathcal{L}$-structures (see 3.2 and 3.3).

The relation between $\mathcal{UL}$-structures and $\mathcal{L}$-structures is investigated in [30] (cf. also [38, 17]). The correspondences between different structures are studied in [34].

1.7. Relations between $\mathscr{L}$-structures and other structures

Let $(X, \mathfrak{L})$ be an $\mathscr{L}$-space. For each $A \subseteqq X$ and each ordinal α we define

$$\lambda^0 A = A,$$
$$\lambda A = \{x \in X \mid \text{there exists } (\langle x_n \rangle, x) \in \mathfrak{L} \text{ such that } \bigcup (x_n) \subseteqq A\},$$
$$\lambda^\alpha A = \bigcup_{\beta<\alpha} \lambda(\lambda^\beta A).$$

For each α the mapping $\lambda^\alpha \colon \mathsf{P}X \to \mathsf{P}X$ satisfies the first three Kuratowski axioms for a closure operator: (i) $\lambda^\alpha \emptyset = \emptyset$, (ii) $A \subseteqq \lambda^\alpha A$, (iii) $\lambda^\alpha(A \cup B) = \lambda^\alpha A \cup \lambda^\alpha B$. The mapping λ^{ω_1} moreover satisfies also the fourth axiom: (iv) $\lambda^{\omega_1}(\lambda^{\omega_1} A) = \lambda^{\omega_1} A$.

The closure spaces (X, λ) are called *sequential convergence spaces* and are studied in detail by J. Novák in [43] and [44]. The topological spaces (X, λ^{ω_1}) are called *sequential spaces* and in the special case of $\lambda = \lambda^{\omega_1}$ they are called *Fréchet spaces* [12] (cf. [11]). If we want to indicate the particular convergence $\mathfrak{L}$ which induced the closure λ resp. λ^{ω_1} we denote them by $(X, \lambda(\mathfrak{L}))$ resp. $(X, \lambda^{\omega_1}(\mathfrak{L}))$.

On the other hand if a closure space (X, u) is given, then

$$\mathfrak{L}(u) = \{(\langle x_n \rangle, x) \in X^N \times X \mid \text{if } U \text{ is any neighborhood of } x, \text{ then } x_n \in U \text{ for all but finitely many } n \in N\}$$

is an $\mathscr{L}$-structure for X.

Another type of space closely related to $\mathscr{L}$-structures is the S-convergence filter space. For $\xi \in X^N$ denote

$$\mathscr{F}(\xi) = \{F \subseteqq X \mid \xi(n) \in F \text{ for all but finitely many } n \in N\},$$

i.e. $\mathscr{F}(\xi)$ is the elementary filter associated with ξ. Define a mapping $q \colon \mathsf{F}X \to \mathsf{P}X$ as follows:

$$x \in q(\mathscr{F}) \quad \text{if there exists} \quad (\xi, x) \in \mathfrak{L} \quad \text{such that} \quad \mathscr{F} \supseteqq \mathscr{F}(\xi) \cap \dot{x}.$$

The mapping $q = q(\mathfrak{L})$ is called an *S-convergence structure.* It satisfies filter convergence structure axioms [33]: (i) $x \in q(\dot{x})$ for each $x \in X$, (ii) $x \in q(\mathscr{F})$ and $\mathscr{G} \supseteqq \mathscr{F}$ imply $x \in q(\mathscr{G})$, (iii) $x \in q(\mathscr{F})$ implies $x \in q(\mathscr{F} \cap \dot{x})$. Some properties of S-convergence structures are discussed in a talk given by R. Frič and D. C. Kent at this summer school.

On the other hand, if a filter convergence space (X, q) is given, then

$$\mathfrak{L}(q) = \{(\xi, x) \in X^N \times X \mid x \in q(\mathscr{F}(\xi))\}$$

is an $\mathscr{L}$-structure.

Mutual relations between $\mathscr{L}$-structures and filter convergence structures are studied in [21].

1.8. Relations between $\mathscr{U}\mathscr{L}$-structures and other structures

Let $(X, \mathfrak{n})$ be a $\mathscr{U}\mathscr{L}$-space. Denote

$$\mathscr{U}(\mathfrak{n}) = \{U \subseteqq X \times X \mid \text{for each } (\xi, \eta) \in \mathfrak{n} \text{ there exists } k \in N \text{ such that for each } n \geqq k \text{ we have } (\xi(n), \eta(n)) \in U\},$$

$$p(\mathfrak{n}) = \{(A, B) \in \mathsf{P}X \times \mathsf{P}X \mid \text{there exists } (\xi, \eta) \in \mathfrak{n} \text{ such that } \bigcup (\xi(n)) \subseteqq A, \bigcup (\eta(n)) \subseteqq B\}.$$

Then $\mathscr{U}(\mathfrak{n})$ is a semi-uniformity and $p(\mathfrak{n})$ is a proximity in the sense of Čech [8]. Conversely, every sequential semi-uniformity $\mathscr{U}$, resp. every sequential proximity p, for X induces a $\mathscr{U}\mathscr{L}$-structure

$$\mathfrak{n}(\mathscr{U}) = \{(\xi, \eta) \in X^N \times X^N \mid \text{for each } U \in \mathscr{U} \text{ there exists } k \in N \text{ such that for each } n \geqq k \text{ we have } (\xi(n), \eta(n)) \in U\},$$

resp.

$$\mathfrak{n}(p) = \left\{(\xi, \eta) \in X^N \times X^N \mid \text{for every } s \text{ we have } \left(\bigcup_{n=1}^{\infty} ((\xi \circ s)(n))\right) p \left(\bigcup_{n=1}^{\infty} ((\eta \circ s)(n))\right)\right\}.$$

The relations between these structures are treated in detail in [31, 49, 17, 34].

1.9. Relations between $\mathscr{C}\mathscr{L}$-structures and other structures

Let (X, L) be a $\mathscr{C}\mathscr{L}$-space. Denote

$$C(L) = \{\mathscr{F} \in \mathsf{F}X \mid \mathscr{F} \supseteqq \mathscr{F}(\xi) \text{ for some } \xi \in L\}.$$

Then, under certain assumptions, $(X, C(L))$ is a Cauchy space (cf. [40, 36, 50]). Conversely, if (X, C) is a Cauchy space, then

$$L(C) = \{\xi \in X^N \mid \mathscr{F}(\xi) \in C\}$$

is a $\mathscr{C}\mathscr{L}$-structure. The properties of structures of the type $C(L)$ are studied in [25]. Completions of this special type of Cauchy spaces are also investigated there.

1.10. $\mathscr{L}$-groups

Definition. Let $(G, +)$ be a (commutative) group. Let $\mathfrak{L}$ be an $\mathscr{L}$-structure for G such that

$$(\mathscr{L}\mathscr{G})\quad (\xi, x) \in \mathfrak{L} \text{ and } (\eta, y) \in \mathfrak{L} \text{ imply } (\langle \xi(n) - \eta(n)\rangle, x - y) \in \mathfrak{L}.$$

Then $(G, \mathfrak{L}, +)$ is said to be a *(commutative) $\mathscr{L}$-group* (cf. [52], [46]).

Every $\mathscr{L}$-group induces in a natural way a $\mathscr{C}\mathscr{L}$-structure and a $\mathscr{U}\mathscr{L}$-structure. For $\mathfrak{L} = \mathfrak{L}^*$ there is a one-to-one correspondence between $\mathscr{L}$-groups and sequential convergence groups of J. Novák. Constructions similar to those for topological groups (e.g. group uniformities, completions, etc.) are possible also for $\mathscr{L}$-groups. However, J. Novák constructed in [46] an $\mathscr{L}$-group satisfying only very "low" separation axioms (e.g. it is not regular), and even the $\mathscr{L}$-group of rational numbers has at least two different completions ([48]).

The relation between $\mathscr{L}$-groups and filter convergence groups is studied in [21]. Further research in this area seems to be promising.

1.11. Concluding remarks

In conclusion we point out that until recently the theories of sequential and filter structures were developed separately. The papers [22, 21] show that methods, constructions, and examples developed in one theory can be exploited in the other. Constructions of the more general filter convergence theory can be applied to sequential structures while "sequential" examples and counterexamples can usually be handled more easily and can be modified so as to be used in the realm of filter spaces.

Similar arguments show that relations between various continuous structures provide a useful tool for the study of a particular structure (cf. [26, 23, 17]).

2. Discussion of axioms for sequential structure

2.1. Axioms for $\mathscr{L}$-structures

The axiom ($\mathscr{L}_0$)

The importance of the axiom ($\mathscr{L}_0$) on uniqueness of limits follows from the validity of a general statement on extension of continuous mappings which we call the Extension of identity principle.

First we give two definitions.

Let $(X, \mathfrak{L})$ be an $\mathscr{L}$-space. A subset A of X is called *sequentially closed* in X if for each sequence $\langle x_n \rangle \in X^N$ such that $\bigcup (x_n) \subseteqq A$ and $(\langle x_n \rangle, x) \in \mathfrak{L}$ we have $x \in A$. A subset X_0 of X is called *sequentially dense* in X if for each sequentially closed subset B of X such that $X_0 \subseteqq B \subseteqq X$ we have $B = X$.

We have seen that to each $\mathscr{U}\mathscr{L}$-structure $\mathfrak{n}$ there corresponds a unique $\mathscr{L}$-structure $\mathfrak{L}(\mathfrak{n})$ and to each $\mathscr{C}\mathscr{L}$-structure L there corresponds a unique $\mathscr{L}$-structure $\mathfrak{L}(L)$. For $\mathscr{N} \in \{\mathscr{L}, \mathscr{U}\mathscr{L}, \mathscr{C}\mathscr{L}\}$ a sequence will be called $\mathscr{N}$-convergent if it is $\mathfrak{L}$-convergent resp. $\mathfrak{L}(\mathfrak{n})$-convergent resp. $\mathfrak{L}(L)$-convergent. Hence we can extend the definitions of sequentially closed and sequentially dense subsets also to $\mathscr{U}\mathscr{L}$- and $\mathscr{C}\mathscr{L}$-structures by refering to the appropriate induced $\mathscr{L}$-structure.

Extension of identity principle (EIP($\mathscr{L}$)). *Let φ and ψ be $\mathscr{L}$-continuous mappings of an $\mathscr{L}$-space $(X, \mathfrak{L})$ into an $\mathscr{L}$-space $(Y, \mathfrak{M})$ such that $\varphi | X_0 = \psi | X_0$ for some sequentially dense subset X_0 of X. If $\mathfrak{M}$ satisfies ($\mathscr{L}_0$), then $\varphi = \psi$.*

Proof. Let $B = \{x \mid \varphi(x) = \psi(x)\}$. *Let* $\langle y_n \rangle \in X^N$, $\bigcup (y_n) \subseteqq B$, and suppose that $(\langle y_n \rangle, y) \in \mathfrak{L}$. By $\mathscr{L}$-continuity of φ and ψ we have $(\langle \varphi(y_n) \rangle, \varphi(y)) \in \mathfrak{M}$ and $(\langle \psi(y_n) \rangle, \psi(y)) \in \mathfrak{M}$. Since $\varphi(y_n) = \psi(y_n)$ for each $n \in N$ and $\mathfrak{M}$ satisfies ($\mathscr{L}_0$) we have $\varphi(y) = \psi(y)$ and hence $y \in B$. Therefore B is sequentially closed in X. Since $X_0 \subseteqq B \subseteqq X$ and X_0 is sequentially dense in X we have $B = X$.

The axiom ($\mathscr{L}_0$) is also a necessary condition for the statement above in the following sense. Let $(Y, \mathfrak{M})$ be an $\mathscr{L}$-space which does not satisfy ($\mathscr{L}_0$). Then there exists an $\mathscr{L}$-space $(X, \mathfrak{L})$, $\mathscr{L}$-continuous mappings $\varphi: X \to Y$ and $\psi: X \to Y$, and a sequentially dense subset X_0 of X such that $\varphi | X_0 = \psi | X_0$ and $\varphi \neq \psi$. Indeed, let $\langle x_n \rangle$ be a sequence in Y such that $(\langle x_n \rangle, x) \in \mathfrak{M}$, $(\langle x_n \rangle, y) \in \mathfrak{M}$, and $x \neq y$. Put $X_0 = \bigcup \left(\frac{1}{n}\right)$ and $X = X_0 \cup (0)$. Let $(\langle z \rangle, z) \in \mathfrak{L}$ for each $z \in X$ and $\left(\left\langle \frac{1}{n_i} \right\rangle, 0\right) \in \mathfrak{L}$ for each subsequence $\langle n_i \rangle$ of $\langle n \rangle$. Define $\varphi\left(\frac{1}{n}\right) = \psi\left(\frac{1}{n}\right) = x_n$, $\varphi(0) = x$, $\psi(0) = y$. Then both $\varphi: X \to Y$ and $\psi: X \to Y$ are $\mathscr{L}$-continuous, $\varphi | X_0 = \psi | X_0$, and $\varphi \neq \psi$.

The axiom ($\mathcal{L}_3$)

The axiom ($\mathcal{L}_3$) is important when we consider the relations between $\mathcal{L}$-structures and topological structures. We have seen that to each $\mathcal{L}$-space $(X, \mathfrak{L})$ there corresponds a unique closure space $(X, \lambda(\mathfrak{L}))$. Denote by $[\mathfrak{L}]$ the set of all $\mathcal{L}$-structures $\mathfrak{M}$ for X for which $\lambda(\mathfrak{M}) = \lambda(\mathfrak{L})$. The following propositions characterize the axiom ($\mathcal{L}_3$).

Proposition ([14, 32, 44]). *Let $(X, \mathfrak{L})$ be an $\mathcal{L}$-space satisfying ($\mathcal{L}_0'$). Then the following statements are equivalent.*

(i) *$(X, \mathfrak{L})$ satisfies ($\mathcal{L}_3$).*
(ii) *If $\mathfrak{M} \in [\mathfrak{L}]$, then $\mathfrak{M} \subseteqq \mathfrak{L}$.*

Proposition ([44]). *Let $(X, \mathfrak{L})$ be an $\mathcal{L}$-space satisfying ($\mathcal{L}_0'$). Then the following statements are equivalent.*

(i) *$(X, \mathfrak{L})$ satisfies ($\mathcal{L}_3$).*
(ii) $\mathfrak{L}(\lambda(\mathfrak{L})) = \mathfrak{L}$.

Proposition ([41]). *Let $(X, \mathfrak{L})$ and $(Y, \mathfrak{M})$ be $\mathcal{L}$-spaces and let $f\colon X \to Y$. If $(Y, \mathfrak{M})$ satisfies ($\mathcal{L}_0'$)* and *($\mathcal{L}_3$), then the following statements are equivalent.*

(i) *$f\colon (X, \mathfrak{L}) \to (Y, \mathfrak{M})$ is $\mathcal{L}$-continuous.*
(ii) *$f\colon (X, \lambda(\mathfrak{L})) \to (Y, \lambda(\mathfrak{M}))$ is continuous.*

For $\mathcal{L}$-spaces satisfying axiom ($\mathcal{L}_0'$) there is a one-to-one correspondence between maximal convergences and closures induced by a convergence. We have also seen that the axiom ($\mathcal{L}_3$) is a necessary and sufficient condition for the relation $\mathfrak{L}(\lambda(\mathfrak{L})) = \mathfrak{L}$ to hold. When we replace the closure λ by the topological closure λ^{ω_1} the statement remains true for single-valued $\mathcal{L}$-structures.

Proposition ([9]). *Let $(X, \mathfrak{L})$ be an $\mathcal{L}$-space satisfying ($\mathcal{L}_0'$). Then the following statements are equivalent.*

(i) *$(X, \mathfrak{L})$ satisfies ($\mathcal{L}_3$).*
(ii) $\mathfrak{L}(\lambda^{\omega_1}(\mathfrak{L})) = \mathfrak{L}$.

The axiom ($\mathcal{F}\mathcal{L}$)

Let q be a filter convergence structure. It follows easily that the induced $\mathcal{L}$-structure $\mathfrak{L}(q)$ satisfies the axiom ($\mathcal{F}\mathcal{L}$). The next proposition (cf. [24]) characterizes this axiom.

Proposition. *Let $(X, \mathfrak{L})$ be an $\mathcal{L}$-space. Then the following statements are equivalent.*

(i) *$(X, \mathfrak{L})$ satisfies ($\mathcal{F}\mathcal{L}$).*
(ii) $\mathfrak{L}(q(\mathfrak{L})) = \mathfrak{L}$.

The axiom ($\mathcal{M}\mathcal{L}$)

The next two propositions show the usefulness of the axiom ($\mathcal{M}\mathcal{L}$) when considering the relations between $\mathcal{L}$-spaces and filter convergence spaces.

Proposition ([21]). *Let (X, q) be a filter convergence space and $\mathfrak{L}(q)$ the induced $\mathcal{L}$-structure. If q satisfies the Limitierung axiom (i.e. $x \in q(\mathcal{F})$ and $x \in q(\mathcal{G})$ imply $x \in q(\mathcal{F} \cap \mathcal{G})$), then $\mathfrak{L}(q)$ satisfies ($\mathcal{M}\mathcal{L}$).*

Proposition ([21]). *Let $(X, \mathfrak{L})$ be an $\mathscr{L}$-space and $q(\mathfrak{L})$ the induced S-convergence structure. If $(X, \mathfrak{L})$ satisfies the axiom $(\mathscr{ML})$, then $q(\mathfrak{L})$ is a Limitierung.*

The axiom $(\mathscr{ML})$ is also important for the investigation of relations between $\mathscr{L}$-structures and $\mathscr{CL}$-structures.

Proposition. *Let (X, L) be a $\mathscr{CL}$-space. Then the induced $\mathscr{L}$-structure $\mathfrak{L}(L)$ satisfies $(\mathscr{ML})$.*

Proof. Let $(\xi, x) \in \mathfrak{L}(L)$ and $(\eta, x) \in \mathfrak{L}(L)$. Then $\xi \wedge \langle x\rangle \in L$, $\eta \wedge \langle x\rangle \in L$. It follows from $(\mathscr{CL}_3)$ that $(\xi \wedge \langle x\rangle) \wedge (\eta \wedge \langle x\rangle) \in L$ and hence, by $(\mathscr{CL}_2)$, we have $(\xi \wedge \eta) \wedge \langle x\rangle \in L$. Thus $(\xi \wedge \eta, x) \in \mathfrak{L}(L)$.

Proposition. *Let $(X, \mathfrak{L})$ be an $\mathscr{L}$-space. If $(X, \mathfrak{L})$ satisfies $(\mathscr{L}_0)$ and $(\mathscr{ML})$, then $L(\mathfrak{L})$ is a $\mathscr{CL}$-structure for X.*

Proof. By $(\mathscr{L}_1)$ we have $(\langle x\rangle, x) \in \mathfrak{L}$ for each $x \in X$ and therefore $\langle x\rangle \in L(\mathfrak{L})$ for each $x \in X$ so that $(\mathscr{CL}_1)$ holds. Let $\xi \in L(\mathfrak{L})$ and let s be given. We have $(\xi, x) \in \mathfrak{L}$ for some $x \in X$ and by $(\mathscr{L}_2)$ we have $(\xi \circ s, x) \in \mathfrak{L}$. Hence $\xi \circ s \in L(\mathfrak{L})$ and $(\mathscr{CL}_2)$ holds. Finally let $\xi, \eta \in L(\mathfrak{L})$ and suppose that $\xi \circ s = \eta \circ t$ for some s and t. We have $(\xi, x) \in \mathfrak{L}$ and $(\eta, y) \in \mathfrak{L}$ for some $x \in X$ and $y \in X$. By $(\mathscr{L}_2)$ we have $(\xi \circ s, x) \in \mathfrak{L}$ and $(\eta \circ t, y) \in \mathfrak{L}$. Since $\eta \circ t = \xi \circ s$ we have also $(\xi \circ s, y) \in \mathfrak{L}$. By $(\mathscr{L}_0)$ $x = y$ and by $(\mathscr{ML})$ we have $(\xi \wedge \eta, x) \in \mathfrak{L}$. Therefore $\xi \wedge \eta \in L(\mathfrak{L})$ and $(\mathscr{CL}_3)$ holds.

Remark. In the preceding statement the axiom $(\mathscr{L}_0)$ can be replaced by the following weaker condition: If $(\xi, x) \in \mathfrak{L}$, $(\eta, y) \in \mathfrak{L}$, and there are s and t such that $\xi \circ s = \eta \circ t$, then $\operatorname{Lim} \xi \cap \operatorname{Lim} \eta \neq \emptyset$. This condition and axiom $(\mathscr{ML})$ are also necessary for the assertion to hold.

Proposition. *Let $(X, \mathfrak{L})$ be an $\mathscr{L}$-space which satisfies $(\mathscr{L}_0)$. Then the following statements are equivalent.*

(i) *$(X, \mathfrak{L})$ satisfies $(\mathscr{ML})$.*

(ii) $\mathfrak{L}(L(\mathfrak{L})) = \mathfrak{L}$.

Proof. I. Let $(X, \mathfrak{L})$ satisfy $(\mathscr{ML})$. By previous statement $L(\mathfrak{L})$ is a $\mathscr{CL}$-structure for X. It remains to prove that $\mathfrak{L}(L(\mathfrak{L})) = \mathfrak{L}$. Let $\mathfrak{L}\text{-lim}\ \xi = x$. Then $\mathfrak{L}\text{-lim}\ (\xi \wedge \langle x\rangle) = x$ by $(\mathscr{ML})$ and hence $\xi \wedge \langle x\rangle \in L(\mathfrak{L})$. It follows that $\mathfrak{L}(L(\mathfrak{L}))\text{-lim}\ \xi = x$. Conversely suppose that $\mathfrak{L}(L(\mathfrak{L}))\text{-lim}\ \xi = x$. Then $\xi \wedge \langle x\rangle \in L(\mathfrak{L})$ and hence $\mathfrak{L}\text{-lim}\ (\xi \wedge \langle x\rangle) = y$ for some $y \in X$. By $(\mathscr{L}_2)$ and $(\mathscr{L}_0)$ it follows that $\mathfrak{L}\text{-lim}\ \xi = x$.

II. As we have proved that for any $\mathscr{CL}$-space (X, L) the induced $\mathscr{L}$-structure $\mathfrak{L}(L)$ satisfies $(\mathscr{ML})$, the converse follows immediately.

2.2. Axioms for $\mathscr{UL}$-structures

The axiom $(\mathscr{UL}_0)$

Let φ be a $\mathscr{UL}$-continuous mapping of a $\mathscr{UL}$-space $(X, \mathfrak{n})$ into a $\mathscr{UL}$-space $(Y, \mathfrak{m})$. Then φ is an $\mathscr{L}$-continuous mapping of the $\mathscr{L}$-space $(X, \mathfrak{L}(\mathfrak{n}))$ into the $\mathscr{L}$-space $(Y, \mathfrak{L}(\mathfrak{m}))$. If $\mathfrak{m}$ satisfies $(\mathscr{UL}_0)$, then $\mathfrak{L}(\mathfrak{m})$ satisfies $(\mathscr{L}_0)$. The following statement is a simple corollary to Proposition $EIP(\mathscr{L})$.

Proposition. *Let φ and ψ be $\mathscr{UL}$-continuous mappings of a $\mathscr{UL}$-space $(X, \mathfrak{n})$ into a $\mathscr{UL}$-space $(Y, \mathfrak{m})$ such that $\varphi \mid X_0 = \psi \mid X_0$ for some sequentially dense subset X_0 of X. If $\mathfrak{m}$ satisfies $(\mathscr{UL}_0)$, then $\varphi = \psi$.*

The axiom $(\mathscr{U}\mathscr{L}_3)$

It is easy to prove [31] that if $\mathscr{U}$ is a semi-uniformity for X, then $\mathfrak{n}(\mathscr{U})$ satisfies the axiom $(\mathscr{U}\mathscr{L}_3)$. The importance of the axiom $(\mathscr{U}\mathscr{L}_3)$ follows from the following statement (cf. [17]).

Proposition. *Let $(X, \mathfrak{n})$ be a separated $\mathscr{U}\mathscr{L}$-space. Then the following statements are equivalent.*

(i) $\mathfrak{n}$ *satisfies* $(\mathscr{U}\mathscr{L}_3)$.
(ii) $\mathfrak{n} = \mathfrak{n}(\mathscr{U}(\mathfrak{n}))$.

The axiom $(\mathscr{U}\mathscr{L}^\square)$

If p is a proximity for X, then $\mathfrak{n}(p)$ satisfies the axiom $(\mathscr{U}\mathscr{L}^\square)$. The axiom $(\mathscr{U}\mathscr{L}^\square)$ is characterized by the following statement (cf. [31]).

Proposition. *Let $(X, \mathfrak{n})$ be a separated $\mathscr{U}\mathscr{L}$-space. Then the following statements are equivalent.*

(i) $\mathfrak{n}$ *satisfies* $(\mathscr{U}\mathscr{L}^\square)$.
(ii) $\mathfrak{n} = \mathfrak{n}(p(\mathfrak{n}))$.

2.3. Axioms for $\mathscr{C}\mathscr{L}$-structures

The axiom $(\mathscr{C}\mathscr{L}_0)$

Let φ be a $\mathscr{C}\mathscr{L}$-continuous mapping of a $\mathscr{C}\mathscr{L}$-space (X, L) into a $\mathscr{C}\mathscr{L}$-space (Y, M). Then φ is an $\mathscr{L}$-continuous mapping of an $\mathscr{L}$-space $(X, \mathfrak{L}(L))$ into an $\mathscr{L}$-space $(Y, \mathfrak{L}(M))$. If M satisfies the axiom $(\mathscr{C}\mathscr{L}_0)$, then $\mathfrak{L}(M)$ satisfies the axiom $(\mathscr{L}_0)$. The following statement follows from the Proposition $EIP(\mathscr{L})$.

Proposition. *Let φ and ψ be $\mathscr{C}\mathscr{L}$-continuous mappings of a $\mathscr{C}\mathscr{L}$-space (X, L) into a $\mathscr{C}\mathscr{L}$-space (Y, M) such that $\varphi \mid X_0 = \psi \mid X_0$ for some sequentially dense subset X_0 of X. If M satisfies $(\mathscr{C}\mathscr{L}_0)$, then $\varphi = \psi$.*

The axiom $(\mathscr{C}\mathscr{L}_4)$

A $\mathscr{C}\mathscr{L}$-space (X, L) is called *complete* if for each $\xi \in L$ there exists $x \in X$ such that $\xi \wedge \langle x \rangle \in L$. Complete $\mathscr{C}\mathscr{L}$-spaces are in fact $\mathscr{L}$-spaces and in that case the axiom $(\mathscr{C}\mathscr{L}_4)$ is equivalent to $(\mathscr{L}_3)$. We shall show that there is a close relationship between the axioms $(\mathscr{C}\mathscr{L}_4)$ and $(\mathscr{L}_3)$ even in the more general case.

Let L be a $\mathscr{C}\mathscr{L}$-structure for X satisfying $(\mathscr{C}\mathscr{L}_0)$. We define an equivalence relation on L as follows: $\xi \sim \eta$ if $\xi \wedge \eta \in L$. We can identify each $x \in X$ with the equivalence class $[\langle x \rangle]$. Then X is a subset of the set Y of all equivalence classes. We introduce a $\mathscr{C}\mathscr{L}$-structure M for y as follows: $\xi \in M$ if $\xi = \langle y \rangle$, $y \in Y$, or $\xi = \langle x_n \rangle$, $\langle x_n \rangle \in L$. Then (Y, M) is a complete $\mathscr{C}\mathscr{L}$-space and hence we can identify the $\mathscr{C}\mathscr{L}$-structure M with the $\mathscr{L}$-structure $\mathfrak{L}(M)$ and the $\mathscr{L}$-structure $\mathfrak{L}(M)^*$ with the $\mathscr{C}\mathscr{L}$-structure M^*.

Proposition. *The following statements are equivalent.*

(i) L *satisfies* $(\mathscr{C}\mathscr{L}_4)$.
(ii) $L = M^* \mid X$.

2.4. Concluding remarks

Throughout this subsection we assume that $\mathcal{N} \in \{\mathcal{L}, \mathcal{UL}, \mathcal{CL}\}$. We can then say that the axiom $(\mathcal{N}_0)$ guarantees that the $\mathcal{N}$-convergent sequences have unique limits. The Extension of identity principle then shows that the function of the axiom $(\mathcal{N}_0)$ is similar to that of Hausdorff separation axiom T_2 for topological spaces. This principle can be expressed in general form in the following way.

Extension of identity principle. *Let φ and ψ be $\mathcal{N}$-continuous mappings of an $\mathcal{N}$-space X into an $\mathcal{N}$-space Y such that $\varphi \mid X_0 = \psi \mid X_0$ for some sequentially dense subset X_0 of X. If Y satisfies the axiom $(\mathcal{N}_0)$, then $\varphi = \psi$.*

A similar statement can be proved for filter structures.

The propositions of preceding subsections show that the axioms $(\mathcal{L}_3)$, $(\mathcal{FL})$, $(\mathcal{ML})$, $(\mathcal{UL}_3)$, $(\mathcal{UL}^\square)$, and $(\mathcal{CL}_4)$ characterize separated $\mathcal{N}$-structures satisfying the equality $\mathfrak{N} = \mathfrak{N}(m(\mathfrak{N}))$ where $m(\mathfrak{N})$ is some continuous structure induced by $\mathfrak{N}$ and $\mathfrak{N}(m(\mathfrak{N}))$ is an $\mathcal{N}$-structure induced by $m(\mathfrak{N})$. These axioms characterize also the extreme members of equivalence classes of separated $\mathcal{N}$-structures inducing the same continuous structure (cf.[43, 44, 17]). In general, a given $\mathcal{N}$-structure $\mathfrak{N}$ can be enlarged to $\mathfrak{N}^*$ using the particular "star" axiom ($\mathfrak{L}$ to $\mathfrak{L}^{\mathcal{F}}$ satisfying $(\mathcal{FL})$ resp. to $\mathfrak{L}^{\mathcal{M}}$ satisfying $(\mathcal{ML})$ resp. to $\mathfrak{L}^*$ satisfying $(\mathcal{L}_3)$, $\mathfrak{u}$ to $\mathfrak{u}^*$ satisfying $(\mathcal{UL}_3)$ resp. to $\mathfrak{u}^\square$ satisfying $(\mathcal{UL}^\square)$, L to L^* satisfying $(\mathcal{CL}_4)$).

3. Categorical approach to sequential structures

In this section we describe some of the properties of the three basic sequential structures as well as their relations to other continuous structures in categorical terms.

We employ the symol $\mathcal{L}$ to denote the category whose objects are $\mathcal{L}$-spaces and whose morphisms are $\mathcal{L}$-continuous mappings. Let $\mathcal{B}$ be a subcategory of a category $\mathcal{A}$. We always assume that (i) $\mathcal{B}$ is full, i.e. whenever A and B are $\mathcal{B}$-objects and $f: A \to B$ is an $\mathcal{A}$-morphism, then f is a $\mathcal{B}$-morphism, (ii) $\mathcal{B}$ is isomorphism closed, i.e. whenever A is a $\mathcal{B}$-object and $f: A \to B$ is an $\mathcal{A}$-isomorphism, then B is a $\mathcal{B}$-object. In particular, we do not distinguish between subcategories of $\mathcal{L}$ and "$\mathcal{L}$-properties", i.e. $(\mathcal{L}_0)$, $(\mathcal{L}_3)$, $(\mathcal{ML})$, $(\mathcal{FL})$. Further, by $\mathcal{L}_0 \cap \mathcal{L}_3$ we mean the subcategory the objects of which satisfy both $(\mathcal{L}_0)$ and $(\mathcal{L}_3)$. The same notational conventions will be used for other subcategories of $\mathcal{L}$ and for the category $\mathcal{UL}$ of all $\mathcal{UL}$-spaces and $\mathcal{UL}$-continuous mappings resp. the category $\mathcal{CL}$ of all $\mathcal{CL}$-spaces and $\mathcal{CL}$-continuous mappings.

3.1. Subcategories of $\mathcal{L}$

Let $f: X \to Y$ be an $\mathcal{L}$-morphism. Then

f is an epimorphism in $\mathcal{L}$	f is onto, i. e. $f[X] = Y$
f is an extremal monomorphism in $\mathcal{L}$	f is an $\mathcal{L}$-embedding

From $EIP(\mathcal{L})$ it follows that

f is an epimorphism in $\mathcal{L}_0$	$f[X]$ is sequentially dense in Y
f is an extremal monomorphism in $\mathcal{L}_0$	f is a closed $\mathcal{L}$-embedding

As mentioned in Section 1, there are natural correspondences between $\mathcal{L}$-structures and other continuous structures. In this way important functors as well as subcategories of $\mathcal{L}$ can be obtained.

The correspondence $\mathfrak{L} \to \lambda^\alpha(\mathfrak{L})$ induces a functor $\mathbf{T}_\alpha$ from $\mathcal{L}$ into the category of all closure spaces and continuous mappings as follows:

$$\text{for} \quad f\colon (X_1, \mathfrak{L}_1) \to (X_2, \mathfrak{L}_2) \quad \text{we put} \quad \mathbf{T}_\alpha f = f\colon \big(X_1, \lambda^\alpha(\mathfrak{L}_1)\big) \to \big(X_2, \lambda^\alpha(\mathfrak{L}_2)\big).$$

Remark. Notice that $\mathbf{T}_1$ is a composition of two functors: one induced by the correspondence $\mathfrak{L} \to q(\mathfrak{L})$ and the other one by $q(\mathfrak{L}) \to k_{q(\mathfrak{L})}$, where $k_{q(\mathfrak{L})}$ is the closure operator derived from the convergence structure $q(\mathfrak{L})$ (cf. [21]).

Similarly, the correspondence $q \to \mathfrak{L}(q)$ induces a functor $\mathbf{L}$ from the category $\mathcal{F}$ of all filter convergence spaces and continuous mappings into $\mathcal{L}$ defined by:

$$\text{for} \quad f\colon (X_1, q_1) \to (X_2, q_2) \quad \text{we put} \quad \mathbf{L}f = f\colon \big(X_1, \mathfrak{L}(q_1)\big) \to \big(X_2, \mathfrak{L}(q_2)\big).$$

Since the categories of all closure, resp. topological, spaces and continuous mappings can be considered as subcategories of $\mathcal{F}$ (cf. [33]), the functor $\mathbf{L}$ can be applied to these categories as well. The functor $\mathbf{T}_{\omega_1}$ and the functor $\mathbf{L}$ defined on the category of all topological spaces and continuous mappings were studied by M. Dolcher in [9].

Composing functors $\mathbf{T}_\alpha$ and $\mathbf{L}$ we can derive isomorphisms between various categories.

Proposition. *Category $\mathcal{L}_0' \cap \mathcal{L}_3$ is isomorphic to the category of all sequential convergence spaces which satisfy the usual separation axiom T_1 for topological spaces.*

Category $\mathcal{F}\mathcal{L}$ is isomorphic to the category of all S-convergence spaces.

Category $\mathcal{F}\mathcal{L} \cap \mathcal{M}\mathcal{L}$ is isomorphic to the category of all S-convergence spaces satisfying the Limitierung axiom.

Remark. In Section 4 isomorphism between the category of all maximal single-valued sequentially regular $\mathcal{L}$-spaces and a certain subcategory of the category of completely regular spaces is established.

3.2. Subcategories of $\mathcal{U}\mathcal{L}$

Let $f\colon X \to Y$ be a $\mathcal{U}\mathcal{L}$-morphism. Then

f is an epimorphism in $\mathcal{U}\mathcal{L}$	f is onto, i.e. $f[X] = Y$
f is an extremal monomorphism in $\mathcal{U}\mathcal{L}$	f is a $\mathcal{U}\mathcal{L}$-embedding

From $EIP(\mathcal{U}\mathcal{L})$ it follows that

f is an epimorphism in $\mathcal{U}\mathcal{L}_0$	$f[X]$ is sequentially dense in Y
f is an extremal monomorphism in $\mathcal{U}\mathcal{L}_0$	f is a sequentially closed $\mathcal{U}\mathcal{L}$-embedding

We list important correspondences between $\mathcal{U}\mathcal{L}$-structures and other continuous structures, as well as the induced functors:

$\mathfrak{n} \to \mathfrak{L}(\mathfrak{n})$, $\mathcal{U}\mathcal{L} \to \mathcal{L}$

$\mathfrak{n} \to p(\mathfrak{n})$, $\mathcal{U}\mathcal{L}_0 \to$ Seq. Prox. (cf. [31])

$\mathfrak{n} \to \mathcal{U}(\mathfrak{n})$, $\mathcal{U}\mathcal{L}_0 \to$ Seq. Semi-unif. (cf. [31])

$p \to \mathfrak{n}(p)$, Seq. Prox. $\to \mathcal{U}\mathcal{L}_0 \cap \mathcal{U}\mathcal{L}^\square$ (cf. [31])

$\mathcal{U} \to \mathfrak{n}(\mathcal{U})$, Seq. Semi-unif. $\to \mathcal{U}\mathcal{L}_0 \cap \mathcal{U}\mathcal{L}_3$ (cf. [31, 17])

Proposition. *Category* $\mathscr{UL}_0 \cap \mathscr{UL}^\square$ *is isomorphic to Seq. Prox.*
Category $\mathscr{UL}_0 \cap \mathscr{UL}_3$ *is isomorphic to Seq. Semi-unif.*

3.3. Subcategories of $\mathscr{CL}$

Let $f: X \to Y$ be a $\mathscr{CL}$-morphism. Then

f is an epimorphism in $\mathscr{CL}$	f is onto, i.e. $f[X] = Y$
f is an extremal monomorphism in $\mathscr{CL}$	f is a $\mathscr{CL}$-embedding

From $EIP(\mathscr{CL})$ if follows that

f is an epimorphism in $\mathscr{CL}_0$	$f[X]$ is sequentially dense in Y
f is an extremal monomorphism in $\mathscr{CL}_0$	f is a sequentially closed $\mathscr{CL}$-embedding

Some of the important correspondences between $\mathscr{CL}$-structures and other continuous structures, as well as the induced functors, are listed below.

$L \to \mathfrak{L}(L)$, $\quad \mathscr{CL} \to \mathscr{ML}$

$L \to \mathfrak{n}(L)$, $\quad \mathscr{CL} \to \mathscr{UL}$

$\mathfrak{n} \to L(\mathfrak{n})$, $\quad \mathscr{UL}_3 \to \mathscr{CL}_4$

$\quad \mathscr{UL}_0 \cap \mathscr{UL}_3 \to \mathscr{CL}_0 \cap \mathscr{CL}_4$

$\mathfrak{L} \to L(\mathfrak{L})$, $\quad \mathscr{L}_0 \cap \mathscr{ML} \to \mathscr{CL}_0$

$\quad \mathscr{L}_0 \cap \mathscr{L}_3 \to \mathscr{CL}_0 \cap \mathscr{CL}_4$

Proposition. *Category* $\mathscr{CL}_0 \cap \mathscr{CL}_4$ *is isomorphic to a subcategory of* $\mathscr{UL}_0 \cap \mathscr{UL}_3$. *Category* $\mathscr{L}_0 \cap \mathscr{ML}$ *and the subcategory of* $\mathscr{CL}$ *the objects of which are separated complete* $\mathscr{CL}$*-spaces are isomorphic. Category* $\mathscr{L}_0 \cap \mathscr{L}_3$ *and the subcategory of* $\mathscr{CL}$ *the objects of which are maximal separated complete* $\mathscr{CL}$*-spaces are isomorphic.*

Remark. Denote by $\mathscr{A}$ the subcategory of $\mathscr{UL}_0 \cap \mathscr{UL}_3$ which is isomorphic to $\mathscr{CL}_0 \cap \mathscr{CL}_4$. It follows from Proposition 1.7 and Proposition 1.10 in [35] that the objects of $\mathscr{A}$ are of the form $(X, (\mathfrak{n}(L^*))^*)$.

3.4. Concluding remarks

In most applications we work with categories $\mathscr{L}_0 \cap \mathscr{L}_3$, $\mathscr{UL}_0 \cap \mathscr{UL}_3$, $\mathscr{CL}_0 \cap \mathscr{CL}_4$. Relations between these categories are especially simple. In this case we can identify $\mathscr{L}_0 \cap \mathscr{L}_3$ with Compl $(\mathscr{CL}_0 \cap \mathscr{CL}_4)$, and $\mathscr{CL}_0 \cap \mathscr{CL}_4$ with the subcategory $\mathscr{A}$ of $\mathscr{UL}_0 \cap \mathscr{UL}_3$ described above.

In Section 4 a completion theory for categories Proj $\mathscr{N}$ of $\mathscr{N}$-spaces ($\mathscr{N} \in \{\mathscr{L}, \mathscr{UL}, \mathscr{CL}\}$) projectively generated by $\mathscr{N}$-continuous functions is discussed. Let Compl Proj $\mathscr{N}$ be the subcategory of Proj $\mathscr{N}$ consisting of $\mathscr{N}$-complete spaces. Then for each X in Proj $\mathscr{N}$ the completion $\sigma X \in$ Compl Proj $\mathscr{N}$ is an epireflection of X into Compl Proj $\mathscr{N}$, i.e. Compl Proj $\mathscr{N}$ is an epireflective subcategory of Proj $\mathscr{N}$.

4. Projectively generated sequential structures

4.1. Introduction

The extension theory for uniform and topological spaces is well developed and has led to such important concepts as the completion of a uniform space and the Čech-Stone compactification. The class of topological spaces involved is the class of completely regular spaces. Here we are concerned with sequential analogues of these notions.

Consider this model example ([45]).

Let X be a set and let $\mathsf{A} \subseteqq \mathsf{P}X$ be an algebra of subsets of X. Each probability measure P on A has a unique extension to a probability measure $\bar{P}$ defined on the generated σ-algebra $\mathsf{S}(\mathsf{A})$.

From the sequential point of view we have the following situation.

First, if X is a set, we can define an $\mathscr{L}$-structure $\mathfrak{L}$ for $\mathsf{P}X$ as follows:

$$(\langle A_n \rangle, A) \in \mathfrak{L} \quad \text{whenever} \quad A = \bigcup_{k \geqq 1} \bigcap_{n \geqq k} A_n = \bigcap_{k \geqq 1} \bigcup_{n \geqq k} A_n .$$

Obviously, $\mathfrak{L}$ is single-valued and maximal. Hence $\mathsf{P}X$ is an $\mathscr{L}$-space and A and $\mathsf{S}(\mathsf{A})$ are its subspaces. Furthermore they have the following important property: If $(\langle A_n \rangle, A) \notin \mathfrak{L}$, then there exists an $\mathscr{L}$-continuous function f, namely $f(B) = c_B(x)$ for a suitable $x \in X$, where $c_B(x)$ is the value of the characteristic function of B at x, such that $\langle f(A_n) \rangle$ does not converge to $f(A)$.

Secondly, each probability measure P on A is $\mathscr{L}$-continuous. Notice, that $f(\cdot) = c_{(\cdot)}(x)$ is a probability measure on A.

Thirdly, the algebra A is a sequentially dense subspace of $\mathsf{S}(\mathsf{A})$ and $\mathsf{S}(\mathsf{A})$ is "complete" in the following sense: If $\langle A_n \rangle \in \mathsf{S}(\mathsf{A})^N$ and $\langle c(x) \rangle$ converges for all $x \in X$, then $\langle A_n \rangle$ converges in $\mathsf{S}(\mathsf{A})$.

So the proposition mentioned above essentially states that certain $\mathscr{L}$-continuous functions defined on certain $\mathscr{L}$-spaces can be uniquely extended to certain "complete" $\mathscr{L}$-spaces which contain the given $\mathscr{L}$-space as a sequentially dense subspace, and the functions in question projectively generate the underlying spaces.

4.2. Projectively generated $\mathscr{N}$-structures

In the sequel $\mathscr{N} \in \{\mathscr{L}, \mathscr{U}\mathscr{L}, \mathscr{C}\mathscr{L}\}$ and all $\mathscr{N}$-structures are assumed to be single-valued, resp. separated, and maximal.

Definition. ([45]). Let $(X, \mathfrak{L})$ be an $\mathscr{L}$-space and let $C_0 \subseteqq C(X)$. We say that $(X, \mathfrak{L})$ is *C_0-sequentially regular* if $\mathfrak{L}$ is projectively generated by C_0, i.e. if, whenever no subsequence of a sequence $\langle x_n \rangle \in X^N$ converges to a point $x \in X$, there exists a function $f \in C_0$ such that the sequence $\langle f(x_n) \rangle$ does not converge to $f(x)$.

If $C_0 = C(X)$, then we simply say that $(X, \mathfrak{L})$ is *sequentially regular*.

Observe that the $\mathscr{L}$-space $\mathsf{P}X$ in our example is C_0-sequentially regular for $C_0 = C(\mathsf{P}X) \cap \{0, 1\}^{\mathsf{P}X}$.

The next two statements show that the sequential regularity is indeed the sequential analogue of complete regularity.

Proposition ([37]). *If (X, u) is a completely regular space, then $(X, \mathfrak{L}(u))$ is sequentially regular.*

Denote by P the class of all completely regular spaces (X, u) such that $C_s(X) = C(X)$. To each sequentially regular $\mathscr{L}$-space $(X, \mathfrak{L})$ there corresponds a unique completely regular space $(X, u) \in \mathsf{P}$ such that $C\big((X, u)\big) = C_s\big((X, u)\big) = C\big((X, \mathfrak{L})\big)$ (u is the weak topology for X relative to $C((X, \mathfrak{L}))$). We denote u by $\bar{\lambda}$ and the space $(X, \bar{\lambda})$ is called the *completely regular modification* of $(X, \mathfrak{L})$ ([37]).

Proposition ([37]). *The category of all sequentially regular $\mathscr{L}$-spaces is isomorphic to* P.

In a similar way projectively generated $\mathscr{U}\mathscr{L}$-structures called $\mathscr{U}$-regular ([30, 18]) and projectively generated $\mathscr{C}\mathscr{L}$-structures called Cauchy-regular ([35]) can be defined.

The class of projectively generated $\mathscr{N}$-structures satisfies two important requirements. It is sufficiently broad to contain the most important structures and at the same time the structures it contains behave reasonably enough to enable us to develop a theory of completions of these structures which permit extension of continuous functions.

4.3. Sequentially complete $\mathscr{N}$-structures

Definition ([26]). Let $(X, \mathfrak{L})$ be an $\mathscr{L}$-space and let $C_0 \subseteqq C(X)$. We say that $(X, \mathfrak{L})$ is *C_0-sequentially complete* if $\langle x_n \rangle$ converges in X whenever $\langle f(x_n) \rangle$ converges in R for each $f \in C_0$.

If $C_0 = C(X)$, then we simply say that $(X, \mathfrak{L})$ is *sequentially complete.*

Observe that in our example the space S(A) is C_0-sequentially complete for $C_0 = C\big(\mathsf{S}(\mathsf{A})\big) \cap \{0, 1\}^{\mathsf{S}(\mathsf{A})}$.

C_0-sequentially complete C_0-sequentially regular $\mathscr{L}$-spaces are characterized by the following statement.

Proposition ([26]). *Let $(X, \mathfrak{L})$ be a C_0-sequentially regular $\mathscr{L}$-space. Then the following statements are equivalent.*

(i) $(X, \mathfrak{L})$ *is C_0-sequentially complete.*

(ii) $(X, \mathfrak{L})$ *is sequentially closed in every $\mathscr{L}$-space Y in which it is C_0-embedded.*

Again, the notion of completeness with similar properties can be defined for $\mathscr{U}\mathscr{L}$-structures ([31]) and for $\mathscr{C}\mathscr{L}$-structures ([35]).

4.4. Completion of $\mathscr{N}$-structures

Definition ([45]). Let $(X, \mathfrak{L})$ be a C_0-sequentially regular $\mathscr{L}$-space. An $\mathscr{L}$-space $(S, \mathfrak{S})$ is said to be a *C_0-sequential envelope* $\sigma_0(X)$ of X if

(e_1) $(X, \mathfrak{L})$ is a sequentially dense C_0-embedded subspace of $(S, \mathfrak{S})$,

(e_2) $(S, \mathfrak{S})$ is $\bar{C}_0(S)$-sequentially regular where

$$\bar{C}_0(S) = \{f \in C(S) \mid f|X \in C_0\}, \quad \text{and}$$

(e_3) S is $\bar{C}_0(S)$-sequentially complete.

If $C_0 = C(X)$, then we say that $(S, \mathfrak{S})$ is a *sequential envelope* $\sigma(X)$ of X.

Observe that in our example S(A) is a C_0-sequential envelope of A for $C_0 = \{P | P$ is a probability measure on A$\}$.

There are several known constructions of $\sigma(X)$. J. Novák originally used the method consisting in embedding X into the product $R^{C(X)}$ of real lines equipped with the pointwise sequential convergence and taking the smallest sequentially closed set in $R^{C(X)}$ containing the image of X. In [22] the sequential envelope is obtained by embedding the given space into the space $C(C(X))$ equipped with the continuous convergence of functions.

Denote by βX, resp. vX, the ČECH-STONE compactification, resp. the HEWITT real-compactification, of the completely regular modification of X. Then ([37, 15]) the sequential envelope of X is the smallest sequentially closed subset of βX, resp. vX, which contains X.

While the sequential envelope is the sequential analogue of βX, resp. vX, there is a fundamental difference between these notions. For example it has been shown in [20] that if $C_0 = C^*(X)$, then $\sigma_0(X) = \sigma(X)$.

The notion of the sequential envelope has become a widely used tool for constructing completions of sequential structures. V. KOUTNÍK pointed out in [38] that the sequential envelope can be considered as a completion for some $\mathcal{UL}$-spaces. R. FRIČ in [18] used the C_0-sequential envelope to develop a general theory of completion for $\mathcal{U}$-regular $\mathcal{UL}$-spaces. In [35] it is shown that this theory can be extended to a yet bigger class of $\mathcal{UL}$-spaces. in [23] the C_0-sequential envelope is used to obtain a D-completion for $\mathcal{CL}$-spaces projectively generated by D. Further research in this area comprises elations to completions of filter structures and general categorical constructions.

4.5. Examples

Example 1. The first example of an $\mathcal{L}$-space $(X, \mathfrak{L})$ for which $\sigma(X) \neq X$ was given by J. NOVÁK ([44]). The space $(X, \mathfrak{L})$ is a modification of the TYCHONOFF plank.

Example 2. V. KOUTNÍK modified Example 1 in [37] to show that there exists a ring of sets E for which $\sigma(\mathsf{E}) \neq \mathsf{E}$.

Example 3. J. NOVÁK has shown ([45]) that for any algebra of sets A the sigma-algebra $\mathsf{S}(\mathsf{A})$ is the C_0-sequential envelope of A where C_0 is the family of all probability measures on A.

Example 4. R. FRIČ has shown ([16]) that the family of BAIRE functions on R is the C_0-sequential envelope of $C(R)$ where C_0 is the set of all finite linear combinations of evaluations on $C(R)$.

Example 5. R. FRIČ has constructed a C_0-sequentially regular $\mathcal{L}$-group G such that $\sigma_0(G)$ cannot be equipped with a group operation compatible with the $\mathcal{L}$-structure of $\sigma_0(G)$.

4.6. Problems

As shown in this talk, there has been considerable development in the theory of sequential structures in recent years due to the effort of many authors. However, many important questions still remain to be answered. Here we mention some of the open problems.

It is known that:

(i) If the completely regular modification $(X, \bar{\lambda})$ of a sequentially regular maximal single-valued $\mathcal{L}$-space $(X, \mathfrak{L})$ is normal, then $\sigma(X) = X$ ([37]);

or, more generally:

(ii) If (X, u) is a normal topological space, then $(X, \mathfrak{L}(u))$ is sequentially complete ([20]);

(iii) There exist $\mathscr{L}$-spaces $(X, \mathfrak{L})$ such that $(X, \bar{\lambda})$ is not normal while $\sigma(X) = X$ ([37]).

The following questions are closely related.

Problem 1. What are the necessary and sufficient conditions for the completely regular topology $\bar{\lambda}$ unter which the equality $\sigma(X) = X$ holds?

Problem 2. Let $(X, \mathfrak{L})$ be a sequentially complete single-valued maximal $\mathscr{L}$-space. Does there exist a normal topological space (X, u) such that $\mathfrak{L} = \mathfrak{L}(u)$?

A positive answer to Problem 2 would give a nice characterization of sequentially complete maximal single-valued $\mathscr{L}$-spaces and show that sequential completeness is a sequential analogue of normality.

Very little is known about sequential envelopes of products (cf. [19]).

Problem 3. Find necessary and sufficient conditions for the equality $\sigma(X \times Y) = \sigma(X) \times \sigma(Y)$.

It is known [46] that if $(G, u, +)$ is a topological group, then $(G, \mathfrak{L}(u), +)$ is a sequentially regular $\mathscr{L}$-group.

Problem 4. Let $(G, \mathfrak{L}^*, +)$ be a sequentially regular $\mathscr{L}$-group and let $(G, \bar{\lambda})$ be the completely regular modification of $(G, \mathfrak{L})$. Is $(G, \bar{\lambda}, +)$ a topological group?

Problem 5. Let $(G, \mathfrak{L}^*, +)$ be a sequentially regular $\mathscr{L}$-group. Is there a completely regular topology u for G such that $\mathfrak{L}^* = \mathfrak{L}(u)$ and $(G, u, +)$ is a topological group?

A positive answer to Problem 4 would solve Problem 5 as well.

References

[1] P. Alexandroff et P. Urysohn, Une condition nécessaire et suffisante pour qu'une classe ($\mathscr{L}$) soit une classe ($\mathscr{D}$), C. R. Acad. Sci. Paris **177** (1923), 1274–1276.

[2] P. Antosik, On a topology of convergence, Colloq. Math. **21** (1970), 205–209.

[3] P. Antosik, Mappings from L-groups into topological groups I, Bull. Acad. Polon. Sci. Sér. Sci. Math. Astronom. Phys. **21** (1973), 145–153.

[4] A. R. Bednarek and J. Mikusiński, Convergence and topology, Bull. Acad. Polon. Sci. Sér. Sci. Math. Astronom. Phys. **17** (1969), 437–441.

[5] N. Bourbaki, Théorie des ensembles, Paris 1957.

[6] E. Čech, Topologické prostory, Časopis Pěst. Mat. Fys. **66** (1937), D225–D264.

[7] E. Čech, Topologické prostory, Praha 1959.

[8] E. Čech, Topological spaces, revised by Z. Frolík and M. Katětov, Praha 1966.

[9] M. Dolcher, Topologie e strutture di convergenza, Ann. Scuola Norm. Sup. Pisa Ser III **14** (1960), 63–92.

[10] R. M. Dudley, On sequential convergence, Trans. Amer. Math. Soc. **112** (1964), 483–507.

[11] R. Engelking, General topology, Warszawa 1977.

[12] S. P. Franklin, Spaces in which sequences suffice, Fund. Math. **57** (1965), 107–115.

[13] M. Fréchet, Sur quelques points du calcul fonctionel, Rend. Circ. Mat. Palermo **22** (1906), 1–74.

[14] M. Fréchet, Les espaces abstraits, Paris 1926.

[15] R. Frič, Sequential envelope and subspaces of the Čech-Stone compactification, General Topology and Its Relations to Modern Analysis and Algebra III, Proc. Third Prague Topological Sympos., 1971, Praha 1972, 123–126.

[16] R. Frič, On a problem of J. Novák, Colloq. Math. Soc. János Bolyai 8. Topics in Topology, Keszthely 1972, 317—319.
[17] R. Frič, On the sequential uniform structures of A. Goetz, Topology and its appl., Budva 1972, Beograd 1973, 90—93.
[18] R. Frič, On the completion of sequential structures, Topology and its appl., Budva 1972, Beograd 1973, 94—96.
[19] R. Frič, On the sequential envelope of products, IV. vědecká konferencia Vysokej školy dopravnej v Žilině, Žilina 1973, 89—92.
[20] R. Frič, On E-sequentially regular spaces, Czechoslovak Math. J. **26** [101] (1976), 604—612.
[21] R. Frič, On convergence spaces and groups, Math. Slovaca **27** (1977), 323—330.
[22] R. Frič, K. McKennon and G. Richardson, Sequential convergence in $C(X)$.
[23] R. Frič and D. C. Kent, Completion of sequential Cauchy spaces, Comment. Math. Univ. Carolinae **18** (1977), 351—361.
[24] R. Frič and D. C. Kent, On c-embedded sequential convergence spaces, these Proceedings, 33–36.
[25] R. Frič and D. C. Kent, On the natural completion functor for Cauchy spaces, Bull. Austral. Math. Soc. **18** (1978), 335—343.
[26] R. Frič and V. Koutník, Sequentially complete spaces, General Topology and Its Relations to Modern Analysis and Algebra IV, Proc. Fourth Prague Topological Sympos., 1976, Part B, Contributed Papers, Praha 1977, 127—131.
[27] R. Frič and V. Koutník, Sequentially complete spaces, Czechoslovak Math. J. **29** (1979), 287—297.
[28] R. J. Gazik and D. C. Kent, Regular completions of Cauchy spaces via function algebras, Bull. Austral. Math. Soc. **11** (1974), 77—88.
[29] A. Goetz, A notion of uniformity for L-spaces of Fréchet, General Topology and Its Relations to Modern Analysis and Algebra, Proc. First Prague Topological Sympos., 1961, Praha 1962, 177—178.
[30] A. Goetz, On a notion of uniformity for L-spaces of Fréchet, Colloq. Math. **9** (1962), 223—231.
[31] A. Goetz, Proximity-like structures in convergence spaces, Topology Conf., 1967, Arizona State University, Tempe 1968, 104—137.
[32] F. Hausdorff, Gestufte Räume, Fund. Math. **25** (1935), 486—502.
[33] B. V. Hearsey and D. C. Kent, Convergence structures, Portugal Math. **31** (1972), 105—118.
[34] J. M. Irwin and D. C. Kent, Sequential Cauchy spaces, Proc. Special Session Convergence Structures, 1976, University of Nevada, Reno 1976, 60—82.
[35] J. M. Irwin and D. C. Kent, Sequential Cauchy spaces, Math. Slovaca **29** (1979), 117—130.
[36] H. H. Keller, Die Limes-Uniformisierbarkeit der Limesräume, Math. Ann. **176** (1968), 334—341.
[37] V. Koutník, On sequentially regular convergence spaces, Czechoslovak Math. J. **17** (92) (1967), 232—247.
[38] V. Koutník, Sequential envelopes and completeness, Proc. I. Internat. Sympos. on Extension Theory of Topological Structures, Berlin, 1967, Berlin 1969, 141—143.
[39] V. Koutník, Convergence in closure spaces, Proc. Internat. Sympos. on Topology and Its Appl., Herceg Novi, 1968, Beograd 1969, 226—230.
[40] H.-J. Kowalsky, Limesräume und Komplettierung, Math. Nachr. **12** (1954), 301—340.
[41] K. Kuratowski, Topologie I, Warszawa 1933.
[42] J. Novák, Sur les espaces ($\mathscr{L}$) et sur les produits cartésiens ($\mathscr{L}$), Publ. Fac. Sci. Univ. Masaryk No. 273, 1939.
[43] J. Novák, On some problems concerning multivalued convergences, Czechoslovak Math. J. **14** (89) (1964), 548—561.
[44] J. Novák, On convergence spaces and their sequential envelopes, Czechoslovak Math. J. **15** (90) (1965), 74—100.

[45] J. Novák, On sequential envelopes defined by means of certain classes of continuous functions, Czechoslovak Math. J. **18** (93) (1968), 450—456.

[46] J. Novák, On convergence groups, Czechoslovak Math. J. **20** (95) (1970), 357—374.

[47] J. Novák, On some problems concerning the convergence spaces and groups, General Topology and Its Relations to Modern Analysis and Algebra, Proc. Kanpur Topological Conf., 1968, Praha 1970, 219—229.

[48] J. Novák, On completion of convergence commutative groups, General Topology and Its Relations to Modern Analysis and Algebra III, Proc. Third Prague Topological Sympos., 1971, Praha 1972, 335—340.

[49] V. Z. Poljakov, O ravnomernych prostranstvach schodimosti, Colloq. Math. **13** (1965), 167—179.

[50] J. F. Ramaley and O. Wyler, Cauchy spaces I. Structure and uniformization theorems, Math. Ann. **187** (1970), 175—186.

[51] G. G. Roussas, Contiguity of probability measures, Cambridge 1972.

[52] O. Schreier, L-Gruppen, Abhandlungen aus dem math. Seminar, 1926, S. 15.

[53] Z. Semadeni, Banach spaces of continuous functions, Warszawa 1971.

[54] P. Urysohn, Sur les classes ($\mathscr{L}$) de M. Fréchet, Enseignement Math. **25** (1926), 77—83.

Anschrift der Verfasser:

Katedra matematiky
Fakulta strojnicka a elektrotechnická
Vysoká škola dopravná
Marxa-Engelsa 25
01088 Žilina
ČSSR

Matematicky Ústav ČSAV
Žitná 25
11 567 Praha 1
ČSSR

Sequential Convergence in $C(X)$

ROMAN FRIČ, Žilina (ČSSR), KELLY MCKENNON, Pullman (U.S.A.), and GARY D. RICHARDSON, Greenville (U.S.A.)

0. Introduction

We study spaces of continuous functions equipped with the continuous convergence of sequences. Modifying the c-embedding technique developed by E. BINZ and others for filter convergence spaces we obtain new results for sequential convergence spaces. We give new representation theorems for sequentially regular and sequentially complete spaces, and a new construction of the sequential envelope introduced by J. NOVÁK. Finally, we investigate relations between a convergence of sequences of functions and the convergence of the corresponding sequences of their extensions.

The notion of a convergence on an arbitrary nonempty set was introduced in terms of sequences by M. FRÉCHET in [7]. The notion was generalized in terms of nets by E. H. MOORE and H. L. SMITH in [17], and in terms of filters by H. CARTAN in [3]. The three concepts have been developed into what is now called the theory of convergence spaces. The reader is referred for details and further references on sequential convergence spaces to [10, 15, 18], on net convergence spaces to [5, 20], and on filter convergence spaces to [2, 14]. Their mutual relations are investigated in [1, 5, 9, 20].

For the reader's convenience we recall some notions used in the paper. Symbols N, R, ω, Ω, stand, respectively, for the natural numbers, the real numbers, the first infinite ordinal, and the first uncountable ordinal. Let X be a nonempty set. By a *sequential convergence* $\mathfrak{L}$ on X we understand one-valued maximal convergence, i.e. $\mathfrak{L} \subseteqq X^N \times X$, and the following axioms are satisfied:

($\mathscr{L}_0$) $(\langle x_n \rangle, x) \in \mathfrak{L}$ and $(\langle x_n \rangle, y) \in \mathfrak{L}$ imply $x = y$,

($\mathscr{L}_1$) $x_n = x$, $n \in N$, implies $(\langle x_n \rangle, x) \in \mathfrak{L}$,

($\mathscr{L}_2$) $(\langle x_n \rangle, x) \in \mathfrak{L}$ implies $(\langle x_n' \rangle, x) \in \mathfrak{L}$ for each subsequence $\langle x_n' \rangle$ of $\langle x_n \rangle$,

($\mathscr{L}_3$) $(\langle x_n \rangle, x) \in \mathfrak{L}$ whenever each subsequence $\langle x_n' \rangle$ of $\langle x_n \rangle$ contains a subsequence $\langle x_n'' \rangle$ such that $(\langle x_n'' \rangle, x) \in \mathfrak{L}$.

If $(\langle x_n \rangle, x) \in \mathfrak{L}$, then we say that $\langle x_n \rangle$ *$\mathfrak{L}$-converges to* x, in symbols $x = \mathfrak{L}\text{-}\lim x_n$; the sequence $\langle x_n \rangle$ is said to be *$\mathfrak{L}$-convergent*. Define an operator λ on 2^X as follows: $\lambda A = \{x \mid x = \mathfrak{L}\text{-}\lim x_n,\ A \supseteqq \bigcup (x_n)\}$. Then λ is a *closure topology* (also a pretopology) for X. The closure space (X, λ) will be called a *sequential convergence space* and denoted by $(X, \mathfrak{L}, \lambda)$ or, when no confusion seems possible, by X. Notice that the pair $(X, \mathfrak{L})$ s an $\mathscr{L}$-space in the sense of M. FRÉCHET; (X, λ) is an L-space in the sense of E. ČECH, a convergence space in the sense of J. Novák, and a pretopological convergence space in the sense of D. C. Kent.

Denote by $C(X)$ the set of all sequentially continuous, more precisely $\mathfrak{L}$-continuous, functions on X, i.e. $f \in C(X)$ whenever $x = \mathfrak{L}\text{-lim } x_n$ implies $f(x) = \lim f(x_n)$. The *continuous sequential convergence* $\mathfrak{C}$ on $C(X)$ is defined as follows: $(\langle f_n \rangle, f) \in \mathfrak{C}$ iff $f(x) = \lim f_n(x_n)$ whenever $(\langle x_n \rangle, x) \in \mathfrak{L}$. The definition of $\mathfrak{C}$ dates back to K. Kuratowski ([16]); however, the notion of a continuous convergence has been fully developed in the filter setting (cf. [2, 4, 14]). Denote by cl the closure induced by $\mathfrak{C}$. Repeating the process, the sequential convergence $\mathfrak{C}_2$ on $C_2(X) = C(C(X), \mathfrak{C}, \text{cl})$ is defined; the corresponding sequential convergence space is denoted by $(C_2(X), \mathfrak{C}_2, \text{cl}_2)$.

A sequential convergence space $(X, \mathfrak{L}, \lambda)$ is called *sequentially regular* if $\mathfrak{L}$ is projectively generated by $C(X)$, i.e. $(\langle x_n \rangle, x) \in \mathfrak{L}$ iff $f(x) = \lim f(x_n)$ for each $f \in C(X)$. Sequentially regular spaces are analogous to completely regular spaces (cf. [10, 15, 18]). Denote by λ^- the topological modification of λ, i.e. λ^- is the finest of all topologies for X coarser than λ. For $A \subseteq X$ we have $\lambda^- A = \lambda^\Omega A$, where for $\xi \leq \Omega$ we define inductively $\lambda^\xi A = \bigcup_{\eta < \xi} \lambda(\lambda^\eta A)$. Denote by $\lambda^\sim$ the completely regular modification of λ, i.e. $\lambda^\sim$ is the finest of all completely regular topologies for X (no separation axiom is assumed) coarser than λ or, equivalently, $\lambda^\sim$ is the $C(X)$-weak topology. The spaces (X, λ), (X, λ^-), $(X, \lambda^\sim)$, are linked together by $\mathfrak{L}$ as follows: $x = \mathfrak{L}\text{-lim } x_n$ iff each λ^*-neighborhood of x contains all but finitely many x_n, where λ^* stands for any of the closure topologies λ and λ^-; if $(X, \mathfrak{L}, \lambda)$ is sequentially regular, then λ^* stands also for $\lambda^\sim$ (cf. [15]). Further, let $(Y, \mathfrak{M}, \mu)$ be a sequential convergence space and f a mapping of X into Y. Then the following are equivalent (cf. [5, 15]):

(a) $(\langle x_n \rangle, x) \in \mathfrak{L}$ implies $(\langle f(x_n) \rangle, f(x)) \in \mathfrak{M}$,

(b) $x \in \lambda A$ implies $f(x) \in \mu f(A)$,

(c) $x \in \lambda^- A$ implies $f(x) \in \mu^- f(A)$;

if $(Y, \mathfrak{M}, \mu)$ is sequentially regular, then (a) is equivalent also to

(d) $x \in \lambda^\sim A$ implies $f(x) \in \mu^\sim f(A)$.

Clearly $C(X) = C(X, \lambda^\sim)$.

1. Continuous convergence in $C(X)$

Let $(X, \mathfrak{L}, \lambda)$ be a sequential convergence space. In this section we study the space $(C(X), \mathfrak{C}, \text{cl})$ and prove a representation theorem for sequentially regular spaces.

Lemma 1. *Let $f_n, f \in C(X)$. Then the following conditions are equivalent:*

(a) $f = \mathfrak{C}\text{-lim } f_n$.

(b) $\lim_{m,n \to \infty} f_n(x_m) = f(x)$ *whenever* $x = \mathfrak{L}\text{-lim } x_m$.

(c) *The sequence $\langle f_n \rangle$ converges to f uniformly on each sequentially compact subset $A \subseteq X$.*

The straightforward proof is omitted.

Lemma 2. *Let $\alpha = (\langle x_m \rangle, x_\omega) \in \mathfrak{L}$. Then the function Φ_α defined on $C(X)$ by $\Phi_\alpha(f) = \max \{f(x_\xi), \xi \leq \omega\}$ is $\mathfrak{C}$-continuous.*

Proof. Let $f = \mathfrak{C}\text{-}\lim f_n$. We are to prove that $\lim \Phi_\alpha(f_n) = \Phi_\alpha(f)$. Obviously: (a) $\lim_{n,m\to\infty} f_n(x_m) = f(x_\omega)$; (b) $\lim f_n(x_\xi) = f(x_\xi)$, $\xi \leqq \omega$; (c) $\Phi_\alpha(f_\xi) \geqq f_\xi(x_\omega)$, $\xi \leqq \omega$. From (a), (b), and (c) it follows immediately

$$\infty > \limsup \Phi_\alpha(f_n) \geqq \liminf \Phi_\alpha(f_n) \geqq f(x_\omega).$$

There are two possibilities: 1. $\Phi_\alpha(f) = f(x_\omega)$. Suppose that, on the contrary, there is an $\varepsilon > 0$ such that $\limsup \Phi_\alpha(f_n) \geqq f(x_\omega) + \varepsilon$. Then there is a subsequence $\langle f_n' \rangle$ of $\langle f_n \rangle$ such that $\limsup \Phi_\alpha(f_n) = \lim \Phi_\alpha(f_n')$. From (a) and (b) it follows that there is a $k \in N$ such that for each $n, \xi > k$, $n \in N$, $\xi \leqq \omega$, we have $|f_n(x_\xi) - f(x_\omega)| < \varepsilon/2$, and hence there is an $m_0 \leqq k$ such that for infinitely many $n \in N$ we have $\Phi_\alpha(f_n') = f_n'(x_{m_0})$. Thus

$$\lim f_n(x_{m_0}) = f(x_{m_0}) \geqq f(x_\omega) + \varepsilon = \Phi_\alpha(f) + \varepsilon,$$

which is a contradiction.

2. $\Phi_\alpha(f) > f(x_\omega) + \varepsilon$, for some $\varepsilon > 0$. Then $\Phi_\alpha(f) = f(x_{m_0})$ for some m_0. From (a) and (b) it follows that there is a $k > m_0$, $k \in N$, such that for each $n \geqq k$ we have $\Phi_\alpha(f_n) = \max \{f_n(x_m), m < k\}$. In this case the rest of the proof is trivial and is omitted.

Theorem 3. *The space* $(C(X), \mathfrak{C}, \mathrm{cl})$ *is sequentially regular.*

Proof. Let $(\langle f_n \rangle, f) \notin \mathfrak{C}$. Then there is a $\beta = (\langle x_m \rangle, x_\omega) \in L$ such that $\lim_{n,m\to\infty} f_n(x_m) = f(x_\omega)$ does not hold. Consequently, there is an $\varepsilon > 0$ such that for each $k \in N$ there are $n_k, m_k > k$ such that $|f_{n_k}(x_{m_k}) - f(x_\omega)| > \varepsilon$. Let m_0 be a natural number such that for each $m > m_0$ we have $|f(x_m) - f(x_\omega)| < \varepsilon$. Denote by $\alpha = (\langle x_{m+m_0} \rangle, x_\omega)$. Then either $\lim \Phi_\alpha(f_n) = \Phi_\alpha(f)$ or $\lim \Psi_\alpha(f_n) = \Psi_\alpha(f)$ does not hold, where $\Psi_\alpha(f) = \Phi_\alpha(-f)$. Since $\Phi_\alpha, \Psi_\alpha \in C_2(X)$, the proof is finished.

We sketch an indirect proof of Theorem 3. Consider the filter convergence space $(X, q(\mathfrak{L}))$ (cf. [9]). It follows from Theorem 3.2 in [14] that $C(X)$ equipped with the continuous filter convergence σ is completely regular. Since $(C(X), \mathfrak{C}, \mathrm{cl})$ is associated with $(C(X), \sigma)$, the assertion follows from Theorem 4 in [15].

Now we shall prove a new representation theorem for sequentially regular spaces. The original one (cf. Theorem 11 in [18]) states that a sequential convergence space $(X, \mathfrak{L}, \lambda)$ is sequentially regular iff it is homeomorphic with a subspace of $R^{C(X)}$ equipped with the pointwise sequential convergence and the induced closure topology.

Let (P, u) be a closure space and $C(P)$ the set of all continuous functions on P. Then the *evaluation mapping* ev_P *of* P *into* $R^{C(P)}$ is defined by $\mathrm{ev}_P(p) = \langle \Phi_p(f) \rangle_{f \in C(P)}$, where $\Phi_p(f) = f(p)$.

Theorem 4. *A sequential convergence space* $(X, \mathfrak{L}, \lambda)$ *is sequentially regular iff it is homeomorphic with a subspace of* $(C_2(X), \mathfrak{C}_2, \mathrm{cl}_2)$.

Proof. *I. Sufficiency.* By Theorem 3 the space $C_2(X)$ is sequentially regular, and sequential regularity is invariant under homeomorphisms and the formation of subspaces (cf. [18]).

II. Necessity. We shall prove that the evaluation mapping ev_X is a homeomorphism of $(X, \mathfrak{L}, \lambda)$ into $(C_2(X), \mathfrak{C}_2, \mathrm{cl}_2)$. Since $C(X)$ separates points, ev_X is one-to-one. Let $(\langle x_n \rangle, x) \in \mathfrak{L}$. Then $(\langle f_n \rangle, f) \in \mathfrak{C}$ implies $\lim f_n(x_n) = f(x)$. Consequently, $(\langle \Phi_{x_n} \rangle, \Phi_x) \in \mathfrak{C}_2$ and ev_X is continuous. Conversely, $(\langle \Phi_{x_n} \rangle, \Phi_x) \in \mathfrak{C}_2$ implies that for each $f \in C(X)$ we have

$\lim \Phi_{x_n}(f) = \Phi_x(f)$, i.e. $\lim f(x_x) = f(x)$. Since X is sequentially regular, we have $(\langle x_n \rangle, x) \in \mathfrak{L}$ and $\mathrm{ev}_X^{\leftarrow}$ is also continuous.

Lemma 5. *Let $f = \mathfrak{C}\text{-}\lim f_n$ in $C(X)$. Then the function g defined on X by $g(x) = \sup \{f_n(x), n \in N\}$ is $\mathfrak{L}$-continuous.*

Proof. Obviously, the function g is well defined. Let $x = \mathfrak{L}\text{-}\lim x_m$. We are to prove that $g(x) = \lim g(x_m)$. For each $\varepsilon > 0$ there is an $n_0 \in N$ such that $f_{n_0}(x) > g(x) - \varepsilon$. Then $\liminf_{m\to\infty} g(x_m) \geqq \liminf_{m\to\infty} f_{n_0}(x_m) = f_{n_0}(x) > g(x) - \varepsilon$ Thus $\liminf_{m\to\infty} g(x_m) \geqq g(x)$. Denote by $a = \limsup_{m\to\infty} g(x_m)$. Now, there are two possibilities:

1. There is an $n_1 \in N$ such that $a \leqq \limsup_{m\to\infty} h(x_m)$, where $h(y) = \sup \{f_k(y), k \leqq n_1\}$. Obviously $h \in C(X)$ and hence $a = \limsup_{m\to\infty} h(x_m) = h(x) \geqq g(x)$.

2. For each $n \in N$ we have $a > \limsup_{m\to\infty} [\sup \{f_k(x_m), k \leqq n\}]$. Then there are subsequences $\langle f_{n_i} \rangle$ of $\langle f_n \rangle$ and $\langle x_{m_i} \rangle$ of $\langle x_m \rangle$ such that $a = \lim f_{n_i}(x_{m_i})$. Consequently, $a = \lim_{m,n\to\infty} f_n(x_m) = f(x) = \lim f(x_m) \leqq g(x)$.

Thus $g(x) \geqq \limsup_{m\to\infty} g(x_m) \geqq \liminf_{m\to\infty} g(x_m) \geqq g(x)$, and the proof is complete.

2. Sequentially complete spaces

The reader is referred to [10] and [11] for a study in this area. We say that a sequentially regular space $(X, \mathfrak{L}, \lambda)$ is *sequentially complete* if X is closed in each sequentially regular space $(Y, \mathfrak{M}, \mu)$ in which it is C-embedded (i.e. X is a subspace of Y and each $f \in C(X)$ has an extension $f^\wedge \in C(Y)$). A sequence $\langle x_n \rangle$ in X is said to be *fundamental* if $\lim f(x_n)$ exists for each $f \in C(X)$. Notice that the following are equivalent:

(a) $(X, \mathfrak{L}, \lambda)$ is sequentially complete,

(b) each fundamental sequence $\langle x_n \rangle$ is $\mathfrak{L}$-convergent in X (cf. [15], where the terms a remarkable sequence and an $\mathscr{L}$-complete space are used),

(c) $(X, \mathfrak{L}, \lambda)$ has the property p (see [8]),

(d) $(X, \mathfrak{L}, \lambda)$ is the sequential envelope of itself.

The category of sequentially complete spaces is closed under the formation of closed subspaces and sequential products; it is an epireflective subcategory of the category of sequentially regular spaces, and the sequential envelope $\sigma(X)$ is the epireflection of $(X, \mathfrak{L}, \lambda)$.

Lemma 6. *Let $(P, \mathfrak{P}, \pi)$ be a sequential convergence space. Let $\langle x_n \rangle$ be a sequence $\mathfrak{P}$-converging to $x \in (P - \bigcup (x_n))$, $\langle f_m{}^\wedge \rangle$ a sequence in $C(P)$, and $f^\wedge \in C(P)$. If $\lim_{m,n\to\infty} f_m{}^\wedge(x_n)$ does not exist, then there are an $\varepsilon > 0$ and subsequences $\langle m_i \rangle$ of $\langle m \rangle$ and $\langle n_i \rangle$ of $\langle n \rangle$ such that the sequence $\langle x_{n_i} \rangle$ is one-to-one and $|f^\wedge_{m_i}(x_{n_i}) - f^\wedge(x_{n_i})| > \varepsilon$ for each $i \in N$.*

The obvious proof is omitted.

Theorem 7. *Let $(X, \mathfrak{L}, \lambda)$ be a sequentially regular space and $\langle x_n \rangle$ a sequence in X. Then the following are equivalent:*

(a) *$\langle x_n \rangle$ is fundamental.*

(b) *$\lim_{m,n\to\infty} f_m(x_n)$ exists whenever $\langle f_m \rangle$ is a $\mathfrak{C}$-convergent sequence.*

Proof. (a) implies (b). Let (P, u) be the HEWITT realcompactification of $(X, \lambda^{\sim})$ and $(P, \mathfrak{P}, \pi)$ the associated sequential convergence space, i.e. $\mathfrak{P}$ is the convergence of sequences in (P, u) and π is the induced closure topology. Then $(P, \mathfrak{P}, \pi)$ is sequentially complete and $(X, \mathfrak{L}, \lambda)$ is C-embedded in $(P, \mathfrak{P}, \pi)$ (cf. Theorem 3 and Theorem 8 in [8]). For each $f \in C(X, \lambda^{\sim}) = C(X, \mathfrak{L}, \lambda)$ denote by $f^{\wedge} \in C(P, u)$ its (unique) continuous extension. From the completeness of $(P, \mathfrak{P}, \pi)$ it follows that there is an $x \in P$ such that $x = \mathfrak{P}\text{-}\lim x_n$. Now suppose that, on the contrary, (b) does not hold. Then there is a sequence $\langle f_m \rangle$ and a function f in $C(X)$ such that $f = \mathfrak{C}\text{-}\lim f_m$ and $\lim_{m,n\to\infty} f_m(x_n)$ does not exist. Thus $x \in P - X$. According to Lemma 6 there are an $\varepsilon > 0$ and subsequences $\langle m_i \rangle$ of $\langle m \rangle$ and $\langle n_i \rangle$ of $\langle n \rangle$ such that the sequence $\langle x_{n_i} \rangle$ is one-to-one and $|f_{m_i}(x_{n_i}) - f(x_{n_i})| > \varepsilon$ for each $i \in N$. Denote by $A = (x) \cup \left(\bigcup_{i=1}^{\infty} (x_{n_{2i-1}}) \right)$. Since A is closed in (P, u), there is for each $x_{n_{2i}}$ a function $h^{\wedge}_{2i} \in C(P, u)$ such that $h^{\wedge}_{2i}(x_{n_{2i}}) = 1$ and $h^{\wedge}_{2i}(A) = 0$. Define a function $g^{\wedge}_{2i}$ on P by $g^{\wedge}_{2i}(y) = |f^{\wedge}_{m_{2i}}(y) - f^{\wedge}(y)| \cdot h^{\wedge}_{2i}(y)$. Then $g^{\wedge}_{2i} \in C(P, u)$, $g^{\wedge}_{2i}(A) = 0$, $g^{\wedge}_{2i}(x_{n_{2i}}) > \varepsilon$, $g_{2i} = g^{\wedge}_{2i}/X \in C(X)$, and $0 = \mathfrak{C}\text{-}\lim g_{2i}$ in $C(X)$. It follows from Lemma 5 that $g = \sup \{g_{2i}, i \in N\} \in C(X)$. Moreover, for each $i \in N$ we have $g(x_{n_{2i}}) \geqq \varepsilon$ and $g(x_{n_{2i-1}}) = 0$. Thus the sequence $\langle x_{n_i} \rangle$ is not fundamental. This contradicts (a).

(b) implies (a). Obvious.

Lemma 8. *Let $(X, \mathfrak{L}, \lambda)$ be a sequential convergence space and $\langle f_m \rangle$ a sequence in $C(X)$ such that*

(*) $\lim_{m,n\to\infty} f_m(x_n)$ *exists whenever $\langle x_n \rangle$ is an $\mathfrak{L}$-convergent sequence.*

Then there is an $f \in C(X)$ such that $f = \mathfrak{C}\text{-}\lim f_m$.

The proof is easy and omitted.

Theorem 9. *Let $(X, \mathfrak{L}, \lambda)$ be a sequential convergence space and $\langle f_m \rangle$ a sequence in $C(X)$. Then the following are equivalent:*

(a) *$\langle f_m \rangle$ is fundamental.*

(b) *$\langle f_m \rangle$ satisfies the condition (*) in Lemma 8.*

(c) *$0 = \mathfrak{C}\text{-}\lim (f_i - f_{m_i})$ for each subsequence $\langle f_{m_i} \rangle$ of $\langle f_m \rangle$.*

Proof. (a) implies (b). Let $\langle f_m \rangle$ be a fundamental sequence and $(\langle x_n \rangle, x) \in \mathfrak{L}$. Then obviously $(\langle \Phi_{x_n} \rangle, \Phi_x) \in \mathfrak{C}_2$ and hence, by Theorem 7, $\lim_{m,n\to\infty} f_m(x_n) = \lim_{m,n\to\infty} \Phi_{x_n}(f_m)$ exists.

(b) implies (c). Let $\langle f_m' \rangle$ be a subsequence of $\langle f_m \rangle$. It follows from Lemma 8 that for some $f \in C(X)$ we have $f = \mathfrak{C}\text{-}\lim f_m = \mathfrak{C}\text{-}\lim f_m'$. Thus $0 = \mathfrak{C}\text{-}\lim (f_m - f_m')$.

(c) implies (a). We shall prove that $\langle f_m \rangle$ is uniformly CAUCHY on each sequentially compact set $A \subseteqq X$, which means that $\langle f_m \rangle$ converges to some $f \in C(X)$ uniformly on each sequentially compact set $A \subseteqq X$. It will follow from Lemma 1 that $f = \mathfrak{C}\text{-}\lim f_m$, and hence that $\langle f_m \rangle$ is fundamental. Now, let $A \subseteqq X$ be a sequentially compact set. Suppose that, on the contrary, $\langle f_m \rangle$ is not uniformly CAUCHY on A. Then there is an $\varepsilon > 0$ such that for each $k \in N$ there are $m, n > k$ and a $y \in A$ such that $|f_m(y) - f_n(y)| \geqq \varepsilon$. Consequently, there is in A a convergent sequence $(\langle x_{2m-1} \rangle, x) \in \mathfrak{L}$ and a subsequence $\langle f_m' \rangle$ of $\langle f_m \rangle$ such that

$$|f'_{2m-1}(x_{2m-1}) - f'_{2m}(x_{2m-1})| \geqq \varepsilon.$$

Define $x_{2m} = x_{2m-1}$, $m \in N$. Then $(\langle x_m \rangle, x) \in \mathfrak{L}$ and $[\limsup |f_m'(x_m) - f'_{m+1}(x_m)|] \geqq \varepsilon$. Since $\langle f_m' \rangle$ also satisfies (c), we have a contradiction.

Corollary 10. *Let $(X, \mathfrak{L}, \lambda)$ be a sequential convergence space. Then the space $(C(X), \mathfrak{C}, \mathrm{cl})$ is sequentially complete.*

Notice that $C(X)$ is a linear convergence space in the sense of R. M. Dudley [6], a *cc*-group in the sense of J. Novák [19], and if we define two sequences $\langle f_n \rangle$ and $\langle g_n \rangle$ to be near whenever $0 = \mathfrak{C}\text{-lim } (f_n - g_n)$, then $C(X)$ is a $\mathcal{UL}^*$-space in the sense of A. Goetz [12]. In all three cases a sequence $\langle f_m \rangle$ satisfying the above condition (c) is called a Cauchy sequence. Since each Cauchy sequence in $C(X)$ is $\mathfrak{C}$-convergent, $C(X)$ is complete in all three senses.

We conclude this section with a new representation theorem for sequentially complete spaces. The original theorem (cf. Theorem 4 in [8]) asserts that a sequential convergence space $(X, \mathfrak{L}, \lambda)$ is sequentially complete iff it is homeomorphic with a closed subspace of a sequential power R^m of real lines.

Theorem 11. *A sequential convergence space $(X, \mathfrak{L}, \lambda)$ is sequentially complete iff it is homeomorphic with a closed subspace of $(C_2(X), \mathfrak{C}_2, \mathrm{cl}_2)$.*

Proof. *I.* The *sufficiency* follows from the fact that $C_2(X)$ is sequentially complete and sequential completeness is invariant under homeomorphisms and the formation of closed subspaces.

II. Necessity. Let $(X, \mathfrak{L}, \lambda)$ be sequentially complete. Since (cf. Theorem 4) ev_X is a homeomorphism of X into $C_2(X)$, it suffices to prove that $\mathrm{ev}_X(X)$ is closed in $C_2(X)$. Let $(\langle \Phi_{x_n} \rangle, \Phi) \in \mathfrak{C}_2$. Then $(\langle f_m \rangle, f) \in \mathfrak{C}$ implies $\Phi(f) = \lim_{m,n\to\infty} \Phi_{x_n}(f_m) = \lim_{m,n\to\infty} f_m(x_n)$. It follows from Theorem 7 that $\langle x_n \rangle$ is a fundamental sequence in X. Thus $(\langle x_n \rangle, x) \in \mathfrak{L}$ for some $x \in X$, and hence $\Phi = \Phi_x \in \mathrm{ev}_X(X)$. Consequently $\mathrm{cl}\,(\mathrm{ev}_X(X)) = \mathrm{ev}_X(X)$ and the proof is complete.

3. Sequential envelope

We say that a sequentially complete space $(Y, \mathfrak{M}, \mu)$ is a *sequential envelope* $\sigma(X)$ of a sequentially regular space $(X, \mathfrak{L}, \lambda)$ if X is C-embedded μ^--dense subspace of Y (cf. Theorem 6 in [8]). Each sequentially regular space $(X, \mathfrak{L}, \lambda)$ possesses a $\sigma(X)$; namely, the smallest sequentially closed subset of $R^{C(X)}$ which contains $\mathrm{ev}_X(X)$, equipped with the pointwise sequential convergence and the induced closure topology. Moreover, $\sigma(X)$ is uniquely determined up to a commuting homeomorphism (cf. [18]). As a rule, $\varphi: (P, u) \to (Q, v)$ will denote a continuous mapping φ of a closure space P into a closure space Q. Consider $\mathrm{ev}_X: (X, \mathfrak{L}, \lambda) \to (C_2(X), \mathfrak{C}_2, \mathrm{cl}_2)$ and denote by $Y = \mathrm{cl}_2\,(\mathrm{ev}_X(X))$, $\mathfrak{M} = \mathfrak{C}_2/_Y$, and $\mu = \mathrm{cl}_2/_Y$ the induced closure topology. We shall prove somewhat surprising result that $\delta(X) = (Y, \mathfrak{M}, \mu)$ is a sequential envelope of $(X, \mathfrak{L}, \lambda)$.

Theorem 12. *Let $(X_1, \mathfrak{L}_1, \lambda_1)$ and $(X_2, \mathfrak{L}_2, \lambda_2)$ be sequentially regular spaces and $\varphi: X_1 \to X_2$. Then there is a uniquely determined mapping $\varphi^\wedge: \delta(X_1) \to \delta(X_2)$ such that the*

diagram

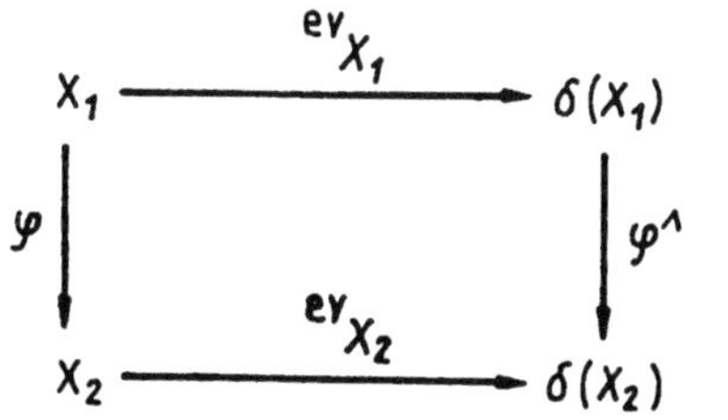

commutes.

Proof. Let $\psi\colon C_2(X_1) \to C_2(X_2)$ be the second conjugate of φ, i.e. $\big(\psi(\Phi)\big)\,(g) = \Phi(\varphi \circ g)$, $\Phi \in C_2(X_1)$, $g \in C(X_2)$. Then

$$\psi\big(\delta(X_1)\big) = \psi\big(\mathrm{cl}_2^{-}(\mathrm{ev}_{X_1}(X_1))\big) \subseteqq \mathrm{cl}_2^{-}\big(\psi(\mathrm{ev}_{X_1}(X_1))\big) \subseteqq \mathrm{cl}_2^{-}\big(\mathrm{ev}_{X_2}(X_2)\big) = \delta(X_2)$$

and the diagram

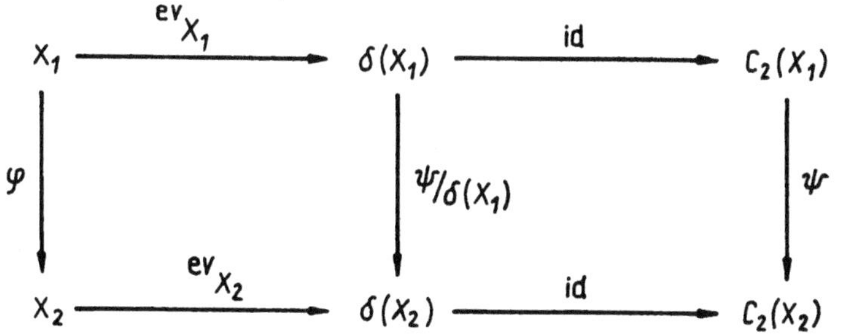

commutes. Put $\varphi^\wedge = \psi/_{\delta(X_1)}$. The uniqueness of $\varphi^\wedge$ follows by a standard topologica argument (cf. Lemma 5 in [18]).

Corollary 13. *$\delta(X)$ is a sequential envelope of X.*

Proof. It follows from the definition of $\delta(X)$ that X can be identified with the cl_2^{-}-dense subspace $\mathrm{ev}_X(X)$ of the sequentially complete space $\delta(X)$. It suffices to prove that X is C-embedded in $\delta(X)$. Put in the above diagram $X_2 = R$. Since R is sequentially complete, R can be identified with $\delta(R)$. Thus each $\varphi \in C(X)$ can be continuously extended to a $\varphi^\wedge \in C\big(\delta(X)\big)$.

Notice that $\delta(X)$ is an epireflection of X in the category of sequentially complete spaces. Since the epireflection is unique up to a commuting isomorphism, Corollary 13 follows from the general category theory.

Lemma 14. *Let $f^\wedge \in C\big(\delta(X)\big)$ be the extension of an $f \in C(X)$. Then $f^\wedge = f^*/_{\delta(X)}$, where $f^*\colon C_2(X) \to R$ is defined by $f^*(\Phi) = \Phi(f)$.*

Proof. Clearly, the diagram

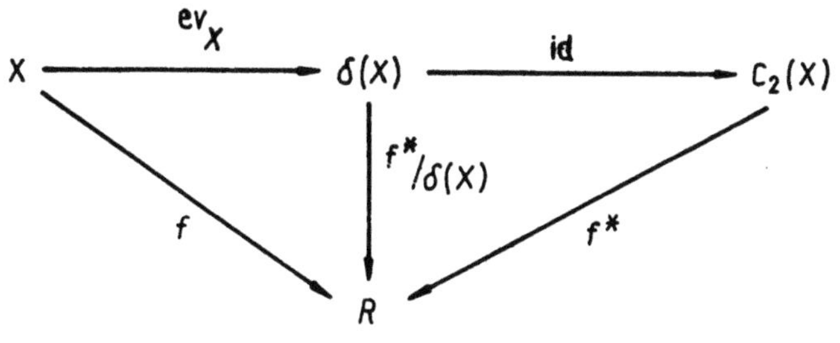

commutes.

Theorem 15. *Let* $f = \mathfrak{C}\text{-}\lim f_n$ *in* $C(X)$. *Then* $f^\wedge = \mathfrak{C}\text{-}\lim f_n^\wedge$ *in* $C(\delta(X))$, *where* $g^\wedge \in C(\delta(X))$ *denotes the unique extension of a* $g \in C(X)$.

Proof. Let $(\langle \Phi_n \rangle, \Phi) \in \mathfrak{C}_2$, $\Phi_n, \Phi \in \delta(X)$. Then $(\langle f_n \rangle, f) \in \mathfrak{C}$ implies $\lim \Phi_n(f_n) = \Phi(f)$, and hence $\lim f_n^\wedge(\Phi_n) = f^\wedge(\Phi)$.

Corollary 16. $C(X)$ *and* $C(\delta(X))$ *are topologically isomorphic.*

Let X be a completely regular topological space and vX the Hewitt realcompactification of X. Then (cf. Theorem 2 in [13]) in $C(X)$ a sequence $\langle f_n \rangle$ converges pointwise on X to f iff in $C(vX)$ the sequence $\langle f_n^\wedge \rangle$ of the extensions $f_n^\wedge$ of f_n converges pointwise on vX to the extension $f^\wedge$ of f. The following question arises. *Is it possible to replace* $\delta(X)$ *in Theorem* 15 *with some larger set, e.g. with* vX?

More generally, let X *be a* C_0*-embedded dense subspace of a completely regular topological space* P, *where* $C_0 \subseteqq C(X)$. *Let* $\mathfrak{K}$ *be a sequential convergence defined for spaces of functions. For which* X, P, C_0, *and* $\mathfrak{K}$, *it is true that* $f = \mathfrak{K}\text{-}\lim f_n$ *in* C_0 *iff* $f^\wedge = \mathfrak{K}\text{-}\lim f_n^\wedge$ *in* $C(P)$?

References

[1] M. Ja. Antonovskiĭ and I. G. Koshevnikova, Fréchet-Urysohn type convergence spaces and generalized metrization (Russian), Mat. Vesnik **24** (1972), 373–378.

[2] E. Binz, Recent results in the functional analytic investigation of convergence spaces, Proc. Third Prague Topological Sympos. 1971, 67–72, Prague 1972.

[3] N. Bourbaki, Topologie générale, Paris 1961.

[4] C. H. Cook and H. R. Fischer, On equicontinuity and continuous convergence, Math, Ann. **159** (1965), 94–104.

[5] E. Čech, Topological spaces, Prague 1966.

[6] R. M. Dudley, On sequential convergence, Trans. Amer. Math. Soc. **112** (1964), 483–507.

[7] M. Fréchet, Sur quelques points du calcul fonctionel, Rend. Circ. Mat. Palermo **22** (1906), 1–74.

[8] R. Frič, Sequential envelope and subspaces of the Čech-Stone compactification, Proc. Third Prague Topological Sympos. 1971, 123–126, Prague 1972.

[9] R. Frič, On convergence spaces and groups, Math. Slovaca **27** (1977), 323–330.

[10] R. Frič, On E-sequentially regular spaces, Czechoslovak Math. J. **26** (1976), 604–612.

[11] R. Frič and V. Koutník, Sequentially complete spaces, Czechoslovak Math. J. **29** (1979), 287–297.

[12] A. Goetz, A notion of uniformity for $\mathscr{L}$-spaces of Fréchet, Proc. First Prague Topological Sympos. 1961, Prague 1962, 177–178.

[13] J. R. Isbell and E. S. Thomas Jr., Mazur's theorem on sequentially continuous functionals, Proc. Amer. Math. Soc. **14** (1963), 644–647.

[14] D. C. Kent, K. McKennon, G. D. Richardson, and M. Schroder, Continuous convergence in $C(X)$, Pacific J. Math. **52** (1974), 271–279.

[15] V. Koutník, On sequentially regular convergence spaces, Czechoslovak Math. J. **17** (1967), 232–247.

[16] K. Kuratowski, Topology, New York 1966.

[17] E. H. Moore and H. L. Smith, A general theory of limits, Amer. J. Math. **44** (1922), 102–121.

[18] J. Novák, On convergence spaces and their sequential envelopes, Czechoslovak Math. J. **15** (1965), 74–100.

[19] J. Novák, On completions of convergence commutative groups, Proc. Third Prague Topological Sympos. 1971, Prague 1972, 335–340.
[20] K. Wichterle, On $\mathfrak{B}$-convergence spaces, Czechoslovak Math. J. 18 (1968), 569–588.

Anschriften der Verfasser:

Fakulta strojnícka a elektrotechnická
Vysoká škola dopravná
Marxa-Engelsa 25
01088 Žilina
ČSSR

Department of Pure and Applied Mathematics
Washington State University
Pullman, Washington 99163
U.S.A.

Department of Mathematics
East Carolina University
Greenville, North Carolina 27834
U.S.A.

Generalized Notions of Differentiability and Their Application in Optimization Theory

SIEGFRIED GÄHLER, Berlin

1. Introduction

In the last years investigations on optimization theory, to avoid unnecessary strong differentiability assumptions, several authors used generalized derivatives. Of special importance are algebraic and topological lower derivatives (or subgradients) and the corresponding upper derivatives [1, 3, 12, 17].

In this paper we study a certain rather weak notion of a topological lower derivative (section 3), two generalizations (section 4), and the corresponding notion of a second topological lower derivative (section 5). These derivatives are introduced for mappings f of a subset Q of a separated pseudo-topological vector space R into an ordered pseudo-topological vector space S and defined on a cone $C_{q,Q}$ of R with 0 as vertex which depends on the set Q and the point q at which the derivatives are to be considered. They are only supposed to have a weakened concavity property, i.e. must neither be homogeneous nor in the case of a convex $C_{q,Q}$ be additive. Concerning results which are obtained for these derivatives, especially we refer to the generalizations of the mean value theorem (Theorems 3 and 9) and of the chain rule (Theorems 5, 7, and 11). These derivatives and the corresponding upper derivatives are useful in vector optimization where we consider local maxima of f at q of the following three kinds: There exists a neighbourhood U of q in R such that

$$f[Q \cap U] - f(q) \subseteq -K, \quad \subseteq S \setminus (K \setminus \{0\}), \quad \text{or} \subseteq S \setminus \underline{K},$$

respectively, where K denotes the positive cone of S. Extending simple classical results, for these three kinds of local maxima the paper gives necessary (section 6 and 8) and sufficient (section 7) optimality conditions.

2. Preliminaries

Let R be a set. By $\mathcal{F}R$ we denote the set of all filters in R (included the improper filter). A non-empty set u of filters in R is called a $\wedge$-*ideal of filters* in R if u has the following properties:

1. $\mathfrak{F}, \mathfrak{F}' \in u \Rightarrow \mathfrak{F} \cap \mathfrak{F}' \in u$,
2. $\mathfrak{F} \in u, \mathfrak{F} \subseteq \mathfrak{F}' \in \mathcal{F}R \Rightarrow \mathfrak{F}' \in u$.

Let $\mathcal{J}R$ denote the set of all $\wedge$-ideals of filters in R. A mapping τ of R into $\mathcal{J}R$ is called a *pseudo-topology of* R if for every $p \in R$, the $\wedge$-ideal $\tau(p)$ contains the principal filter $[p] = \{A; p \in A \subseteq R\}$ and the filters $F \in \tau(p)$ are said to be *convergent* to p (with respect to τ). A net $(p_i)_{i\in I}$ of points $p_i \in R$ is said to be convergent to p (with respect to τ) if with respect to τ the filter in R having $\{\{p_j; j \geqq i\}; i \in I\}$ as base converges to p. Let σ and τ be pseudo-topologies of R. σ is called *finer* than τ and τ *coarser* than σ if $\sigma(p) \subseteqq \tau(p)$ for all $p \in R$. In this case we write $\sigma \leqq \tau$. A pseudo-topological space R and also its pseudo-topology τ are said to be *separated* if for all points p, $p' \in R$ with $p \neq p'$, $\tau(p) \cap \tau(p')$ is the improper filter. A point p of a pseudo-topological space R is called an *adherent point* of a filter $\mathfrak{F}$ in R if there is a proper filter containing $\mathfrak{F}$ and converging to p. The set of all adherent points of the principal filter $[A] = \{B; A \subseteqq B \subseteqq R\}$ is called *adherence of* A and is denoted by $\bar{A}$ (or $\bar{A}^\tau$ if necessary). In the case $A = \bar{A}$, the set A is said to be *closed*. The closed hull $\bar{A}^{t(\tau)}$ of A with respect to the finest topology $t(\tau)$ of R coarser than τ is the smallest closed set containing A. The set $R \setminus \overline{R \setminus A}$ is called interior of A and denoted by $\underline{A}$ (or $\underline{A}_\tau$ if necessary). In the case $A = \underline{A}$, the set A is said to be *open*. For every $p \in R$, the filter $\mathfrak{N}(p) = \bigcap_{\mathfrak{F}\in\tau(p)} \mathfrak{F}$ is called *neighbourhood filter of* p and every set $N \in \mathfrak{N}(p)$ is called *neighbourhood of* p. A set A is a neighbourhood of p if and only if $p \in \underline{A}$. A mapping f of a pseudo-topological space R into a pseudo-topological space S is called *continuous at a point* $p \in R$ if for every filter $\mathfrak{F}$ in R converging to p, the filter $f(\mathfrak{F})$ in S (with $\{f[F]; F \in \mathfrak{F}\}$ as a base) converges to $f(p)$. f is called *continuous* if f is continuous at each point of R.

A pseudo-topology τ of a vector space R over $\boldsymbol{R}$ is called *compatible* (*with the linear structure of* R) or a *vector pseudo-topology* and R equipped with τ is called *pseudo-topological vector space* if the vector addition and the scalar multiplication are continuous operations. τ is a compatible pseudo-topology of R if and only if $\tau(p) = p + \tau(0)$ for every $p \in R$ and $\tau(0)$ has the following properties:

1. $\tau(0) + \tau(0) \subseteqq \tau(0)$,
2. $\alpha\tau(0) \subseteqq \tau(0)$ for every $\alpha \in \boldsymbol{R}$,
3. $\boldsymbol{V}\tau(0) \subseteqq \tau(0)$ where $\boldsymbol{V}$ denotes the neighbourhood filter of 0 in $\boldsymbol{R}$,
4. $\boldsymbol{V}p \subseteqq \tau(0)$ for every $p \in R$.

In the case $\dim R < \infty$, every separated vector pseudo-topology on R is a vector topology. For every subset A of a vector space R, by cA we denote the convex hull of A. A pseudo-topological vector space R is said to be *locally convex* if for every filter $\mathfrak{F} \in \tau(0)$, the filter with $\{cF; F \in \mathfrak{F}\}$ as base belongs to $\tau(0)$. A subset A of a pseudo-topological vector space R is said to be *bounded* if $\boldsymbol{V}A$ converges to 0. If every filter of $\tau(0)$ contains a bounded set, the pseudo-topological vector space R is called *locally bounded*.

A vector space R is said to be *ordered* if it is equipped with a partial ordering $\leqq$ which is compatible with the linear structure of R, i.e. such that

1. $p \leqq p'$ implies $p + q \leqq p' + q$ for all $p, p', q \in R$

and

2. $p \leqq p'$ implies $\alpha p \leqq \alpha p'$ for all $p, p' \in R$ and $\alpha \geqq 0$.

$K = \{p \in R; p \geqq 0\}$ is a cone in R with 0 as vertex, the so-called *positive cone*, which is proper, that means is convex and fulfils $K \cap (-K) = \{0\}$. Conversely, R being a vector space and K a proper cone in R, R may be ordered by $p \leqq p' \Leftrightarrow p' - p \in K$ where as positive cone one gets K. A pseudo-topological vector space R is said to be *ordered* if the underlying vector space is ordered and the positive cone K is closed. A subset A of an ordered vector space R is said to be *order-convex* if $(A + K) \cap (A - K)$

$= A$. For every subset A of R, there exists an *order-convex hull* co A, i.e. a smallest order-convex subset of R containing A. An ordered pseudo-topological vector space R is called *locally order-convex* if for every filter $\mathfrak{F} \in \tau(0)$, the filter with $\{\text{co } F; F \in \mathfrak{F}\}$ as base belongs to $\tau(0)$. Every ordered pseudo-topological vector space of finite dimension is locally order-convex. The set $\boldsymbol{R}^n$ becomes an ordered topological vector space if we equip it with the natural topology and take $K = \{(\sigma_1, \ldots, \sigma_n); \sigma_i \geqq 0 \text{ for } i = 1, \ldots, n\}$ as positive cone. $\boldsymbol{R}^n$ will be considered always to be this ordered topological vector space. R being a pseudo-topological vector space and K a cone in R with 0 as vertex, by K^* we denote the algebraically dual cone of K, i.e. the set of all linear functionals ϕ on R such that $\phi(p) \geqq 0$ whenever $p \in K$. If $\underline{K} \neq \emptyset$, then every $\phi \in K^*$ is continuous.

Let R be a vector space, A a convex cone in R with 0 as vertex, S an ordered vector space, and K its positive cone. A mapping f of A into S is called *subadditive* if

$$f(p) + f(p') - f(p + p') \in K$$

for every $p, p' \in A$, and *convex* if

$$\alpha f(p) + (1 - \alpha) f(p') - f\big(\alpha p + (1 - \alpha) p'\big) \in K$$

for every $\alpha \in (0, 1)$ and $p, p' \in A$. In the vector space of all mappings of A into S, the set of all subadditive or of all convex mappings is a convex cone containing the zero-mapping 0 which is its vertex. A subadditive mapping f is convex if $\alpha f(p) - f(\alpha p) \in K$ whenever $\alpha \in (0, 1)$ and $p \in A$. A convex mapping is subadditive if $f(\alpha p) - \alpha f(p) \in K$ whenever $\alpha \in (0, 1)$ and $p \in A$. If $-f$ is subadditive or convex, then f is called *super-additive* or *concave*, respectively. f is additive if and only if f is subadditive as well as superadditive. A mapping f of a pseudo-topological space R with the pseudo-topology τ_R into an ordered pseudo-topological vector space S with the vector pseudo-topology τ_S and the positive cone K is called *lower semicontinuous at a point* $p \in R$ if for every filter $\mathfrak{F} \in \tau_R(p)$, there exists a filter $\mathfrak{G} \in \tau_S\big(f(p)\big)$ and for every $G \in \mathfrak{G}$, there exists an $F \in \mathfrak{F}$ such that $f[F] \subseteqq G + K$. f is said to be *upper semicontinuous at* p if $-f$ is lower semicontinuous at p. From the continuity of f at p, there follows the lower and the upper semicontinuity of f at p. In the case of f being locally order-convex, f is continuous at p if and only if f is lower as well as upper semicontinuous at p. f is called *lower* or *upper semicontinuous* if at every point $p \in R$, f is lower or upper semicontinuous, respectively. Now, let A be a subset of a pseudo-topological vector space R with the vector pseudo-topology τ_R and S an ordered pseudo-topological vector space with the vector pseudo-topology τ_S and the positive cone K. For every $p' \in R$, let $\mathcal{N}_{p'}$ denote the set of all filters $\boldsymbol{V}^+(p' + \mathfrak{F})$ where $\boldsymbol{V}^+$ is the neighbourhood filter of the point 0 in $\boldsymbol{R}^+ = [0, \infty)$ and $\mathfrak{F} \in \tau_R(0)$. A mapping f of A into S is called *lower semicontinuous at a point* $p \in A$ *in a direction* $p' \in R$ if for every filter $\mathfrak{N} \in \mathcal{N}_{p'}$, there exists a filter $\mathfrak{G} \in \tau_S\big(f(p)\big)$ and for every $G \in \mathfrak{G}$, there exists an $N \in \mathfrak{N}$ such that $f[(p + N) \cap A] \subseteqq G + K$. If f is lower semicontinuous at p, f is lower semicontinuous at p in every direction $p' \in R$. In the case $\mathcal{N}_0 = \tau_R(0)$, f is lower semicontinuous at p if and only if f is lower semicontinuous at p in the direction 0. f is said to be *upper semicontinuous at* p *in the direction* p' if $-f$ is lower semicontinuous at p in the direction p'.

3. l-Differentiability

Let R be a separated pseudo-topological vector space with the vector pseudo-topology τ_R, Q a subset of R, and q a point of Q. Let $C_{q,Q} = \{p \in R; \forall \mathfrak{N} \in \mathcal{N}_p \exists N \in \mathfrak{N}: q + N \subseteq Q\}$. $C_{q,Q}$ is a cone in R with 0 as vertex. In the case of Q being convex, $C_{q,Q}$ is convex, too. If Q is a convex cone with q as vertex, $C_{q,Q} = \underline{Q} - q$. $q \in \underline{Q}$ implies $0 \in C_{q,Q}$ which is equivalent to $C_{q,Q} = R$. Assume $C_{q,Q} \neq \emptyset$. Let S be an ordered pseudo-topological vector space with the vector pseudo-topology τ_S and the positive cone K. For every subset A of R, let $\mathfrak{A}(A, S)$ denote the linear space of all mappings of A into S.

A mapping $f \in \mathfrak{A}(Q, S)$ is said to be *l-differentiable* (*lower differentiable*) *at* q if there exists a mapping $\phi \in \mathfrak{A}(C_{q,Q}, S)$ which has the following properties:

1. $\phi(\alpha p) - \alpha\phi(p) \in K$ whenever $\alpha \in (0, 1)$ and $p \in C_{q,Q}$.
2. For every point $p \in C_{q,Q}$ and every filter $\mathfrak{F} \in \tau_R(0)$, there exists a filter $\mathfrak{G} \in \tau_S(0)$ and for every $G \in \mathfrak{G}$, there exists an $F \in \mathfrak{F}$ and an $\varepsilon > 0$ such that

$$\frac{f(q + \varepsilon'(p + p')) - f(q)}{\varepsilon'} \in \phi(p) + G + K$$

whenever $\varepsilon' \in (0, \varepsilon]$, $p' \in F$, and $q + \varepsilon'(p + p') \in Q$.

ϕ is called *l-derivative* (*lower derivative*) of f at q. In the case $0 \in C_{q,Q}$, from properties 1 and 2, there follows $\phi(0) = 0$. The set of all l-derivatives of f at q will be denoted by $D_l f(q)$. $D_l f(q)$ is a subset of the linear space $\mathfrak{A}(C_{q,Q}, S)$.

Theorem 1. *$D_l f(q)$ is convex.*

Proof. Let $\phi, \phi' \in D_l f(q)$, $\beta \in (0, 1)$, and $\phi^* = \beta\phi + (1 - \beta)\phi'$. Trivially, $\phi^*(\alpha p) - \alpha\phi^*(p) \in K$ whenever $\alpha \in (0, 1)$ and $p \in C_{q,Q}$. For every point $p \in C_{q,Q}$ and every filter $\mathfrak{F} \in \tau_R(0)$, there exists a filter $\mathfrak{G} \in \tau_S(0)$ and for every $G \in \mathfrak{G}$, there exists an $F \in \mathfrak{F}$ and an $\varepsilon > 0$ such that

$$\frac{f(q + \varepsilon'(p + p')) - f(q)}{\varepsilon'} \in \big(\phi(p) + G + K\big) \cap \big(\phi'(p) + G + K\big)$$

whenever $\varepsilon' \in (0, \varepsilon]$, $p' \in F$, and $q + \varepsilon'(p + p') \in Q$. For every such ε' and p', we get

$$\frac{f(q + \varepsilon'(p + p')) - f(q)}{\varepsilon'} \in \beta\phi(p) + \beta G + \beta K + (1 - \beta)\phi'(p) + (1 - \beta) G + (1 - \beta) K$$
$$\subseteq \phi^*(p) + \big(\beta G + (1 - \beta) G\big) + K.$$

As $\mathfrak{G}^* = \beta\mathfrak{G} + (1 - \beta)\mathfrak{G}$ belongs to $\tau_S(0)$, the theorem is proven.

Theorem 2. *If $D_l f(q) \neq \emptyset$, the mapping f is lower semicontinuous at q in every direction $p \in C_{q,Q}$.*

Proof. Let $\phi \in D_l f(q)$ and $p \in C_{q,Q}$. For every filter $\mathfrak{F} \in \tau_R(0)$, there exists a filter $\mathfrak{G} \in \tau_S(0)$ and for every $G \in \mathfrak{G}$, there exists an $F = F(G) \in \mathfrak{F}$ and an $\varepsilon = \varepsilon(G) > 0$ such that

$$\frac{f(q + \varepsilon'(p + p')) - f(q)}{\varepsilon'} \in \phi(p) + G + K$$

whenever $\varepsilon' \in (0, \varepsilon(G)]$, $p' \in F(G)$, and $q + \varepsilon'(p + p') \in Q$. We may assume $q + [0, \varepsilon(G)] \times (p + F(G)) \subseteqq Q$. $\mathfrak{N} = V^+(p + \mathfrak{F})$ belongs to $\mathcal{N}_p$ and $\mathfrak{G}' = f(q) + V^+(\phi(p) + \mathfrak{G})$ belongs to $\tau_S(f(q))$. Let $G' = f(q) + [0, \varepsilon^*](\phi(p) + G)$ where $\varepsilon^* > 0$ and $G \in \mathfrak{G}$, and let $N = [0, \min(\varepsilon^*, \varepsilon(G))](p + F(G))$. Then

$$f[q + N] \subseteqq f(q) + [0, \varepsilon^*](\phi(p) + G + K)$$
$$\subseteqq G' + K$$

which proves the theorem.

If, especially, we have $Q = R$, $q = 0$, and $f(q) = 0$, if f is lower semicontinuous at q and superadditive, and if $f(\alpha p) - \alpha f(p) \in K$ whenever $\alpha \in (0, 1)$ and $p \in C_{q,Q}$, then $f \in D_1 f(q)$. This is true as for every filter $\mathfrak{F} \in \tau_R(0)$, there exists a filter $\mathfrak{G} \in \tau_S(0)$ and for every $G \in \mathfrak{G}$, there exists an $F \in \mathfrak{F}$ such that $f[F] \subseteqq G + K$, and hence

$$\frac{f(q + \varepsilon(p + p')) - f(q)}{\varepsilon} \in f(p + p') + K$$
$$\subseteqq f(p) + f(p') + K$$
$$\subseteqq f(p) + G + K$$

whenever $\varepsilon > 0$, $p \in R$, and $p' \in F$.

For every $\alpha \geqq 0$ and $\phi \in D_1 f(q)$, $\alpha\phi \in D_1(\alpha f)(q)$. If $\phi \in D_1 f(q)$ and $\gamma \in D_1 g(q)$, where $g \in \mathfrak{A}(Q, S)$, then $\phi + \gamma \in D_1(f + g)(q)$. In the case $\underline{K} \neq \emptyset$, for every $\psi \in K^*$ and $\phi \in D_1 f(q)$, $\psi \circ \phi \in D_1(\psi \circ f)(q)$.

Theorem 3 (*Generalized mean value theorem*). *Let p be a point $\neq 0$ of $C_{q,Q}$ such that $q + [0, 1]\, p \subseteqq Q$ and such that $p \in C_{q+\varepsilon p,Q}$ for every $\varepsilon \in (0, 1)$. Let $f|_{q+[0,1]p}$ be lower semicontinuous at q and $f|_{q+[0,\varepsilon]p}$ be upper semicontinuous at $q + \varepsilon p$ whenever $\varepsilon \in (0, 1]$. Assume for $\varepsilon \in (0, 1)$, there exists an l-derivative $\phi_{q+\varepsilon p}$ of f at $q + \varepsilon p$. Then,*

$$f(q + p) - f(q) \in \overline{c\{\phi_{q+\varepsilon p}(p); \varepsilon \in (0, 1)\} + K}^{t^\wedge(\tau_S)}$$

where $t^\wedge(\tau_S)$ denotes the finest locally convex vector topology on S coarser than τ_S.

Proof. Let V be a convex $t^\wedge(\tau_S)$-neighbourhood of the point 0 in S, n a natural number > 1, and A the set of all $\varepsilon \in \left[0, 1 - \frac{1}{n}\right]$ such that

$$f\left(q + \left(\frac{1}{n} + \varepsilon'\right)p\right) - f\left(q + \frac{1}{n}p\right) \in \overline{\varepsilon' c(\{\phi_{q+\varepsilon p}(p); \varepsilon \in (0, 1)\} + V) + K}^{t^\wedge(\tau_S)}$$

for every $\varepsilon' \in [0, \varepsilon)$. Trivially, $0 \in A$. For every $\varepsilon\ (> 0) \in A$, $[0, \varepsilon) \in A$. Let $\varepsilon_s = \sup_{\varepsilon \in A} \varepsilon$. We shall prove that

$$f\left(q + \left(\frac{1}{n} + \varepsilon_s\right)p\right) - f\left(q + \frac{1}{n}p\right) \in \overline{\varepsilon_s c(\{\phi_{q+\varepsilon p}(p); \varepsilon \in (0, 1)\} + V) + K}^{t^\wedge(\tau_S)}. \qquad (*)$$

In the case $\varepsilon_s = 0$, (*) is evident. So we may assume $\varepsilon_s > 0$. Let $\mathfrak{B}$ be the set of all $t^\wedge(\tau_S)$-neighbourhoods of 0 in S, N the set of all natural numbers $> \frac{1}{\varepsilon_s}$, and I the set

$\mathfrak{B} \times N$ which is directed by $(W, m) \prec (W', m') \Leftrightarrow W' \subseteq W$, $m \leqq m'$. As $f|_{q+[0,1/n+\varepsilon_s]p}$ is upper semicontinuous at $q + \left(\frac{1}{n} + \varepsilon_s\right)p$, for every $i = (W, m) \in I$ there exists a real number $\mu_i \in \left(\varepsilon_s - \frac{1}{m}, \varepsilon_s\right)$ such that

$$f\left(q + \left(\frac{1}{n} + \varepsilon_s\right)p\right) - f\left(q + \left(\frac{1}{n} + \mu_i\right)p\right) \in W + K.$$

Hence

$$\begin{aligned} &f\left(q + \left(\frac{1}{n} + \varepsilon_s\right)p\right) - f\left(q + \frac{1}{n}p\right) \\ &= \left(f\left(q + \left(\frac{1}{n} + \varepsilon_s\right)p\right) - f\left(q + \left(\frac{1}{n} + \mu_i\right)p\right)\right) \\ &\quad + \left(f\left(q + \left(\frac{1}{n} + \mu_i\right)p\right) - f\left(q + \frac{1}{n}p\right)\right) \\ &\in W + K + \mu_i \overline{c\left(\{\phi_{q+\varepsilon p}(p); \varepsilon \in (0, 1)\} + V\right) + K}^{t^\wedge(\tau_s)} \\ &\subseteq \mu_i \overline{c\left(\{\phi_{q+\varepsilon p}(p); \varepsilon \in (0, 1)\} + V\right) + K}^{t^\wedge(\tau_s)} + W. \end{aligned}$$

This implies (*) which we wanted to demonstrate. Now we shall prove that $\frac{1}{n} + \varepsilon_s = 1$. In the contrary, let us assume $\frac{1}{n} + \varepsilon_s < 1$. As $p \in C_{q+(1/n+\varepsilon_s)p,Q}$, there exists an $\varepsilon \in \left(0, 1 - \frac{1}{n} - \varepsilon_s\right)$ such that

$$\frac{f\left(q + \left(\frac{1}{n} + \varepsilon_s + \varepsilon'\right)p\right) - f\left(q + \left(\frac{1}{n} + \varepsilon_s\right)p\right)}{\varepsilon'} \in \phi_{q+(1/n+\varepsilon_s)p}(p) + V + K$$

whenever $\varepsilon' \in (0, \varepsilon]$. Using this and (*), we get

$$\begin{aligned} &f\left(q + \left(\frac{1}{n} + \varepsilon_s + \varepsilon'\right)p\right) - f\left(q + \frac{1}{n}p\right) \\ &= \left(f\left(q + \left(\frac{1}{n} + \varepsilon_s + \varepsilon'\right)p\right) - f\left(q + \left(\frac{1}{n} + \varepsilon_s\right)p\right)\right) \\ &\quad + \left(f\left(q + \left(\frac{1}{n} + \varepsilon_s\right)p\right) - f\left(q + \frac{1}{n}p\right)\right) \\ &\in \varepsilon'\left(\phi_{q+(1/n+\varepsilon_s)p}(p) + V + K\right) + \varepsilon_s \overline{c\left(\{\phi_{q+\varepsilon p}(p); \varepsilon \in (0, 1)\} + V\right) + K}^{t^\wedge(\tau_s)} \\ &\subseteq (\varepsilon_s + \varepsilon')\, \overline{c\left(\{\phi_{q+\varepsilon p}(p); \varepsilon \in (0, 1)\} + V\right) + K}^{t^\wedge(\tau_s)} \end{aligned}$$

and hence $\varepsilon_s + \varepsilon \in A$ which is a contradiction. Consequently, $\frac{1}{n} + \varepsilon_s = 1$. As V was arbitrary, from (*) we get

$$f(q + p) - f\left(q + \frac{1}{n}p\right) \in \left(1 - \frac{1}{n}\right) \overline{c\{\phi_{q+\varepsilon p}(p); \varepsilon \in (0, 1)\} + K}^{t^\wedge(\tau_s)}$$

Taking into consideration the lower semicontinuity of $f|_{q+[0,1]p}$ at q, we get

$$f(q+p) - f(q) \in \overline{c\{\phi_{q+\varepsilon p}(p);\ \varepsilon \in (0,1)\} + K}^{t^\wedge(\tau_S)}.$$

Thus, the theorem is proven.

f is said to be *u-differentiable* (*upper differentiable*) *at* q if $-f$ is l-differentiable at q. Every mapping of $D_\mathrm{u}f(q) = -D_\mathrm{l}(-f)(q)$ is called *u-derivative* (*upper derivative*) *of* f *at* q.

Theorem 4. *For every* $\phi \in D_\mathrm{l}f(q)$ *and* $\phi' \in D_\mathrm{u}f(q)$,

$$(\phi - \phi')[C_{q,Q}] \subseteqq -K.$$

Proof. Let ϕ and ϕ' be given as in the theorem. For every $p \in C_{q,Q}$ and every filter $\mathfrak{F} \in \tau_R(0)$, there exists a filter $\mathfrak{G} \in \tau_S(0)$ and for every $G \in \mathfrak{G}$, there exists an $F \in \mathfrak{F}$ and an $\varepsilon > 0$ such that

$$\frac{f\big(q + \varepsilon'(p+p')\big) - f(q)}{\varepsilon'} \in \big(\phi(p) + G + K\big) \cap \big(\phi'(p) + G - K\big)$$

whenever $\varepsilon' \in (0, \varepsilon]$, $p' \in F$, and $q + \varepsilon'(p + p') \in Q$. Thus, there are nets $(s_i)_{i\in I}$ and $(s_i')_{i\in I}$ of points of S converging to 0 and nets $(t_i)_{i\in I}$ and $(t_i')_{i\in I}$ of points of K such that

$$\phi(p) + s_i + t_i = \phi'(p) + s_i' - t_i'$$

for all $i \in I$. From this, we get

$$(\phi - \phi')(p) \in s_i' - s_i - K$$

and consequently, as K is closed,

$$(\phi - \phi')(p) \in -K.$$

This was to be proven.

Corollary. $Df(q) = D_\mathrm{l}f(q) \cap D_\mathrm{u}(f(q)$ *contains at most one element.*

Proof. By Theorem 4, for every $\phi, \phi' \in Df(q)$, we have

$$(\phi - \phi')[C_{q,Q}] \subseteqq K \cap (-K)$$

from which $\phi = \phi'$.

In the case $Df(q) \neq \emptyset$, f is said to be *differentiable at* q and the unique mapping $\phi \in Df(q)$ is called *derivative of* f *at* q. ϕ is positive homogeneous. If S is locally order-convex, then f is differentiable at q if and only if there is a positive homogeneous mapping $\phi \in \mathfrak{A}(C_{q,Q}, S)$ which has the following property: For every point $p \in C_{q,Q}$ and every filter $\mathfrak{F} \in \tau_R(0)$, there exists a filter $\mathfrak{G} \in \tau_S(0)$ and for every $G \in \mathfrak{G}$, there exists an $F \in \mathfrak{F}$ and an $\varepsilon > 0$ such that

$$\frac{f\big(q + \varepsilon'(p+p')\big) - f(q)}{\varepsilon'} \in \phi(p) + G$$

whenever $\varepsilon' \in (0, \varepsilon]$, $p' \in F$, and $q + \varepsilon'(p + p') \in Q$. Thus, in the case of S being locally order-convex and $q \in \underline{Q}$, under the additional assumption of the additivity and continuity of ϕ, our notion of differentiability agrees with the notion of Michal-Bastiani-differentiability [2].

Without difficulty, there can be found mappings which are l- and u-differentiable but not differentiable. For example, let $R = Q = \boldsymbol{R}$, $q = 0$ (hence $C_{q,Q} = R$), $S = \boldsymbol{R}$, and

$$f(\sigma) = \begin{cases} \sigma \sin \dfrac{1}{\sigma} + 2\sigma & \text{if } \sigma > 0, \\ 0 & \text{if } \sigma \leqq 0. \end{cases}$$

Every positive homogeneous mapping of $\mathfrak{A}(C_{q,Q}, S)$ transforming 0 in 0 is given by

$$\phi_{\alpha,\beta}(\sigma) = \begin{cases} \alpha\sigma & \text{if } \sigma \geqq 0, \\ \beta\sigma & \text{if } \sigma \leqq 0 \end{cases}$$

where α and β are suitable real numbers. Let $P = \{\phi_{\alpha,\beta};\, \alpha, \beta \in (-\infty, +\infty)\}$. Then we have $D_l f(q) \cap P = \{\phi_{\alpha,\beta};\, (\alpha, \beta) \in (-\infty, 1] \times [0, +\infty)\}$, $D_u f(q) \cap P = \{\phi_{\alpha,\beta};\, (\alpha, \beta) \in [3, +\infty) \times (-\infty, 0]\}$, and $Df(q) = \emptyset$.

By an easy modification of the proof of Theorem 3, we get the following

Corollary to Theorem 3 (*Mean value theorem*). *Assume S is locally order-convex. Let p be a point $\neq 0$ of $C_{q,Q}$ such that $q + [0, 1]\, p \subseteqq Q$ and such that $p \in C_{q+\varepsilon p,Q}$ for every $\varepsilon \in (0, 1)$. Let $f|_{q+[0,1]p}$ be continuous at q and $f|_{q+[0,\varepsilon]p}$ be continuous at $q + \varepsilon p$ whenever $\varepsilon \in (0, 1]$. Assume for $\varepsilon \in (0, 1)$, there exists the derivative $\phi_{q+\varepsilon p}$ of f at $q + \varepsilon p$. Then,*

$$f(q + p) - f(q) \in \overline{c\{\phi_{q+\varepsilon p}(p);\, \varepsilon \in (0, 1)\}}^{\,t^{\wedge}(\tau_S)}.$$

Besides S let be given another ordered pseudo-topological vector space S' with the vector pseudo-topology $\tau_{S'}$ and let us denote the positive cone of S' by K'. Let f' be a mapping of an open subset Q' of S into S' for which $f[Q] \subseteqq Q'$.

Theorem 5 (*Generalized chain rule*). *Assume S is locally order-convex. Let f be differentiable at q and ϕ denote the derivative. Then,*

$$\{\phi' \circ \phi;\, \phi' \in D_l f'(f(q))\} \subseqq D_l(f' \circ f)\,(q).$$

Proof. Let $\phi' \in D_l f'(f(q))$. Then, $\phi' \circ \phi(\alpha p) - \alpha\phi' \circ \phi(p) \in K$ whenever $\alpha \in (0, 1)$ and $p \in C_{q,Q}$. For every point $p \in C_{q,Q}$ and every filter $\mathfrak{F} \in \tau_R(0)$, there exist filters $\mathfrak{G} \in \tau_S(0)$ and $\mathfrak{G}' \in \tau_{S'}(0)$ and for every $G' \in \mathfrak{G}'$, there exists a $G \in \mathfrak{G}$, an $F \in \mathfrak{F}$ and an $\varepsilon > 0$ such that $f(q) + [0, \varepsilon]\, (\phi(p) + G) \subseteqq Q'$, and

$$\frac{f'\big(f(q) + \varepsilon'(\phi(p) + P')\big) - f'(f(q))}{\varepsilon'} \in \phi' \circ \phi(p) + G' + K'$$

whenever $\varepsilon' \in (0, \varepsilon]$ and $P' \in G$ and

$$\frac{f\big(q + \varepsilon'(p + p')\big) - f(q)}{\varepsilon'} \in \phi(p) + G$$

whenever $\varepsilon' \in (0, \varepsilon]$, $p' \in F$, and $q + \varepsilon'(p + p') \in Q$. Hence, for every such $\varepsilon' \in (0, \varepsilon]$ and $p' \in F$

$$\frac{f' \circ f(q + \varepsilon'(p + p')) - f' \circ f(q)}{\varepsilon'} \in \frac{f'[f(q) + \varepsilon'(\phi(p) + G)] - f'(f(q))}{\varepsilon'}$$

$$\subseteqq \phi' \circ \phi(p) + G' + K'.$$

Thus, the theorem is proven.

Corollary (*Chain rule*). *Assume S is locally order-convex. Let f be differentiable at q and ϕ denote the derivative. Let f' be differentiable at $f(q)$ and ϕ' denote the derivative. Then $f' \circ f$ is differentiable at q and has $\phi' \circ \phi$ as derivative.*

4. The sets $D_{l_1}f(q)$ and $D_{l_2}f(q)$

For certain applications, two generalizations of the notion of l-differentiability are of interest which will be studied in this section. As above, let Q be a subset of R, q a point of Q such that $C_{q,Q} \neq \emptyset$, and f a mapping of $\mathfrak{A}(Q, S)$. Assume $\underline{K} \neq \emptyset$. Let $\phi \in \mathfrak{A}(C_{q,Q}, S)$ be such that $\phi(\alpha p) - \alpha\phi(p) \in K$ whenever $\alpha \in (0, 1)$ and $p \in C_{q,Q}$. ϕ is called *l_1-derivative of f at q* if for every point $p \in C_{q,Q}$ for which $\phi(p) \in \underline{K}$ and for every $\mathfrak{N} \in \mathcal{N}_p$, there exists an $N \in \mathfrak{N}$ such that $q + N \subseteqq Q$ and

$$f[(q + N) \setminus \{q\}] - f(q) \subseteqq \underline{K}$$

([5], page 132, and [6]). ϕ is called *l_2-derivative of f at q* if for every point $p \in C_{q,Q}$ for which $\phi(p) \in S \setminus (-K)$ and for every $\mathfrak{N} \in \mathcal{N}_p$, there exists an $N \subseteqq \mathfrak{N}$ such that $q + N \subseteqq Q$ and

$$f[(q + N) \setminus \{q\}] - f(q) \subseteqq S \setminus (-K).$$

The set of all l_1- or l_2-derivatives of f at q will be denoted by $D_{l_1}f(q)$ or $D_{l_2}f(q)$, respectively. $D_{l_1}f(q)$ and $D_{l_2}f(q)$ are subsets of the linear space $\mathfrak{A}(C_{q,Q}, S)$.

Theorem 6. $D_l f(q) \subseteqq D_{l_1}f(q) \cap D_{l_2}f(q)$.

Proof. 1. Let be $\phi \in D_l f(q)$ and $p \in C_{q,Q}$ such that $\phi(p) \in \underline{K}$. For every filter $\mathfrak{F} \in \tau_R(0)$, there exists a filter $\mathfrak{G} \in \tau_S(0)$ and for every $G \in \mathfrak{G}$, there exists an $F \in \mathfrak{F}$ and an $\varepsilon > 0$ such that

$$\frac{f(q + \varepsilon'(p + p')) - f(q)}{\varepsilon'} \in \phi(p) + G + K$$

whenever $\varepsilon' \in (0, \varepsilon]$, $p' \in F$, and $q + \varepsilon'(p + p') \in Q$. By a suitable choice of G, $\phi(p) + G \subseteqq \underline{K}$. $\mathfrak{N} = V^+(p + \mathfrak{F})$ belongs to $\mathcal{N}_p$ and there exists an $N = [0, \varepsilon]\,(p + F)$ ($\varepsilon > 0$, $F \in \mathfrak{F}$) in $\mathfrak{N}$ such that $q + N \subseteqq Q$. Hence,

$$f[(q + N) \setminus \{q\}] - f(q) = f[(q + (0, \varepsilon]\,(p + F)) \setminus \{q\}] - f(q)$$

$$\subseteqq (0, \varepsilon]\,(\phi(p) + G + K) \subseteqq (0, \varepsilon]\,(\underline{K} + K) = \underline{K}$$

which we wanted to prove.

2. The proof of $D_l f(q) \subseteqq D_{l_2}f(q)$ is to be given analogously.

Evidently, $0 \in D_{1_1}f(q) \cap D_{1_2}f(q)$. For every $\alpha \geqq 0$, $\beta > 0$, and $\phi \in D_{1_i}f(q)$ $(i = 1, 2)$, $\alpha\phi \in D_{1_i}(\beta f)(q)$. $Df_{1_2}(q)$ is convex. f is lower semicontinuous at q in every direction $p \in C_{q,Q}$ for which there exists a $\phi \in D_{1_1}f(q)$ such that $\phi(p) \in \underline{K}$.

Let S', K', f' be given as before Theorem 5 and assume $\underline{K}' \neq \emptyset$.

Theorem 7 (*Generalized chain rule*). *Assume S is locally order-convex. Let f be differentiable at q and let ϕ denote the derivative. Then, for $i = 1, 2$,*

$$\left\{\phi' \circ \phi ; \phi' \in D_{1_i}f'\big(f(q)\big)\ \phi'(0) = 0\right\} \subseteqq D_{1_i}(f' \circ f)(q).$$

Proof. 1. Let $\phi' \in D_{1_1}f'\big(f(q)\big)$. Then, $\phi' \circ \phi(\alpha p) - \alpha\phi' \circ \phi(p) \in K$ whenever $\alpha \in (0, 1)$ and $p \in C_{q,Q}$. Let $p \in C_{q,Q}$ be such that $\phi' \circ \phi(p) \in \underline{K}'$. For every filter $\mathfrak{F} \in \tau_R(0)$, there exists a filter $\mathfrak{G} \in \tau_S(0)$ and for every $G \in \mathfrak{G}$, there exists an $F \in \mathfrak{F}$ and an $\varepsilon > 0$ such that

$$\frac{f\big(q + \varepsilon'(p + p')\big) - f(q)}{\varepsilon'} \in \phi(p) + G$$

whenever $\varepsilon' \in (0, \varepsilon]$, $p' \in F$, and $q + \varepsilon'(p + p') \in Q$. $\mathfrak{N} = V^+(p + \mathfrak{F})$ belongs to $\mathcal{N}_p$. Let $N = [0, \varepsilon](p + F)$. We may assume $q + N \subseteqq Q$. As $V^+\big(\phi(p) + \mathfrak{G}\big)$ belongs to $\mathcal{N}_{\phi(p)}$ and $\phi(p) \neq 0$, we may choose ε and G such that $f(q) + [0, \varepsilon]\big(\phi(p) + G\big) \subseteqq Q'$, $-\phi(p) \notin G$, and

$$f'\big[\big(f(q) + [0, \varepsilon](\phi(p) + G)\big) \setminus \{f(q)\}\big] - f'\big(f(q)\big) \subseteqq \underline{K}'.$$

Using

$$\begin{aligned} f[(q + N) \setminus \{q\}] &= f\big[\big(q + (0, \varepsilon](p + F)\big) \setminus \{q\}\big] \\ &\subseteqq f(q) + (0, \varepsilon]\big(\phi(p) + G\big), \end{aligned}$$

we get

$$f' \circ f[(q + N) \setminus \{q\}] - f' \circ f(q) \subseteqq f'\big[f(q) + (0, \varepsilon]\big(\phi(p) + G\big)\big] - f'\big(f(q)\big) \subseteqq \underline{K}'$$

which we wanted to prove.

2. In the case $i = 2$, the proof is to be given analogously.

The mappings of $D_{u_i}f(q) = -D_{1_i}(-f)(q)$ $(i = 1, 2)$ are generalized u-derivatives of f at q which are corresponding to the 1_i-derivatives of f at q. They are called *u_i-derivatives of f at q.*

5. Second l-derivatives

Q being a subset of R and q a point of Q such that $C_{q,Q} \neq \emptyset$, let $\mathfrak{S}(C_{q,Q}, S)$ denote the linear space of all positive homogeneous continuous mappings of $\mathfrak{A}(C_{q,Q}, S)$. Let τ^b be the separated vector pseudotopology on $\mathfrak{S}(C_{q,Q}, S)$ of uniform convergence on bounded sets. With respect to τ^b, a filter $\mathfrak{F}$ of $\mathfrak{S}(C_{q,Q}, S)$ converges to 0 if and only if for every bounded set $B \subseteqq C_{q,Q}$, the filter in S with $\{F[B]; F \in \mathfrak{F}\}$ as base converges to 0. Let be

$$\mathfrak{K} = \{\phi \in \mathfrak{S}(C_{q,Q}, S); \phi\,[C_{q,Q}] \subseteqq K\}.$$

Theorem 8. *With respect to τ^b, $\mathfrak{K}$ is a closed proper cone in $\mathfrak{S}(C_{q,Q}, S)$.*

Proof. Evidently, $\mathfrak{K}$ is a proper cone in $\mathfrak{S}(C_{q,Q}, S)$. For the proof of $\mathfrak{K}$ to be closed with respect to τ^b, let $\mathfrak{F}$ be a filter in $\mathfrak{S}(C_{q,Q}, S)$ containing $[\mathfrak{K}]$ and τ^b-converging to an element ϕ. For every $p \in C_{q,Q}$, the filter in S with $\{F(p); F \in \mathfrak{F}\}$ as base converges to $\phi(p)$. As $\mathfrak{F} \ni [\mathfrak{K}]$ and K is closed, $\phi(p) \in K$. Therefore $\phi \in \mathfrak{K}$, that is, $\mathfrak{K}$ is closed. Thus, the theorem is proven.

In what follows, let us assume that $\mathfrak{S}(C_{q,Q}, S)$ is endowed with τ^b and $\mathfrak{K}$ and hence is an ordered pseudo-topological vector space. If S is locally convex or locally order-convex, then $\mathfrak{S}(C_{q,Q}, S)$ also has this property.

A mapping $f \in \mathfrak{A}(Q, S)$ is said to be *twice l-differentiable at* q if in R, there exists a neighbourhood U of q with the following properties:

1. $C_{q,Q} \subseteqq C_{p,Q}$ for every $p \in Q \cap U$.

2. f admits a continuous derivative ϕ_q at q and a positive homogeneous continuous l-derivative ϕ_p at every point $p(\neq q) \in Q \cap U$.

3. The mapping $\phi^1 \in \mathfrak{A}(Q \cap U, \mathfrak{S}(C_{q,Q}, S))$ given by $\phi^1(p) = \phi_p|_{C_{p,Q}}$ is l-differentiable at q.

Every l-derivative of every such mapping ϕ^1 is called *second l-derivative of f at q*. The set of all second l-derivatives of f at q will be denoted by $D_1^2f(q)$. As $C_{q,Q} = C_{q,Q \cap U}$, $D_1^2f(q)$ is a subset of the linear space $\mathfrak{A}(C_{q,Q}, \mathfrak{S}(C_{q,Q}, S))$ and can be considered also to be a subset of the linear space $\mathfrak{A}(C_{q,Q} \times C_{q,Q}, S)$. For $\phi^2 \in D_1^2f(q)$ and $p, p' \in C_{q,Q}$, instead of $\phi^2(p)(p')$, therefore we write $\phi^2(p, p')$, too.

Analogously as Theorem 1 and using Theorem 1, one can prove that $D_1^2f(q)$ is convex. By Theorem 2, it follows that in the case $D_1^2f(q) \neq \emptyset$, every above mapping ϕ^1 is lower semicontinuous at q in every direction $p \in C_{q,Q}$. For every $\alpha \geqq 0$ and $\phi^2 \in D_1^2f(q)$, $\alpha\phi^2 \in D_1^2(\alpha f)(q)$. If $\phi^2 \in D_1^2f(q)$ and $\gamma^2 \in D_1^2g(q)$ where $g \in \mathfrak{A}(Q, S)$, then $\phi^2 + \gamma^2 \in D_1^2(f + g)(q)$. In the case $\underline{K} \neq \emptyset$, for every $\psi \in K^*$ and $\phi^2 \in D_1^2f(q)$, $\psi \circ \phi^2 \in D_1^2(\psi \circ f)(q)$.

Theorem 9 (*Generalized second order mean value theorem*). *Let p be a point $\neq 0$ of $C_{q,Q}$ such that $q + [0, 1]\, p \subseteqq Q$ and such that $C_{q,Q} \cup \{-p\} \subseteqq C_{q+\varepsilon p,Q}$ for every $\varepsilon \in [0, 1]$. Assume for $\varepsilon \in [0, 1]$, there exists the derivative $\phi_{q+\varepsilon p}$ of f at $q + \varepsilon p$ and denote by ϕ the mapping of $\mathfrak{A}(q + [0, 1]\, p, \mathfrak{S}(C_{q,Q}, S))$ which is given by $\phi(p') = \phi_{p'}|_{C_{q,Q}}$. Let ϕ be lower semicontinuous at q and $\phi|_{q+[0,\varepsilon]p}$ be upper semicontinuous at $q + \varepsilon p$ whenever $\varepsilon \in (0, 1]$. Assume for $\varepsilon \in (0, 1)$, there exists a second l-derivative $\phi^2_{q+\varepsilon p}$ of f at $q + \varepsilon p$. Then,*

$$f(q + p) - f(q) \in \phi_q(p) + \overline{c\Big(\bigcup_{\varepsilon \in (0,1)} (0, 1)\, \phi^2_{q+\varepsilon p}(p, p)\Big) + K}^{t^\wedge(\tau_S)}.$$

Proof. By Theorem 2, $f|_{q+[0,1]p}$ is lower semicontinuous at q and $f|_{q+[0,\varepsilon]p}$ is upper semicontinuous at $q + \varepsilon p$ whenever $\varepsilon \in (0, 1]$. Using Theorem 3, therefore

$$\begin{aligned} f(q + p) - f(q) &\in \overline{c\{\phi_{q+\varepsilon p}(p); \varepsilon \in (0, 1)\} + K}^{t^\wedge(\tau_S)} \\ &= \phi_q(p) + \overline{c\{\phi_{q+\varepsilon p}(p) - \phi_q(p); \varepsilon \in (0, 1)\} + K}^{t^\wedge(\tau_S)}. \end{aligned}$$

For every $\varepsilon \in (0, 1)$, using Theorem 3, we get

$$\phi_{q+\varepsilon p}(p) - \phi_q(p) \in \overline{c\{\phi^2_{q+\varepsilon' p}(\varepsilon p, p); \varepsilon' \in (0, \varepsilon)\} + K}^{t^\wedge(\tau_S)}.$$

Taking into consideration that $\phi^2_{q+\varepsilon' p}(\varepsilon p, p) - \varepsilon\phi^2_{q+\varepsilon' p}(p, p) \in K$ whenever $\varepsilon \in (0, 1)$ and $p \in C_{q,Q}$, hence

$$\phi_{q+\varepsilon p}(p) - \phi_q(p) \in \overline{c\Big(\bigcup_{\varepsilon'\in(0,1)} (0, 1)\, \phi^2_{q+\varepsilon' p}(p, p)\Big) + K}^{t^\wedge(\tau_S)}.$$

Consequently,

$$f(q + p) - f(q) \in \phi_q(p) + \overline{c\Big(\overline{c\Big(\bigcup_{\varepsilon\in(0,1)} (0, 1)\, \phi^2_{q+\varepsilon p}(p, p) + K\Big)}^{t^\wedge(\tau_S)}\Big) + K}^{t^\wedge(\tau_S)}$$

$$= \phi_q(p) + \overline{c\Big(\bigcup_{\varepsilon\in(0,1)} (0, 1)\, \phi^2_{q+\varepsilon p}(p, p)\Big) + K}^{t^\wedge(\tau_S)}$$

which proves the theorem.

If $-f$ is twice l-differentiable at q, f is said to be *twice u-differentiable at q*. Let $D_u{}^2f(q) = -D_l{}^2(-f)\,(q)$. Every mapping of $D_u{}^2f(q)$ is called *second u-derivative of f at q*.

Theorem 10. *For every* $\phi^2 \in D_l{}^2f(q)$ *and* $\phi'^2 \in D_u{}^2f(q)$,

$$(\phi^2 - \phi'^2)\,[C_{q,Q}] \subseteqq -\mathfrak{K}.$$

Proof. Let ϕ^2 and ϕ'^2 be given as in the theorem. Then in R, there exists a neighbourhood U of q with the following properties:

1. $C_{q,Q} \subseteqq C_{p,Q}$ for every $p \in Q \cap U$.

2. f admits a continuous derivative ϕ at q and a positive homogeneous continuous l-derivative ϕ_p as well as a positive homogeneous continuous u-derivative ϕ_p' at every point $p(\neq q) \in Q \cap U$ such that

3. for every point p and p' of $C_{q,Q}$, there exists a filter $\mathfrak{G} \in \tau_S(0)$ and for every $G \in \mathfrak{G}$ there exists an $\varepsilon > 0$ such that

$$\frac{\phi_{q+\varepsilon' p}(p') - \phi(p')}{\varepsilon'} \in \phi^2(p, p') + G + K$$

and

$$\frac{\phi'_{q+\varepsilon' p}(p') - \phi(p')}{\varepsilon'} \in \phi'^2(p, p') + G - K$$

whenever $\varepsilon' \in (0, \varepsilon]$ and $q + \varepsilon' p \in Q \cap U$.

By Theorem 4,

$$\phi_{q+\varepsilon' p}(p') - \phi'_{q+\varepsilon' p}(p') \in -K.$$

Thus, there exist nets $(s_i)_{i\in I}$ and $(s_i')_{i\in I}$ of points of S converging to 0 for which

$$\phi^2(p, p') - \phi'^2(p, p') \in s_i' - s_i - K.$$

As K is closed, consequently

$$\phi^2(p, p') - \phi'^2(p, p') \in -K.$$

This was to be proven.

Corollary. $D^2f(q) = D_l{}^2f(q) \cap D_u{}^2f(q)$ *contains at most one element.*

Proof. By Theorem 10, for every $\phi^2, \phi'^2 \in D^2f(q)$, we have

$$(\phi^2 - \phi'^2)\,[C_{q,Q}] \subseteqq \mathfrak{K} \cap (-\mathfrak{K})$$

from which $\phi^2 = \phi'^2$.

In the case $D^2f(q) \neq \emptyset$, f is said to be *twice differentiable at* q and the unique mapping $\phi^2 \in D^2f(q)$ is called *second derivative of* f *at* q. ϕ^2 is positive homogeneous in both arguments.

By an easy modification of the proof of Theorem 9, using the Corollary to Theorem 3, we get the following

Corollary to Theorem 9 (*Second order mean value theorem*). *Assume S is locally order-convex. Let p be a point $\neq 0$ of $C_{q,Q}$ such that $q + [0, 1]\,p \subseteqq Q$ and such that $C_{q,Q} \cup \{-p\} \subseteqq C_{q+\varepsilon p,Q}$ for every $\varepsilon \in (0, 1]$. Assume for $\varepsilon \in [0, 1]$, there exists the derivative $\phi_{q+\varepsilon p}$ of f at $q + \varepsilon p$ and denote by ϕ the mapping of $\mathfrak{A}(q + [0, 1]\,p, \mathfrak{S}(C_{q,Q}, S))$ which is given by $\phi(p') = \phi_{p'}|_{C_{q,Q}}$. Let ϕ be continuous at q and $\phi|_{q+[0,\varepsilon]p}$ be continuous at $q + \varepsilon p$ whenever $\varepsilon \in (0, 1]$. Assume for $\varepsilon \in (0, 1)$, there exists the second derivative $\phi^2_{q+\varepsilon p}$ of f at $q + \varepsilon p$. Then,*

$$f(q + p) - f(q) \in \phi_q(p) + \overline{c\left(\bigcup_{\varepsilon \in (0,1)} (0, 1)\, \phi^2_{q+\varepsilon p}(p, p)\right)}^{\,t^{\wedge}(\tau_S)}.$$

Let S', K', f' ge given as before Theorem 5.

Theorem 11 (*Generalized second order chain rule*). *Assume S is locally bounded and locally order-convex. Let f be continuous and twice differentiable at q and ϕ and ϕ^2 denote the first or second derivative of f at q, respectively. For every p $(\neq q) \in Q$ belonging to a certain neighbourhood of q let there exist a continuous derivative of f at p. Let f' be differentiable at $f(q)$ and assume the derivative ϕ' of f' at $f(q)$ is superadditive. Then*

$$\{\phi'^2(\phi, \phi) + \phi' \circ \phi^2;\ \phi'^2 \in D_1^2 f'(f(q))\} \subseteqq D_1^2(f' \circ f)\,(q).$$

Proof. Because of $f(q) \in Q' = \underline{Q}'$, $C_{f(q),Q'} = S$. Let be $\mathfrak{S} = \mathfrak{S}(C_{q,Q}, S)$ and let $\tau_{\mathfrak{S}}$ denote the vector pseudo-topology of $\mathfrak{S}$. Moreover, we need the (evidently defined) ordered pseudo-topological vector spaces $\mathfrak{S}' = \mathfrak{S}(S, S')$ and $\mathfrak{S}^+ = \mathfrak{S}(C_{q,Q}, S')$. The corresponding vector pseudo-topologies are denoted by $\tau_{\mathfrak{S}'}$ and $\tau_{\mathfrak{S}^+}$ and the positive cones by $\mathfrak{K}'$ or $\mathfrak{K}^+$, respectively. Let U be a neighbourhood of q with the following properties:

1. $C_{q,Q} \subseteqq C_{p,Q}$ for every $p \in Q \cap U$.
2. f admits a continuous derivative ϕ_p at every p $(\neq q) \in Q \cap U$ such that
3. the mapping $\phi^1 \in \mathfrak{A}(Q \cap U, \mathfrak{S})$ given by $\phi^1(p) = \tilde{\phi}_p = \phi_p|_{C_{q,Q}}$ where $\phi_q = \phi$, has ϕ^2 as derivative at q.

Assume ϕ'^2 is a mapping of $D_1^2 f'(f(q))$. Let V be a neighbourhood of $f(q)$ with the following properties:

1. $V \subseteqq Q'$, which implies $C_{s,Q'} = S$ for every $s \in V$.
2. f' admits a positive homogeneous continuous 1-derivative ϕ_s' at every s $(\neq f(q)) \in V$ such that
3. the mapping $\phi'^1 \in \mathfrak{A}(V, \mathfrak{S}')$ defined by $\phi'^1(s) = \phi_s'$ where $\phi'_{f(q)} = \phi'$, has ϕ'^2 as 1-derivative at $f(q)$.

We may assume $f[Q \cap U] \subseteqq V$. Using Theorem 5, we get that the positive homogeneous continuous mapping $\phi' \circ \phi$ is the derivative of $f' \circ f$ at q and that for every $p\ (\neq q) \in Q \cap U$, the positive homogeneous continuous mapping $\phi'_{f(p)} \circ \phi_p$ is an l-derivative of $f' \circ f$ at p. $\psi = \phi'^2(\phi, \phi) + \phi' \circ \phi^2$ is a mapping of $\mathfrak{A}(C_{q,Q}, \mathfrak{S}^+)$ for which $\psi(\alpha p) - \alpha\psi(p) \in \mathfrak{K}^+$ whenever $\alpha \in (0, 1)$ and $p \in C_{q,Q}$.

Let p be a point of $C_{q,Q}$ and $\mathfrak{F}$ a filter of $\tau_R(0)$. Then there exist filters $\mathfrak{G} \in \tau_S(0)$, $\mathfrak{H} \in \tau_{\mathfrak{S}}(0)$, and $\mathfrak{H}' \in \tau_{\mathfrak{S}'}(0)$ such that for every $\boldsymbol{H} \in \mathfrak{H}$ and $\boldsymbol{H}' \in \mathfrak{H}'$, there exist an $F \in \mathfrak{F}$, a $G \in \mathfrak{G}$, and an $\varepsilon > 0$ such that $q + [0, \varepsilon]\,(p + F) \subseteqq Q \cap U$, $f(q) + [0, \varepsilon]\,\big(\phi(p) + G\big) \subseteqq V$ and the following holds:

$$\frac{f\big(q + \varepsilon'(p + p')\big) - f(q)}{\varepsilon'} \in \phi(p) + G$$

and

$$\frac{\tilde{\phi}_{q+\varepsilon'(p+p')} - \phi}{\varepsilon'} \in \phi^2(p) + \boldsymbol{H}$$

whenever $\varepsilon' \in (0, \varepsilon]$, $p' \in F$, and

$$\frac{\phi'_{f(q)+\varepsilon'(\phi(p)+s)} - \phi'}{\varepsilon'} \in \phi'^2\big(\phi(p)\big) + \boldsymbol{H}' + \mathfrak{K}'$$

whenever $\varepsilon' \in (0, \varepsilon]$, $s \in G$. The filter $\mathfrak{H}^+$ in $\mathfrak{S}^+$ with

$$\phi' \circ \mathfrak{H} + \phi'^2\big(\phi(p), \phi + \boldsymbol{V}^+(\phi^2(p) + \mathfrak{H})\big) - \phi'^2\big(\phi\,(p), \phi\big) + \mathfrak{H}' \circ \big(\phi + \boldsymbol{V}^+(\phi^2(p) + \mathfrak{H})\big)$$

as base belongs to $\tau_{\mathfrak{S}^+}(0)$.

$$\boldsymbol{H}^+ = \phi'[\boldsymbol{H}] + \phi'^2\big(\phi(p), \phi + [0, \varepsilon']\,(\phi^2(p) + \boldsymbol{H})\big) - \phi'^2\big(\phi(p), \phi\big) + \boldsymbol{H}' \circ \big(\phi + [0, \varepsilon']\,(\phi^2(p) + \boldsymbol{H})\big)$$

where $\varepsilon' \in (0, \varepsilon]$, is a set of the filter $\mathfrak{H}^+$. For every $\varepsilon^+ \in (0, \varepsilon']$, $p' \in F$, and suitable $s \in G$, we get

$$\frac{\phi'_{f(q+\varepsilon^+(p+p'))} \circ \tilde{\phi}_{q+\varepsilon^+(p+p')} - \phi' \circ \phi}{\varepsilon^+}$$

$$\in \frac{\phi'_{f(q)+\varepsilon^+(\phi(p)+s)} \circ \big(\phi + \varepsilon^+\phi^2(p) + \varepsilon^+\boldsymbol{H}\big) - \phi' \circ \phi}{\varepsilon^+}$$

$$\subseteqq \frac{\big(\phi' + \varepsilon^+\phi'^2(\phi(p)) + \varepsilon^+\boldsymbol{H}' + \varepsilon^+\mathfrak{K}'\big) \circ \big(\phi + \varepsilon^+\phi^2(p) + \varepsilon^+\boldsymbol{H}\big) - \phi' \circ \phi}{\varepsilon^+}$$

and from this without difficulty

$$\frac{\phi'_{f(q+\varepsilon^+(p+p'))} \circ \tilde{\phi}_{q+\varepsilon^+(p+p')} - \phi' \circ \phi}{\varepsilon^+} \in \psi(p) + \boldsymbol{H}^+ + \mathfrak{K}^+.$$

Thus, the theorem is proven.

Corollary (*Second order chain rule*). *Assume S is locally bounded and locally order-convex. Let f be continuous and twice differentiable at q and ϕ and ϕ^2 denote the first or second derivative of f at q, respectively. For every $p\ (\neq q) \in Q$ belonging to a certain neighbourhood of q let there exist a continuous derivative of f at q. Let f' be twice differentiable at $f(q)$ ϕ' and ϕ'^2 denote the first or second derivative of f' at $f(q)$, respectively. Assume ϕ' is additive. Then $f' \circ f$ is twice differentiable at q and has $\phi'^2(\phi, \phi) + \phi' \circ \phi^2$ as second derivative.*

For the second l-differentiation, analogously to the l_1- and l_2-differentiation, certain generalizations can be introduced which, however, shall not be studied here. Also the l-differentiation of order $n > 2$ and its generalizations are not considered in this paper.

6. Necessary optimality conditions

As above, let Q be a subset of R, q a point of Q for which $C_{q,Q} \neq \emptyset$, and f a mapping of $\mathfrak{A}(Q, S)$. Assume $\underline{K} \neq \emptyset$. f is said to have a *local maximum at* q and q to be a *local maximizing point of f of the first, second,* or *third kind* if there exists a neighbourhood U of q such that

$$f[Q \cap U] - f(q) \subseteqq -K, \quad \subseteqq S \setminus (K \setminus \{0\}), \quad \text{or} \subseteqq S \setminus \underline{K},$$

respectively ([5], page 127/128). Every local maximum of the first kind is a local maximum of the second kind and every local maximum of the second kind is a local maximum of the third kind. In the case $S = \boldsymbol{R}^n$, $f = (f_1, \ldots, f_n)$ has a local maximum at q

1. of the first kind if and only if there exists a neighbourhood U of q such that $f_i(p) \leqq f_i(q)$ whenever $p \in Q \cap U$ and $i \in \{1, \ldots, n\}$,
2. of the second kind if and only if there exists a neighbourhood U of q such that $f_i(q) \leqq f_i(p)$, $i = 1, \ldots, n$, implies $f_i(q) = f_i(p)$, $i = 1, \ldots, n$, where $p \in Q \cap U$, and
3. of the third kind if and only if there exists a neighbourhood U of q such that for every $p \in Q \cap U$, $f_i(p) \leqq f_i(q)$ for at least one $i \in \{1, \ldots, n\}$.

In the case $S = \boldsymbol{R}$, the three kinds of local maxima are the same.

Theorem 12. 1. *Let q be a local maximizing point of f of the third (especially of the first or of the second) kind. Then*

$$D_{1_1} f(q) \, [C_{q,Q}] \subseteqq S \setminus \underline{K}.$$

If $C_{q,Q}$ is convex, then for every superadditive (and hence concave) $\phi \in D_{1_1} f(q)$, there exists a ψ $(\neq 0) \in K^*$ for which

$$\psi \circ \phi[C_{q,Q}] \subseteqq (-\infty, 0].$$

2. *Let q be a local maximizing point of f of the first kind. Then,*

$$D_{1_2} f(q) \, [C_{q,Q}] \subseteqq -K$$

and consequently

$$K^* \circ D_{1_2} f(q) \, [C_{q,Q}] \subseteqq (-\infty, 0].$$

Proof. 1. Let q be a local maximizing point of f of the third kind. Then, there exists a neighbourhood U of 0 in R for which

$$f[(q + U) \cap Q] - f(q) \subseteqq S \setminus \underline{K}.$$

Let $\phi \in D_{1_1} f(q)$. If $\phi(p) \in \underline{K}$ where $p \in C_{q,Q}$, then for every $\mathfrak{N} \in \mathcal{N}_p$, there exists an $N \in \mathfrak{N}$ such that $q + N \subseteqq Q$ and

$$f[(q + N) \setminus \{q\}] - f(q) \subseteqq \underline{K}.$$

This is impossible. Hence $D_{1_1}f(q)\,[C_{q,Q}] \subseteq S \setminus \underline{K}$. Now, let $C_{q,Q}$ be convex and let ϕ be a superadditive mapping of $D_{1_1}f(q)$. Let $M = c\phi[C_{q,Q}]$. For arbitrary points $p_1, \ldots, p_n \in C_{q,Q}$ and arbitrary real numbers $\alpha_1, \ldots, \alpha_n > 0$ such that $\sum_{i=1}^{n} \alpha_i = 1$, we have

$$\sum_{i=1}^{n} \alpha_i \phi(p_i) - \phi\left(\sum_{i=1}^{n} \alpha_i p_i\right) \in -K.$$

In the case $\sum_{i=1}^{n} \alpha_i \phi(p_i) \in \underline{K}$, we would get $\phi\left(\sum_{i=1}^{n} \alpha_i p_i\right) \in \underline{K}$ which is impossible. Therefore, we have $M \cap \underline{K} = \emptyset$. Using the separation theorem for convex sets ([10], page 190), we get a continuous linear functional $\psi \neq 0$ on S for which $\psi \in (-M^*) \cap K^*$ where M^* denotes the set of all linear functionals on S being $\geqq 0$ on M. Thus, $\psi \circ \phi[C_{q,Q}] \subseteq (-\infty, 0]$.

2. The proof of the second statement of the theorem is to be given analogously as the proof of the first part of the first statement.

7. Sufficient optimality conditions

f is said to have a *strong local maximum at* q and q to be a *strong local maximizing point of* f *of the first, second,* or *third kind* if q is a local maximizing point of f of the first, second, or third kind, respectively, and if there exists a neighbourhood U of q such that $f(p) \neq f(q)$ for every $p\ (\neq q) \in Q \cap U$.

Theorem 13. *Assume the dimension of R is finite. Let $f' : Q' \to S$ be an extension of f where $Q' \supseteqq Q$ is a subset of R for which there exists a neighbourhood U of q such that $\overline{[0, \infty)\,(Q \cap U - q)} \subseteq C_{q,Q'} \cup \{0\}$.*

1. *Let there exist a $\phi' \in D_{u_1}f'(q)$ for which*

$$\phi'[C_{q,Q'} \setminus \{0\}] \subseteq -\underline{K}. \tag{1}$$

Then q is a strong local maximizing point of f of the first (and hence the second and third) kind.

2. *Let there exist a $\phi' \in D_{u_2}f'(q)$ for which*

$$\phi'[C_{q,Q'} \setminus \{0\}] \subseteq S \setminus K. \tag{2}$$

Then q is a strong local maximizing point of f of the second (and hence the third) kind.

Proof. 1. Under the assumptions of statement 1, assume that q is not a strong local maximizing point of f of the first kind. Then, there exists a sequence of points $q_1, q_2, \ldots$ of $Q \cap U$ different from q and converging to q for which

$$f(q_i) - f(q) \subset S \setminus (-K \setminus \{0\}).$$

The q_i can be choosen such that $q_i = q + \varepsilon_i(p + p_i)$ where ε_i are positive real numbers for which $\lim_{i\to\infty} \varepsilon_i = 0$, p is a point of $C_{q,Q'}$ different from 0, and $p_1, p_2, \ldots$ is a sequence of points of R converging to 0. Thus, we get

$$f\big(q + \varepsilon_i(p + p_i)\big) - f(q) \in S \setminus (-K \setminus \{0\}).$$

On the other hand, from $\phi' \in D_{u_1}f'(q)$ and $\phi'(p) \in -\underline{K}$, for almost all i, we have

$$f(q + \varepsilon_i(p + p_i)) - f(q) = f'(q + \varepsilon_i(p + p_i)) - f'(q) \in -\underline{K},$$

which is a contradiction.

2. The proof of the second statement of the theorem is to be given analogously.

Corollary. *Condition* (1) *is true if*

$$(K^* \setminus \{0\}) \circ \phi'[C_{q,Q'} \setminus \{0\}] \subseteqq (-\infty, 0).$$

Condition (2) *is true if there exists a* $\psi \in K^*$ *such that*

$$\psi \circ \phi'[C_{q,Q'} \setminus \{0\}] \subseteqq (-\infty, 0).$$

Proof. Using the separation theorem for convex sets, the first part of the Corollary becomes clear. The second part is immediate.

Theorem 14. *Let there exist a neighbourhood U of such that $U - q$ is equilibrated and $(q + C_{q,Q}) \cap U \subseteqq Q \cap U \subseteqq (q + \overline{C_{q,Q}}) \cap U$. Let f be lower semicontinuous at every point of $(Q \cap U) \setminus ((q + C_{q,Q}) \cap U)$. For $p(\neq 0) \in C_{q,Q} \cap (U - q)$, assume $f|_{q+[\varepsilon,1]p}$, $\varepsilon \in [0, 1)$, is upper semicontinuous at $q + \varepsilon p$ and lower semicontinuous at $q + p$, there exists a positive homogeneous u-derivative ϕ_{q+p} of f at $q + p$, and $p \in C_{q+p,Q}$.*

1. *For every point p $(\neq 0) \in C_{q,Q} \cap (U - q)$, let*

$$K^* \circ \phi_{q+p}(p) \subseteqq (-\infty, 0]. \tag{3}$$

Then q is a local maximizing point of f of the first (and hence second and third) kind. Moreover, if especially $(q + C_{q,Q}) \cap U = Q \cap U$ and if there exists a $\psi \in K^*$ such that

$$\psi \circ \phi_{q+p}(p) \in (-\infty, 0) \tag{4}$$

whenever $p(\neq 0) \in C_{q,Q} \cap (U - q)$, then q is a strong local maximizing point of f of the first (and hence second and third) kind.

2. *Let there exist a ψ $(\neq 0) \in K^*$ such that*

$$\psi \circ \phi_{q+p}(p) \in (-\infty, 0] \tag{5}$$

whenever p $(\neq 0) \in C_{q,Q} \cap (U - q)$. Then q is a local maximizing point of f of the third kind. Moreover, if especially $(q + C_{q,Q}) \cap U = Q \cap U$ and instead of (5), (4) *holds whenever p $(\neq 0) \in C_{q,Q} \cap (U - q)$, then q is a strong local maximizing point of f of the third kind.*

Proof. 1. Let $p(\neq 0) \in C_{q,Q} \cap (U - q)$. Then, $[0, 1]\, p \in (C_{q,Q} \cup \{0\}) \cap (U - q)$ and hence, $q + [0, 1]\, p \subseteqq Q$. For every $\psi \in K^*$ and every $\varepsilon \in (0, 1)$, $\psi \circ \phi_{q+\varepsilon p}$ is a u-derivative of $\psi \circ f$ at $q + \varepsilon p$, and $\psi \circ \phi_{q+\varepsilon p}(p) \in (-\infty, 0]$. Thus, using Theorem 3, we get

$$\psi \circ f(q + p) - \psi \circ f(q) \in \overline{c\{\psi \circ \phi_{q+\varepsilon p}(p);\, \varepsilon \in (0, 1)\} + (-\infty, 0]} \subseteqq (-\infty, 0]. \tag{*}$$

As $(q + C_{q,Q}) \cap U \subseteqq Q \cap U \subseteqq (q + \overline{C_{q,Q}}) \cap U$ and as $\psi \circ f$ is lower semicontinuous at every point of $(Q \cap U) \setminus ((q + C_{q,Q}) \cap U)$, this implies

$$\psi[f[Q \cap U] - f(q)] \subseteqq (-\infty, 0]$$

from which, using the separation theorem for convex sets, we get

$$f[Q \cap U] - f(q) \subseteqq -K.$$

Consequently, q is a local maximizing point of f of the first kind.

Now, let $(q + C_{q,Q}) \cap U = Q \cap U$ and let there exist a $\psi \in K^*$ such that (4) holds whenever $p(\neq 0) \in C_{q,Q} \cap (U - q)$. We shall prove that $f(q+p) \neq f(q)$ for every point $p(\neq 0) \in Q \cap U - q$ which implies that q is a strong local maximizing point of f of the first kind. Let us assume in the contrary, there exists a point p $(\neq 0) \in Q \cap U - q$ such that $f(q+p) = f(q)$. Analogously as (*), for every $\varepsilon \in (0, 1)$ we can prove that

$$\psi \circ f(q + \varepsilon p) - \psi \circ f(q) \in (-\infty, 0] \quad \text{and} \quad \psi \circ f(q+p) - \psi \circ f(q + \varepsilon p) \in (-\infty, 0]$$

from which

$$\psi \circ f(q + \varepsilon p) = \psi \circ f(q).$$

Hence

$$\psi \circ f[q + (0, 1)\, p] = \{\psi \circ f(q)\}.$$

For every $\varepsilon \in (0, 1)$, consequently

$$\psi \circ \phi_{q+\varepsilon p}(p) \in [0, \infty)$$

which is a contradiction. Thus, statement 1 of the theorem is proven.

2. The proof of the second statement of the theorem is to be given analogously.

Corollary. *Conditions* (3) *and* (5) *are true if*

$$\phi_{q+p}(p) \in -K.$$

Condition (4) *is true if*

$$\phi_{q+p}(p) \in -\underline{K}.$$

Theorem 15. *Assume the dimension of R is finite. Let $f' : Q' \to S$ be an extension of f where Q' is a subset of R for which there exists a neighbourhood U of q with $\overline{[0, \infty)\,(Q \cap U - q)} \subseteqq C_{q,Q'} \cup \{0\}$ and such that $(Q' \cap U) \setminus \{q\}$ is open. For every point p $(\neq 0)$ for which $q + [0, 1]\, p \subseteqq Q' \cap U$, assume $f'|_{q+[0,1]p}$ is lower semicontinuous at $q + p$. Let f' be twice u-differentiable at q and ϕ' and ϕ'^2 denote the derivative or second u-derivative of f' at q, respectively. For every point $p(\neq 0) \in C_{q,Q'}$, let*

$$\phi'(p) \in S \setminus \underline{K}$$

and in the case $\phi'(p) \in K \setminus \underline{K}$, let there exist a $\psi \in K^*$ and a neighbourhood W of 0 in R for which

$$\psi \circ \phi'[p + W] \in (-\infty, 0] \quad \textit{and} \quad \psi \circ \phi'^2(p, p) < 0.$$

Then q is a strong local maximizing point of f of the second (and hence the third) kind.

Proof. Assume that q is not a strong local maximizing point of f of the second kind. Then there is a sequence of points $q_1, q_2, \ldots$ of $Q \cap U$ different from q and converging to q for which

$$f(q_i) - f(q) \in K.$$

The q_i can be choosen such that $q_i = q + \varepsilon_i(p + p_i)$ where ε_i are positive real numbers for which $\lim_{i\to\infty} \varepsilon_i = 0$, p is a point $\neq 0$ of $C_{q,Q'}$, and $p_1, p_2, \ldots$ is a sequence of points of R converging to 0.

$$f'(q + \varepsilon_i(p + p_i)) - f'(q) = f(q + \varepsilon_i(p + p_i)) - f(q) \in K$$

implies $\phi'(p) \in K$. As also $\phi'(p) \in S \setminus \underline{K}$, we have $\phi'(p) \in K \setminus \underline{K}$ and thus, there exists a $\psi \in K^*$ and a bounded neighbourhood W of 0 in R for which $\psi \circ \phi'[p + W] \subseteqq (-\infty, 0]$ and $-\alpha = \psi \circ \phi'^2(p, p) < 0$. We can find a real number $\varepsilon > 0$ and choose W such that $-p \notin W$, $q + [0, \varepsilon](p + W) \subseteqq Q' \cap U$, $p + W \in C_{q,Q'}$, and $\psi \circ \phi'^2(p)\,[p + W] \subseteqq \left(-\infty, -\frac{3\alpha}{4}\right]$. As $(Q' \cap U) \setminus \{q\}$ is open, we have

1. $C_{q',Q'} = R$ for every point $q'(\neq q) \in Q' \cap U$.

By a suitable choice of ε, U, and W, the following becomes true:

2. f' admits a positive homogeneous continuous u-derivative $\phi'_{q'}$ at every point $q'(\neq q) \in Q' \cap U$ such that
3. for every $\varepsilon' \in (0, \varepsilon]$ and $p' \in W$,

$$\frac{\psi \circ \phi'_{q+\varepsilon'(p+p')}(p + p') - \psi \circ \phi'(p + p')}{\varepsilon'} \in \psi \circ \phi'^2(p)\,(p + p') + \left(-\frac{\alpha}{4}, \frac{\alpha}{4}\right) + (-\infty, 0)$$

and hence

$$\psi \circ \phi'_{q+\varepsilon'(p+p')}\,(p + p') \in \left(-\infty, -\frac{\varepsilon'\alpha}{2}\right).$$

From this, using Theorem 2 and Theorem 3, we get

$$\begin{aligned}
&\psi \circ f'(q + \varepsilon'(p + p')) - \psi \circ f'(q) \\
&= \left(\psi \circ f'(q + \varepsilon'(p + p')) - \psi \circ f'\left(q + \frac{\varepsilon'}{2}(p + p')\right)\right) \\
&\quad + \left(\psi \circ f'\left(q + \frac{\varepsilon'}{2}(p + p')\right) - \psi \circ f'(q)\right) \\
&\in \overline{c\left\{\psi \circ \phi'_{q+(\varepsilon'/2)(1+\varepsilon^*)(p+p')}\left(\frac{\varepsilon'}{2}(p + p')\right);\ \varepsilon^* \in (0, 1)\right\} + (-\infty, 0]} \\
&\quad + \overline{c\left\{\psi \circ \phi'_{q+(\varepsilon'\varepsilon^*/2)(p+p')}\left(\frac{\varepsilon'}{2}(p + p')\right);\ \varepsilon^* \in (0, 1)\right) + (-\infty, 0]} \\
&\subseteqq \left(-\infty, -\frac{\varepsilon'^2\alpha}{8}\right] + (-\infty, 0] \subseteqq (-\infty, 0).
\end{aligned}$$

This implies

$$f'(q + \varepsilon'(p + p')) - f'(q) \in S \setminus K$$

whenever $\varepsilon' \in (0, \varepsilon]$ and $p' \in W$ which is a contradiction. Thus, the theorem is proven.

As a special case, Theorem 15 contains the sufficient optimality condition given in the Main Theorem of [13].

8. Optimality problems with constraints

To simplify the following considerations, we denote S, K, and f by S_1, K_1, and f_1, respectively. Let $C_{q,Q}$ be convex. Let S_2 be an ordered pseudo-topological vector space with the positive cone K_2 for which $\underline{K}_2 \neq \emptyset$, and let S_3 be a separated topological vector space of finite dimension. For $i = 2, 3$, let f_i be a mapping of Q into S_i such that $f_2(q) \in K_2$ and $f_3(q) = 0$. Then q belongs to $\tilde{Q} = \{p \in Q; f_2(p) \in K_2, f_3(p) = 0\}$. For every finite dimensional linear subspace R' of R, let $f_3|_{R' \cap Q}$ be continuous. Assume f_3 is differentiable at q and its derivative ϕ_3 is the restriction onto $C_{q,Q}$ of a linear mapping of R into S_3.

Theorem 16 (*Generalized Lagrange multiplier rule*). 1. *Let q be a local maximizing point of $f|_{\tilde{Q}}$ of the third (especially of the first or of the second) kind. Then for superadditive (and hence concave) $\phi_i \in D_{1_1}f_i(q)$, $i = 1, 2$, there exist $\psi_i \in K_i^*$, $i = 1, 2$, and a linear functional ψ_3 on S_3, not all 0, such that*

$$\sum_{i=1}^{3} \psi_i \circ \phi_i[C_{q,Q}] \subseteqq (-\infty, 0].$$

2. *Let q be a local maximizing point of $f|_{\tilde{Q}}$ of the first kind. Then for superadditive (and hence concave) $\phi_1 \in D_{1_1}f_1(q)$ and $\phi_2 \in D_{1_1}f_2(q)$, there exist $\psi_i \in K_i^*$, $i = 1, 2$, and a linear functional ψ_3 on S_3, not all 0, such that*

$$\sum_{i=1}^{3} \psi_i \circ \phi_i[C_{q,Q}] \subseteqq (-\infty, 0].$$

Proof. 1. Let q be a local maximizing point of $f|_{\tilde{Q}}$ of the third kind. If the linear subspace $l\phi_3[C_{q,Q}]$ generated by the vectors of $\phi_3[C_{q,Q}]$ is different from S_3, we may find a linear functional $\psi_3 \neq 0$ on S_3 such that $\psi_3[l\phi_3[C_{q,Q}]] = \{0\}$. Taking $\psi_i = 0$ for $i = 1, 2$, then for $\phi_i \in D_{1_1}f_i(q)$, $i = 1, 2$, we get $\sum_{i=1}^{3} \psi_i \circ \phi_i = 0$. In what follows, therefore we may assume $l\phi_3[C_{q,Q}] = S_3$.

At first, we shall prove that

$$\left(D_{1_1}f_1(q), D_{1_1}f_2(q), \phi_3\right) [C_{q,Q} \setminus \{0\}] \subseteqq (S_1 \times S_2 \times S_3) \setminus (\underline{K}_1 \times \underline{K}_2 \times \{0\}). \quad (*)$$

Let us assume that this is not true, i.e., there exist a point $p\ (\neq 0) \in C_{q,Q}$ and $\phi_i \in D_{1_1}f_i(q)$, $i = 1, 2$, such that $\phi_1(p) \in \underline{K}_1$, $\phi_2(p) \in \underline{K}_2$, and $\phi_3(p) = 0$. Let $p_1, \ldots, p_n$ be points $\neq 0$ of $C_{q,Q}$ such that $\{\psi_3(p_1), \ldots, \phi_3(p_n)\}$ is a base of S_3 and denote by R' the linear subspace of R which is generated by the vectors of $\{q, p, p_1, \ldots, p_n\}$. The restriction of the vector pseudo-topology τ_R on R' is the natural vector topology $\tau_{R'}$ of R'. There exists an $\varepsilon > 0$ and a $\tau_{R'}$-neighbourhood V of the point 0 in R' such that $-p \notin V$ and $q + [0, \varepsilon]\ (p + V) \subseteqq Q$. Let ψ denote the mapping of $\{(\alpha, \varrho_1, \ldots, \varrho_n) \in \boldsymbol{R}^{n+1};\ \alpha p + \sum_{i=1}^{n} \varrho_i p_i \in [0, \varepsilon]\,(p + V)\}$ into S_3 which is given by $\psi(\alpha, \varrho_1, \ldots, \varrho_n) = f_3\left(q + \alpha p + \sum_{i=1}^{n} \varrho_i p_i\right)$. By a modification of the proof of [12], Theorem 4.7 (for the restricted case $\lambda > 0$ and $\|x\| \leqq c\lambda$ where c is a suitable real number > 0), we find an $\varepsilon' \in (0, \varepsilon]$ and for every $\alpha \in (0, \varepsilon']$ real numbers $\varrho_i(\alpha)$, $i = 1, \ldots, n$, such that $\lim_{\alpha \to 0} \frac{\varrho_i(\alpha)}{\alpha} = 0$, $\sum_{i=1}^{n} \frac{\varrho_i(\alpha)}{\alpha} p_i \in V$, and $\psi(\alpha, \varrho_1(\alpha), \ldots, \varrho_n(\alpha)) = 0$. Let $q_\alpha = q + \alpha p + \sum_{i=1}^{n} \varrho_i(\alpha) p_i$ whenever $\alpha \in (0, \varepsilon']$. Then, $q_\alpha \neq q$, $q_\alpha \in q + [0, \varepsilon] (p + V) \subseteqq Q$, the filter $\mathfrak{Q}$ in R with $\{\{q_\alpha; \alpha \in (0, \varepsilon']\}; \varepsilon' \in (0, \varepsilon]\}$ as a base converges to q, and

$$f_3(q_\alpha) = 0.$$

Let $\mathfrak{F}$ denote the filter in R which has the set of all $\tau_{R'}$-neighbourhoods of the point 0 in R' as base. Trivially $\mathfrak{F} \in \tau_R(0)$. Because of $\phi_i(p) \in \underline{K}_i$ and $\phi_i \in D_{1_1} f_i(q)$, $i = 1, 2$, using the filter $\mathfrak{N} = V^+(p + \mathfrak{F})$ of $\mathcal{N}_p$, we get an $\varepsilon^+ \in (0, \varepsilon']$ such that

$$f_1(q_\alpha) - f_1(q) \in \underline{K}_1$$

and

$$f_2(q_\alpha) \in f_2(q) + \underline{K}_2 \subseteqq K_2$$

whenever $\alpha \in (0, \varepsilon^+]$. For every such α, $f_2(q_\alpha) \in K_2$ and $f_3(q_\alpha) = 0$ imply $q_\alpha \in \tilde{Q}$. $f_1(q_\alpha) - f_1(q) \in \underline{K}_1$, $q_\alpha \in \tilde{Q}$, and $\mathfrak{Q} \in \tau_R(q)$ are in contradiction to the assumption that q is a local maximizing point of $f|_{\tilde{Q}}$ of the third kind. Consequently, (*) is proved.

Now, for $i = 1, 2$, let ϕ_i be a superadditive mapping belonging to $D_{1_1} f_i(q)$. $M = (0, \infty)(\phi_1, \phi_2, \phi_3) [C_{q,Q} \setminus \{0\}] - K_1 \times K_2 \times \{0\}$ is a cone in $S_1 \times S_2 \times S_3$ which has the point 0 as vertex and contains algebraically inner points. As for $i = 1, 2$, and $\alpha_1, \alpha_2 > 0$, and $\alpha \in (0, 1)$, and $p_1, p_2 \in C_{q,Q} \setminus \{0\}$, we have

$$\begin{aligned}&(\alpha\alpha_1 + (1 - \alpha) \alpha_2) \phi_i \left(\frac{\alpha\alpha_1}{\alpha\alpha_1 + (1 - \alpha) \alpha_2} p_1 + \frac{(1 - \alpha) \alpha_2}{\alpha\alpha_1 + (1 - \alpha) \alpha_2} p_2\right) \\ &\in \alpha\alpha_1\phi_i(p_1) + (1 - \alpha) \alpha_2\phi_i(p_2) + K_i\end{aligned}$$

and as ϕ_3 is the restriction of a linear mapping of R into S_3, M is convex. Also $N = \underline{K}_1 \times \underline{K}_2 \times \{0\}$ is a convex cone in $S_1 \times S_2 \times S_3$ with 0 as vertex. We shall prove that $M \cap N = \emptyset$. Otherwise, there exist a real number $\alpha > 0$, a point $p \in C_{q,Q} \setminus \{0\}$ and points $q_i \in K_i$, $i = 1, 2$, such that

$$\alpha\phi_i(p) - q_i \in \underline{K}_i, \quad i = 1, 2, \quad \text{and} \quad \phi_3(p) = 0.$$

From this, we get

$$\phi_i(p) \in \underline{K}_i, \quad i = 1, 2, \quad \text{and} \quad \phi_3(p) = 0$$

which by (*) is impossible. Hence $M \cap N = \emptyset$. Using the separation theorem for convex sets, we get a linear functional $\psi (\neq 0)$ on $S_1 \times S_2 \times S_3$ for which $\psi \in (-M^*) \cap N^*$. The spaces $S_1 \times \{0\} \times \{0\}$, $\{0\} \times S_2 \times \{0\}$, and $\{0\} \times \{0\} \times S_3$ will be identified with S_1, S_2, and S_3, respectively. For $i = 1, 2, 3$, let $\psi_i = \psi|_{S_i}$. From $\psi \in N^*$, we get $\psi_i \in K_i^*$, $i = 1, 2$. As $\psi \in -M^*$, we have

$$\sum_{i=1}^{3} \psi_i \circ \phi_i[C_{q,Q} \setminus \{0\}] \subseteqq (-\infty, 0]$$

and from this, we get

$$\sum_{i=1}^{3} \psi_i \circ \phi_i[C_{q,Q}] \subseteqq (-\infty, 0].$$

Thus, statement 1 of the theorem is proven.

2. The proof of the second statement of the theorem is to be given analogously. Instead of (*), we have to take

$$\left(\psi_1 \circ D_{1_1} f_1(q), D_{1_1} f_2(q), \phi_3\right) [C_{q,Q} \setminus \{0\}] \subseteq (\boldsymbol{R} \times S_2 \times S_3) \setminus \left((0, \infty) \times K_2 \times \{0\}\right)$$

where $\psi_1 \in K_1{}^*$.

References

[1] M. A. BAZARAA, J. J. GOODE, and M. Z. NASHED, On the Cones of Tangents with Applications to Mathematical Programming, JOTA **13** (1974), 389–426.

[2] A. BASTIANI, Applications différentiables et varietés différentiables de dimension infinie, J. Analyse math. **13** (1964), 1–114.

[3] K. H. ELSTER und R. NEHSE, Konjugierte Operatoren und Subdifferentiale, Math. Operationsforschung und Statistik **6** (1975), 641–657.

[4] J. EKELAND et G. LEBOURG, Generic Fréchet-differentiability and perturbed optimization problems in Banach-spaces, Transact. Amer. Math. Soc. **224** (1976), 193–216.

[5] S. GÄHLER, Eine Verallgemeinerung des Satzes von Dubovickij-Miljutin, Math. Nachr. **72** (1977), 117–138.

[6] S. GÄHLER, Optimality Conditions in Polyoptimization, in: Theory of Non-Linear Operators, Proceedings of the fifth international summer school, Abh. Akad. Wiss. DDR, Jahrgang 1978, Nr. 6N, 353–355.

[7] W. GÄHLER, Grundstrukturen der Analysis. I, II, Berlin/Basel und Stuttgart 1977, 1978.

[8] A. GITTLEMAN, A General Multiplier Rule, JOTA **7** (1976), 29–38.

[9] K. H. HOFFMANN und I. KOLUMBÁN, Verallgemeinerte Differenzierbarkeitsbegriffe und ihre Anwendung in der Optimierungstheorie, Computing **12** (1974), 17–41.

[10] G. KÖTHE, Topologische lineare Räume, Berlin–Göttingen–Heidelberg 1960.

[11] I. LAZAREVIC and I. KOLUMBÁN, On the rule of multipliers in optimal problems, Mathematica Balkanica **4** (1974), 399–407.

[12] B. N. PSENICNYJ, Notwendige Optimalitätsbedingungen, Leipzig 1972.

[13] H. H. SCHAEFER, Topological Vector Spaces, New York–Heidelberg–Berlin 1970.

[14] S. SMALE, Sufficient Conditions for an Optimum, in: Dynamical Systems–Warwick 1974, Lecture Notes in Mathematics **468**, Springer-Verlag, Berlin–Heidelberg–New York 1975, 287–292.

[15] M. VALADIER, Sous-différentiabilité de fonctions convexes á valeur dans un espace vectoriel ordonné, Math. Scand. **30** (1972), 65–74.

[16] YAU-CHUEN WONG and KUNG-FU NG, Partially Ordered Topological Vector Spaces, Oxford 1973.

[17] J. ZOWE, Subdifferentiability of Convex Functions with Values in an Ordered Vector Space, Math. Scand. **34** (1974), 69–83.

Anschrift des Verfassers:

Akademie der Wissenschaften der DDR
Zentralinstitut für Mathematik und Mechanik
108 Berlin
Mohrenstraße 39
DDR

Generalized Marinescu Convergence

WERNER GÄHLER, Berlin

1. Introduction

For arbitrary separated locally convex topological vector spaces X and Y, H. H. KELLER [8] introduced in $L^n(X, Y)$ a useful pseudo-topology, the pseudo-topology of MARINESCU convergence. We generalize this pseudo-topology to the case that X and Y are pseudo-topological linear groups. Besides this generalized MARINESCU convergence we consider two modifications. By means of certain kinds of convergence we define notions of differentiation. We introduce a modified notion of continuity, the strong continuity, and investigate differentiability classes defined with the strong continuity. In one case for mappings of open connected subsets of separated locally convex topological vector spaces into such spaces the introduced differentiability classes are equivalent to the differentiability classes defined by GUTKNECHT [7].

2. The functors u, u' and d

Let $\boldsymbol{K}$ be equal to $\boldsymbol{R}$ or $\boldsymbol{C}$ and let $\boldsymbol{V}$ be the neighbourhood filter of 0 on $\boldsymbol{K}$. Let M be a set. A dual filter $\mathfrak{F}$ on M is a set of subsets of M such that $\{M \setminus F \mid F \in \mathfrak{F}\}$ is a filter on M. A non-empty set $\mathfrak{u}$ of filters (dual filters) on M is called an $\wedge$-*ideal* if for arbitrary $\mathfrak{F}, \mathfrak{G} \in \mathfrak{u}$ also $\mathfrak{F} \cap \mathfrak{G}$ belongs to $\mathfrak{u}$ and for arbitrary $\mathfrak{F} \in \mathfrak{u}$ each filter (dual filter) $\mathfrak{G}$ on X with $\mathfrak{F} \subseteqq \mathfrak{G}$ belongs to $\mathfrak{u}$, too. A *convergence structure* of M is a mapping τ which has M as domain and whose values $\tau(x)$ are $\wedge$-ideals of filters on M. By a *pseudo-topology* of M we understand a convergence structure τ of M such that for each $x \in M$, the filter $[x] = \{A \subseteqq M \mid x \in A\}$ belongs to $\tau(x)$ (see [3] and [9]. A *pseudo-topological* $\boldsymbol{K}$*-vector space* is a $\boldsymbol{K}$-vector space E equipped with a vector pseudo-topology, i.e. with a pseudo-topology of E such that the vector addition and the scalar multiplication are continuous mappings. For each pseudo-topological $\boldsymbol{K}$-vector space $X = (E, \tau)$ let τ^u be the coarsest pseudo-topology of E which is finer than τ and such that $X^u = (E, \tau^u)$ is a *pseudo-topological union* (of topological vector spaces), i.e., X^u is an inductive limit $\bigcup_{i \in I} X_i$ of an embedding spectrum of topological vector spaces X_i ($i \in I$) where I is a non-empty directed set; the inductive limit is to be understood in the category PVEC of all pseudo-topological $\boldsymbol{K}$-vector spaces. KUTZLER [10] proved the following characterization of pseudo-topological unions.

Theorem 2.1. *A pseudo-topological* **K**-*vector space* $X = (E, \tau)$ *is a pseudo-topological union if and only the* $\wedge$-ideal $\tau(0)$ *has a base of filters* $\mathfrak{F}$ *with* $\bigcap_{F \in \mathfrak{F}} \boldsymbol{K}F \in \mathfrak{F} = \mathfrak{F} + \mathfrak{F} = \boldsymbol{V}\mathfrak{F}$.

As an easy consequence we get the

Corollary. *For each pseudo-topological vector space* $X = (E, \tau)$, $\tau^u(0)$ *is the* $\wedge$-*ideal having as base* $\left\{\mathfrak{F} \in \tau(0) \mid \bigcap_{F \in \mathfrak{F}} \boldsymbol{K}F \in \mathfrak{F} = \mathfrak{F} + \mathfrak{F} = \boldsymbol{V}\mathfrak{F}\right\}$.

Let PVEC^u be the full subcategory of PVEC of all pseudo-topological unions. We get a functor $\boldsymbol{u}$: PVEC $\to$ PVEC^u if for each object X of PVEC we define $\boldsymbol{u}(X)$ to be X^u and for each morphism $f: X \to Y$ of PVEC we define $\boldsymbol{u}(f)$ to be the mapping $f: X^u \to Y^u$.

According to [1] by a *pseudo-topological* **K**-*linear group* we understand a **K**-vector space E equipped with a pseudo-topology τ such that the vector addition is continuous and the scalar multiplication $(\alpha, x) \mapsto \alpha x$ is partially continuous at the second argument and is continuous at $(0, 0)$. Of course, a *topological linear group* is a pseudo-topological linear group whose pseudo-topology is a topology. For each pseudo-topological **K**-linear group $X = (E, \tau)$ let $\tau^{u'}$ be the coarsest pseudo-topology of E which is finer than τ and such that $X^{u'} = (E, \tau^{u'})$ is a *pseudo-topological union in the wide sense* (of topological linear groups), i.e., $X^{u'}$ is an inductive limit $\bigcup_{i \in I} X_i$ of an embedding spectrum of topological linear groups X_i $(i \in I)$ where I is a non-empty directed set and the inductive limit is to be understood in the category PLING of all pseudo-topological linear groups. Analogously as Theorem 2.1 we can prove:

Theorem 2.2. *A pseudo-topological* **K**-linear *group* $X = (E, \tau)$ *is a pseudo-topological union in the wide sense if and only in the* $\wedge$-*ideal* $\tau(0)$ *has a base of filters* $\mathfrak{F}$ *with* $\mathfrak{F} = \mathfrak{F} + \mathfrak{F} = \boldsymbol{V}\mathfrak{F}$.

In consequence of this theorem we get the

Corollary. *For each pseudo-topological linear group* $X = (E, \tau)$, $\tau^{u'}(0)$ *is the* $\wedge$-*ideal with base* $\{\mathfrak{F} \in \tau(0) \mid \mathfrak{F} = \mathfrak{F} + \mathfrak{F} = \boldsymbol{V}\mathfrak{F}\}$.

Let PLING$^{u'}$ be the full subcategory of PLING of all pseudo-topological unions in the wide sense. There exists a functor $\boldsymbol{u}'$: PLING $\to$ PLING$^{u'}$ such that for each object X of PLING, $\boldsymbol{u}'(X)$ is equal to $X^{u'}$ and for each morphism $f: X \to Y$ of PLING, $\boldsymbol{u}'(f)$ is equal to $f: X^{u'} \to Y^{u'}$.

A pseudo-topological linear group X is called *locally convex* if for each zero-filter $\mathfrak{F}$ on X the filter $\mathfrak{F}^\wedge$ generated by the convex hulls of the sets $F \in \mathfrak{F}$ is a zero-filter, too. For each subcategory $\mathfrak{C}^\wedge$ of PLING we denote by $\mathfrak{C}^\wedge$ the full subcategory of $\mathfrak{C}$ of all locally convex objects of $\mathfrak{C}$. It is easy to prove that PVEC$^{u\wedge}$ is the full subcategory of PVEC of all pseudo-topological unions of locally convex topological vector spaces, i.e. of all *Marinescu spaces,* and that PLING$^{u'\wedge}$ is the full subcategory of PLING of all pseudo-topological unions in the wide sense of locally convex topological linear groups. By restriction of $\boldsymbol{u}$ and $\boldsymbol{u}'$ we get functors $\boldsymbol{u}^\wedge$: PVEC$^\wedge$ $\to$ PVEC$^{u\wedge}$ and $\boldsymbol{u}'^\wedge$: PLING$^\wedge$ $\to$ PLING$^{u'\wedge}$, respectively. Obiously the functors $\boldsymbol{u}$, $\boldsymbol{u}'$, $\boldsymbol{u}^\wedge$ and $\boldsymbol{u}'^\wedge$ are bicoreflectors.

Of course, each pseudo-topological union in the wide sense of topological vectors spaces is a pseudo-topological union. But a pseudo-topological union in the wide sense which is a pseudo-topological vector space is not a pseudo-topological union in general even if further we suppose that this pseudo-topological vector space is locally convex. We show this by an example.

Let E be a $\boldsymbol{K}$-vector space with countable-infinite dimension and let $e_1, e_2, \ldots$ be the vectors of a linear base of E. By $\mathfrak{N}$ we denote the filter on E which has as a base the set of all subspaces of E of finite codimensions. There exists a locally convex vector pseudo-topology τ on E such that $\{\mathfrak{N} + \boldsymbol{V}e_1 + \cdots + \boldsymbol{V}e_n \mid n = 1, 2, \ldots\}$ is a base of $\tau(0)$. By Theorem 2.2 it follows that $X = (E, \tau)$ is a pseudo-topological union in the wide sense. For each filter $\mathfrak{F} = \mathfrak{N} + \boldsymbol{V}e_1 + \cdots + \boldsymbol{V}e_n$, $\bigcap_{\mathfrak{F} \in F} \boldsymbol{K}F = \boldsymbol{K}e + \cdots + \boldsymbol{K}e_n$ does not belong to $\mathfrak{F}$. Hence by Theorem 2.1 we can easily prove that X is not a pseudo-topological union.

From this example it follows that $\boldsymbol{u}$ is not a restriction of $\boldsymbol{u}'$ and $\boldsymbol{u}^\wedge$ is not a restriction of $\boldsymbol{u}'^\wedge$.

A pseudo-topological linear group $X = (E, \tau)$ is called *locally bounded* if each zero-filter $\mathfrak{F}$ on X contains a *bounded* subset of X, i.e. a subset B such that the filter $\boldsymbol{V}B$ with the base $\{VB \mid V \in \boldsymbol{V}\}$ is a zero-filter. For each pseudo-topological linear group $X = (E, \tau)$ let τ^d be the coarsest pseudo-topology of E finer than τ such that $X^{\boldsymbol{d}} = (E, \tau^d)$ is a pseudo-topological union in the wide sense of locally bounded topological linear groups. $\tau^d(0)$ is the $\wedge$-ideal with the base $\{\boldsymbol{V}B \in \tau(0) \mid \boldsymbol{V}B = \boldsymbol{V}B + \boldsymbol{V}B\}$. Let PLING^d be the full subcategory of PLING of all pseudo-topological unions in the wide sense of locally bounded topological linear groups. There exists a functor $\boldsymbol{d}$: PLING $\to$ PLING^d such that for each object X of PLING, $\boldsymbol{d}(X)$ is equal to $X^{\boldsymbol{d}}$ and for each morphism $f\colon X \to Y$ of PLING, $\boldsymbol{d}(f)$ is equal to $f\colon X^{\boldsymbol{d}} \to Y^{\boldsymbol{d}}$. Let PVEC^d be the full subcategory of PVEC of all pseudo-topological unions of locally bounded topological vector spaces. Taking into account that for each subset B of a $\boldsymbol{K}$-vector space it follows $\bigcap_{F \in \boldsymbol{V}B} \boldsymbol{K}F = \boldsymbol{K}B \in \boldsymbol{V}B$, by Theorem 2.1 we get that the restriction of $\boldsymbol{d}$ to PVEC is a functor of PVEC onto PVEC^d. By means of 5.19.15 of [6], vol. II, we can prove that PVEC$^{d\wedge}$ is the full subcategory of PVEC of all pseudo-topological unions of semi-normable topological vector spaces and that PLING$^{d\wedge}$ is the full subcategory of PLING of all pseudo-topological unions in the wide sense of pseudo-normable topological linear groups. Of course, *semi-normable* topological vector spaces are topological vector spaces whose topologies are generated by semi-norms, and *pseudo-normable* topological linear groups are topological linear groups whose topologies are generated by pseudo-norms where *pseudo-norms* are defined analogously as semi-norms, but the values of pseudo-norms may be infinite. Obviously, $\boldsymbol{d}$ and the mentioned restrictions of $\boldsymbol{d}$ are bicoreflectors.

Theorem 2.3. *For each subspace Y of a pseudo-topological vector space X, $Y^{\boldsymbol{u}}$ is a subspace of $X^{\boldsymbol{u}}$. For each pseudo-topological linear subgroup Y of a pseudo-topological linear group X, $Y^{\boldsymbol{u}'}$ is a pseudo-topological linear subgroup of $X^{\boldsymbol{u}'}$ and $Y^{\boldsymbol{d}}$ is a pseudo-topological linear subgroup of $X^{\boldsymbol{d}}$.*

Proof. Let $Y = (T, \sigma)$ be a subspace of a pseudo-topological $\boldsymbol{K}$-vector space $X = (E, \tau)$. Let $\mathfrak{F}$ be a filter on E with $T \in \mathfrak{F}$ and let $\mathfrak{F}_T$ be the trace $\{F \cap T \mid F \in \mathfrak{F}\}$ of $\mathfrak{F}$ on T. Because of $VF \cap T = V(F \cap T)$ and $F \cap T \supseteqq (G \cap T) + (G \cap T)$ where $V \in \boldsymbol{V}$ and $F, G \subseteqq E$ with $F \supseteqq G + G$, from $\mathfrak{F} = \boldsymbol{V}\mathfrak{F} = \mathfrak{F} + \mathfrak{F}$ it follows $\mathfrak{F}_T = \boldsymbol{V}\mathfrak{F}_T$ and $\mathfrak{F}_T \subseteqq \mathfrak{F}_T + \mathfrak{F}_T$. Taking into account that $\boldsymbol{V}\mathfrak{F}_T \subseteqq [0]$ if $\mathfrak{F}$ is a proper filter, we get $\mathfrak{F}_T = \mathfrak{F}_T + \mathfrak{F}_T$. If $\bigcap_{F \in \mathfrak{F}} \boldsymbol{K}F \in \mathfrak{F}$, we get $\bigcap_{F \in \mathfrak{F}} \boldsymbol{K}(F \cap T) = \left(\bigcap_{F \in \mathfrak{F}} \boldsymbol{K}F\right) \cap T \in \mathfrak{F}_T$. Hence the induced pseudo-topology $(\tau^{\boldsymbol{u}})_T$ of $\tau^{\boldsymbol{u}}$ on T is finer than $\sigma^{\boldsymbol{u}}$. Because of the continuity of the identical mapping ι of Y into X, as mapping of Y^u into X^u, ι is continuous, too. Therefore we have $(\tau^u)_T = \sigma^u$, i.e., the first part of the theorem is proved. The second part of the theorem can be proved analogously.

To get another property of $\boldsymbol{u}$, $\boldsymbol{u}'$ and $\boldsymbol{d}$ we use the following

Theorem 2.4. *Let $\mathfrak{C}$ be an arbitrary subcategory of* PLING *and let S be a functor of $\mathfrak{C}$ into itself such that for each object L of $\mathfrak{C}$, L and $S(L)$ have the same underlying vector space and for each morphism f of $\mathfrak{C}$, f and $S(f)$ are the same linear mappings. Let $X_1, \ldots, X_n$ be objects of $\mathfrak{C}$. Then we have $S(X_1 \times \cdots \times X_n) = S(X_1) \times \cdots \times S(X_n)$.*

Proof. The projections $\pi_i: X \to X_i$ $(X = X_1 \times \cdots \times X_n)$ and therefore the mappings $S(\pi_i): S(X) \to S(X_i)$ are continuous. Hence the identical mapping of $S(X)$ onto $S(X_1) \times \cdots \times S(X_n)$ is continuous, too. Taking into account that the canonical embeddings $\iota_i: X_i \to X$ and therefore also the mappings $S(\iota_i): S(X_i) \to S(X)$ are continuous, the identical mapping of (the direct sum) $S(X_1) \times \cdots \times S(X_n)$ onto $S(X)$ is continuous, too. Hence $S(X) = S(X_1) \times \cdots \times S(X_n)$.

Theorem 2.5. *If $X_1, \ldots, X_n$ and Y are pseudo-topological vector spaces, then $X_1^{\boldsymbol{u}} \times \cdots \times X_n^{\boldsymbol{u}} = (X_1 \times \cdots \times X_n)^{\boldsymbol{u}}$ and each continuous multilinear mapping $l: X_1 \times \cdots \times X_n \to Y$ is continuous as mapping of $X_1^{\boldsymbol{u}} \times \cdots \times X_n^{\boldsymbol{u}}$ into $Y^{\boldsymbol{u}}$. If $X_1, \ldots, X_n$ and Y are pseudo-topological linear groups, then $X_1^{\boldsymbol{u}'} \times \cdots \times X_n^{\boldsymbol{u}'} = (X_1 \times \cdots \times X_n)^{\boldsymbol{u}'}$ and $X_1^{\boldsymbol{d}} \times \cdots \times X_n^{\boldsymbol{d}} = (X_1 \times \cdots \times X_n)^{\boldsymbol{d}}$ and each continuous multilinear mapping $l: X_1 \times \cdots \times X_n \to Y$ is a continuous mapping of $X_1^{\boldsymbol{u}'} \times \cdots \times X_n^{\boldsymbol{u}'}$ into $Y^{\boldsymbol{u}'}$ and of $X_1^{\boldsymbol{d}} \times \cdots \times X_n^{\boldsymbol{d}}$ into $Y^{\boldsymbol{d}}$.*

Proof. The validity of the equations follows from Theorem 2.4. Now let $X_1, \ldots, X_n$ and Y be pseudo-topological $\boldsymbol{K}$-vector spaces with the pseudo-topologies $\tau_1, \ldots, \tau_n$ and σ, respectively, and let l be a continuous multilinear mapping of $X_1 \times \cdots \times X_n$ into Y. For each $i = 1, 2, \ldots$ let $\mathfrak{F}_i$ be a proper filter of $\tau_i(0)$ with $\bigcap_{F \in \mathfrak{F}_i} \boldsymbol{K}F \in \mathfrak{F}_i = \mathfrak{F}_i + \mathfrak{F}_i = \boldsymbol{V}\mathfrak{F}_i$. $\mathfrak{F} = l(\mathfrak{F}_1 \times \cdots \times \mathfrak{F}_n)$ belongs to $\sigma(0)$. Of course, $\bigcap_{F \in \mathfrak{F}} \boldsymbol{K}F \supseteq l\left[\bigcap_{F \in \mathfrak{F}_1} \boldsymbol{K}F \times \cdots \times \bigcap_{F \in \mathfrak{F}_n} \boldsymbol{K}F\right] \in \mathfrak{F} = \boldsymbol{V}\mathfrak{F}$. For arbitrary sets $F_i \subseteq X_i$ with $0 \in F_i (i = 1, \ldots, n)$ we have

$$l[(F_1 + F_1) \times \cdots \times (F_n + F_n)] \supseteq l[F_1 \times \cdots \times F_n] + l[F_1 \times \cdots \times F_n].$$

Therefore $\mathfrak{F} \subseteq \mathfrak{F} + \mathfrak{F}$ and because of $\mathfrak{F} \subseteq [0]$, $\mathfrak{F} = \mathfrak{F} + \mathfrak{F}$. Hence $l: X_1^{\boldsymbol{u}} \times \cdots \times X_n^{\boldsymbol{u}} \to Y^{\boldsymbol{u}}$ is continuous at 0 and therefore at each point of $X_1^{\boldsymbol{u}} \times \cdots \times X_n^{\boldsymbol{u}}$ (see 5.12.2 of [6], vol. II). If $X_1, \ldots, X_n$ and Y are pseudo-topological linear groups we can prove analogously that each continuous multilinear mapping $l: X_1 \times \cdots \times X_n \to Y$ is continuous as mapping of $X_1^{\boldsymbol{u}'} \times \cdots \times X_n^{\boldsymbol{u}'}$ into $Y^{\boldsymbol{u}'}$ and as mapping of $X_1^{\boldsymbol{d}} \times \cdots \times X_n^{\boldsymbol{d}}$ into $Y^{\boldsymbol{d}}$.

3. Generalized Marinescu Convergence and other kinds of convergence

Let $X = (E, \tau)$ be a pseudo-topological linear group and let $\mathfrak{X}$ be a set of subsets of X. Let $\mathfrak{L}$ be a filter on $\mathfrak{X}$, let $\mathfrak{L}^*$ be the *grill* $\{M \subseteq \mathfrak{X} \mid \forall L \in \mathfrak{L} (L \cap M \neq \emptyset)\}$ of $\mathfrak{L}$ and let A be a set belonging to $\mathfrak{X}$. A is called a *simple limit of* $\mathfrak{L}$ if $A = \bigcap_{L \in \mathfrak{L}} \bigcup_{C \in L} C = \bigcup_{L \in \mathfrak{L}} \bigcap_{C \in L} C$, and A is said to be an *open limit* of $\mathfrak{L}$ if for each point $x \in X$

$$x \in M \Leftrightarrow \forall \mathfrak{F} \in \tau(x) \exists L \in \mathfrak{L} \left(\bigcap_{C \in L} C \in \mathfrak{F} \right)$$
$$\Leftrightarrow \forall \mathfrak{F} \in \tau(x) \exists L \in \mathfrak{L}^* \left(\bigcap_{C \in L} C \in \mathfrak{F} \right).$$

By $\mathfrak{S}_q$ we denote the set of all *quasi-bounded* filters on X, i.e. of all filters $\mathfrak{F}$ on X with $V\mathfrak{F} \in \tau(0)$. A is called an $\mathfrak{S}_q$*-limit* of $\mathfrak{L}$ if for each filter $\mathfrak{F} \in \mathfrak{S}_q$ we have

$$A \in \mathfrak{F} \Leftrightarrow \exists L \in \mathfrak{L} \left(\bigcap_{C \in L} C \in \mathfrak{F} \right)$$
$$\Leftrightarrow \exists L \in \mathfrak{L}^* \left(\bigcap_{C \in L} C \in \mathfrak{F} \right).$$

By $\mathfrak{B}_h$ we denote the set of all *hypo-bounded* $\wedge$-ideals of dual filters on X, i.e. of all $\wedge$-ideals $\mathfrak{u}$ of dual filters on X with $\bigvee_{\mathfrak{F} \in \mathfrak{u}} \bigcap_{F \in \mathfrak{F}} VF \in \tau(0)$ where the union is to be understood in the filter lattice of X. We say that A is a $\mathfrak{B}_h$*-limit* of $\mathfrak{L}$ if for each $\mathfrak{u} \in \mathfrak{B}_h$

$$(A) \in \mathfrak{u} \Leftrightarrow \exists \mathfrak{F} \in \mathfrak{u} \forall F \in \mathfrak{F} \exists L \in \mathfrak{L} \left(F \subseteqq \bigcap_{C \in L} C \right)$$
$$\Leftrightarrow \exists \mathfrak{F} \in \mathfrak{u} \forall F \in \mathfrak{F} \exists L \in \mathfrak{L}^* \left(F \subseteqq \bigcap_{C \in L} C \right),$$

where (A) is the dual filter $\{B \mid B \subseteqq A\}$ on X.

Furthermore let $Y = (F, \sigma)$ be a pseudo-topolotical linear group. Let $\mathfrak{A}$ be a set of mappings of subsets of X into Y. For each $g \in \mathfrak{A}$ by db$g$ we denote the domain of g and for each filter $\mathfrak{K}$ on $\mathfrak{A}$ by db$\mathfrak{K}$ we denote the filter on db$\mathfrak{A} = \{\text{db}g \mid g \in \mathfrak{A}\}$ with the base $\{\{\text{db}g \mid g \in K\} \mid K \in \mathfrak{K}\}$. We use the pseudo-topology s of *pointwise convergence* in $\mathfrak{A}$, the convergence structure c of *continuous convergence* in $\mathfrak{A}$, the pseudo-topology q of *quasi-bounded convergence* in $\mathfrak{A}$ and the pseudo-topology h of *hypobounded convergence* in $\mathfrak{A}$ (see [6], vol. II). These kinds of convergence structures are defined as follows: Let $\mathfrak{K}$ be a filter on $\mathfrak{A}$ and let f be a mapping belonging to $\mathfrak{A}$.

$\mathfrak{K}$ *converges pointwise to* f if db$f$ is a simple limit of db$\mathfrak{K}$ and for each $x \in$ db$f$ the filter $\mathfrak{K}(x)$ with the base $\{\{g(x) \mid g \in K, x \in \text{db}g\} \mid K \in \mathfrak{K}\}$ converges to $f(x)$,

$\mathfrak{K}$ *converges continuously to* f if db$f$ is an open limit of db$\mathfrak{K}$ and for each $x \in$ db$f$ and $\mathfrak{F} \in \tau(x)$ the filter $\mathfrak{K}(\mathfrak{F})$, having as base $\{K[F] \mid K \in \mathfrak{K}, F \in \mathfrak{F}\}$ with $K[F] = \{g(y) \mid g \in K, y \in F \cap \text{db}g\}$, belongs to $\sigma(f(x))$,

$\mathfrak{K}$ *converges to* f *with respect to* q if db$f$ is an $\mathfrak{S}_q$-limit of db$\mathfrak{K}$ and for each $\mathfrak{F} \in \mathfrak{S}_q$ the filter $(\mathfrak{K} - f)(\mathfrak{F})$ belongs to $\sigma(0)$, and

$\mathfrak{K}$ *converges to* f *with respect to* h if db$f$ is a $\mathfrak{B}_h$-limit of db$\mathfrak{K}$ and for each $\mathfrak{u} \in \mathfrak{B}_h$, the filter $\bigvee_{\mathfrak{F} \in \mathfrak{u}} \bigcap_{F \in \mathfrak{F}} (\mathfrak{K} - f)[F]$ belongs to $\sigma(0)$, where $(\mathfrak{K} - f)[F]$ are the filters with the bases $\{(K - f)[F] \mid K \in \mathfrak{K}\}$.

If $\mathfrak{A}$ consists of continuous mappings with open sets as domains, c is a pseudo-topology, too. For each filter $\mathfrak{K}$ on $\mathfrak{A}$ and each mapping f with db$f = X$ the implications

$$\mathfrak{K} \underset{\mathsf{h}}{\rightarrow} f \Rightarrow \mathfrak{K} \underset{\mathsf{q}}{\rightarrow} f \Rightarrow^* \mathfrak{K} \underset{\mathsf{c}}{\rightarrow} f \Rightarrow \mathfrak{K} \underset{\mathsf{s}}{\rightarrow} f \tag{1}$$

hold where in the case marked by a star, X is supposed to be a pseudo-topological vector space.

Now let especially X and Y be locally convex topological vector spaces. Following H. H. KELLER [8] for arbitrary $n = 1, 2, \ldots$ we introduce a pseudo-topology M of the vector space $L^n(X, Y)$ of all continuous n-linear mappings of $X^n = \underbrace{X \times \cdots \times X}_{n}$ into Y such that $L_{\mathsf{M}}{}^n(X, Y)$, i.e. $L^n(X, Y)$ equipped with M, is a MARINESCU space. In [8] M is denoted by Λ. Let Γ_X and Γ_Y be the sets of all continuous semi-norms on X and Y, respectively. For each $\alpha \in \Gamma_x$, $\beta \in \Gamma_Y$ and $l \in L^n(X, Y)$ by $|l|_{\beta\alpha}$ we denote the supremum of $\{\beta l(x_1, \ldots, x_n) \mid \alpha(x_1) \leqq 1, \ldots, \alpha(x_n) \leqq 1\}$. Let Φ be the set of all mappings of Γ_Y into Γ_X, directed by $\varphi \lesssim \psi \Leftrightarrow \varphi(\beta) \leqq \psi(\beta)$ $(\beta \in \Gamma_Y)$. For each $\varphi \in \Phi$, $L_\varphi{}^n(X, Y) = \{l \in L^n(X, Y) \mid |l|_{\beta\varphi(\beta)} < \infty$ for each $\beta \in \Gamma_Y\}$ is a subspace of $L^n(X, Y)$. Let $L_\varphi{}^n(X, Y)$ be equipped with the topology generated by the family of semi-norms $l \mapsto |l|_{\beta\varphi(\beta)}$ $\big(l \in L_\varphi{}^n(X, Y)\big)$ $(\beta \in \Gamma_Y)$. By $L^n_{\mathsf{M}'}(X, Y)$ we understand the pseudo-topological union of the locally convex topological vector spaces $L_\varphi{}^n(X, Y)$ and, of course, by M we denote the pseudo-topology of this space.

There exists an interesting modification $L^n_{\mathsf{M}'}(X, Y)$ of $L_{\mathsf{M}}{}^n(X, Y)$, defined as follows: For each $\varphi \in \Phi$, $L^n(X, Y)$, equipped with the topology generated by the family of pseudo-norms $l \mapsto |l|_{\beta\varphi(\beta)}$ $\big(l \in L^n(X, Y)\big)$ $(\beta \in \Gamma_Y)$, is a locally convex topological linear group $L_\varphi{}'^n(X, Y)$. By $L^n_{\mathsf{M}'}(X, Y)$ we understand the pseudo-topological union in the wide sense of the locally convex topological linear groups $L_\varphi{}'^n(X, Y)$ and, of course, by M' we denote the pseudo-topology of $L^n_{\mathsf{M}'}(X, Y)$.

For arbitrary pseudo-topological linear groups X and Y by $L_{\mathsf{s}}{}^n(X, Y)$, $L_{\mathsf{c}}{}^n(X, Y)$, $L_{\mathsf{q}}{}^n(X, Y)$ and $L_{\mathsf{h}}{}^n(X, Y)$. we denote the vector space $L^n(X, Y)$ equipped with the pseudo-topology s, c, q, or h on $L^n(X, Y)$, respectively.

Theorem 3.1 (see [6], vol. II). *Let X and Y be locally convex topological vector spaces. Then $L_{\mathsf{M}}{}^n(X, Y) = L_{\mathsf{h}}{}^n(X, Y)^{\boldsymbol{u}}$ and $L_{\mathsf{M}}{}^n(X, Y) = L_{\mathsf{h}}{}^n(X, Y)^{\boldsymbol{u}'}$.*

By means of this theorem we can generalize M and M'. Let X and Y be pseudo-topological linear groups. Because of the fact $L_{\mathsf{h}}{}^n(X, Y)$ is a pseudo-topological vector space, $L_{\mathsf{h}}{}^n(X, Y)^{\boldsymbol{u}}$ exists. $L_{\mathsf{h}}{}^n(X, Y)^{\boldsymbol{u}}$ is equal to $L_{\mathsf{q}}{}^n(X, Y)^{\boldsymbol{u}}$, and if X is a pseudo-topological vector space, $L_{\mathsf{h}}{}^n(X, Y)^{\boldsymbol{u}}$ is also equal to $L_{\mathsf{c}}{}^n(X, Y)^{\boldsymbol{u}}$. This result remains true if we replace $\boldsymbol{u}$ by $\boldsymbol{u}'$ and by $\boldsymbol{d}$. We denote the pseudo-topologies of the spaces $L_{\mathsf{h}}{}^n(X, Y)^{\boldsymbol{u}}$, $L_{\mathsf{h}}{}^n(X, Y)^{\boldsymbol{u}'}$ and $L_{\mathsf{h}}{}^n(X, Y)^{\boldsymbol{d}}$ by u, u' and d, respectively, and these spaces also by $L_{\mathsf{u}}{}^n(X, Y)$, $L_{\mathsf{u}}{}^n(X, Y)$ and $L_{\mathsf{d}}{}^n(X, Y)$, respectively.

In the case that Y is locally convex, $L_{\mathsf{h}}{}^n(X, Y)$ and therefore also $L_{\mathsf{u}}{}^n(X, Y)$, $L^n_{\mathsf{u}'}(X, Y)$ and $L_{\mathsf{d}}{}^n(X, Y)$ are locally convex. In this case sometimes we use the same denotations as those of the first mentioned special case, i.e. we denote u and u' by M and M', respectively. Obviously we have

$$\mathsf{d} \leqq \mathsf{u} \leqq \mathsf{u}' \leqq \mathsf{h}. \tag{2}$$

If Y is locally convex, u' is equal to h.

Theorem 3.2. *Let X, Y and Z be pseudo-topological linear groups and let a be equal to c, q, h, u', u or d. Then the composition mapping $(f, g) \mapsto g \circ f$ of $L_a{}^n(X, Y) \times L_a(Y, Z)$ into $L_a{}^n(X, Z)$ is continuous.*

Proof. Concerning the cases $a = \mathsf{c}$, q and h we refer to [6], vol. II. In the other cases the proof is a consequence of the case $a = \mathsf{h}$ and of Theorem 2.4.

Let X and Y be pseudo-topological linear groups. For each $n = 1, 2 \ldots$ and $a = \mathsf{s}, \mathsf{c}, \mathsf{q}, \mathsf{h}, \mathsf{u}', \mathsf{u}$ and d we define recursively pseudo-topological linear groups $H_a{}^n(X, Y)$ by $H_a{}^1(X, Y) = L_a(X, Y)$ and $H_a{}^n(X, Y) = L_a\big(X, H_a{}^{n-1}(X, Y)\big)$ $(n > 1)$ and identify each $l \in H_a{}^n(X, Y)$ with the multilinear mapping $(h_1, \ldots, h_n) \mapsto l(h_1) \cdots (h_n)$ $(h_1, \ldots, h_n \in X)$. A filter $\mathfrak{F}$ on a vector space is called *equable* if $\mathfrak{F}$ is equal to $V\mathfrak{F}$, and a pseudo-topological linear group X is called *equable* (see [4]) if for each zero-filter $\mathfrak{F}$ on X there exists an equable zero-filter $\mathfrak{G}$ with $\mathfrak{G} \subseteq \mathfrak{F}$.

Theorem 3.3. *Let X and Y be pseudo-topological linear groups and let n be a positive integer. Then*

$$H_{\mathsf{c}}{}^n(X, Y) = L_{\mathsf{c}}{}^n(X, Y), \tag{3}$$

and if X is an equable pseudo-topological vector space, then

$$H_{\mathsf{q}}{}^n(X, Y) = L_{\mathsf{q}}{}^n(X, Y) \quad \text{and} \quad H_{\mathsf{h}}{}^n(X, Y) = L_{\mathsf{h}}{}^n(X, Y). \tag{4}$$

If we suppose that X is a pseudo-topological union in the wide sense and a pseudo-topological vector space, then

$$H^n_{\mathsf{u}'}(X, Y) = L^n_{\mathsf{u}'}(X, Y), \tag{5}$$

if X is a pseudo-topological union, then

$$H_{\mathsf{u}}{}^n(X, Y) = L_{\mathsf{u}}{}^n(X, Y), \tag{6}$$

and if especially X is a pseudo-topological union of locally bounded topological vector spaces, then

$$H_{\mathsf{d}}{}^n(X, Y) = L_{\mathsf{d}}{}^n(X, Y). \tag{7}$$

Proof. Concerning the equations (3) and (4) we refer to [6], vol. II. Now let X be a pseudo-topological union. Of course, (6) is true if $n = 1$. In the following we suppose that (6) is true for any n. For each $l \in L^{n+1}(X, Y)$ let $\varphi(l)$ be the mapping of X into $L_{\mathsf{q}}{}^n(X, Y)$ with $\varphi(l)\,(x)\colon (h_1, \ldots, h_n) \mapsto l(x, h_1, \ldots, h_n)$. Because of $\varphi(l)$ being continuous, $\varphi(l)$ is also continuous as mapping of X^u into $L_{\mathsf{u}}{}^n(X, Y) = H_{\mathsf{u}}{}^n(X, Y)$. Therefore $\varphi\colon l \mapsto \varphi(l)$ is a mapping of $L^{n+1}(X, Y)$ into $L\big(X, H_{\mathsf{u}}{}^n(X, Y)\big)$. Taking into account that each mapping $f \in L\big(X, H_a{}^n(X, Y)\big)$ is continuous as mapping of X into $L_{\mathsf{c}}{}^n(X, Y)$, it follows easily that φ is a bijection of $L^{n+1}(X, Y)$ onto $L\big(X, H_{\mathsf{u}}{}^n(X, Y)\big)$. $(l, x) \mapsto \varphi(l)\,(x)$ is a continuous bilinear mapping of $L_{\mathsf{q}}{}^{n+1}(X, Y) \times X$ into $L_{\mathsf{q}}{}^n(X, Y)$ and by Theorem 2.4 therefore of $L_{\mathsf{u}}{}^{n+1}(X, Y) \times X$ into $H_{\mathsf{u}}{}^n(X, Y)$. For each equable zero-filter $\mathfrak{K}$ on $L_{\mathsf{u}}{}^{n+1}(X, Y)$ and each quasi-bounded filter $\mathfrak{F}$ on X, $\varphi(\mathfrak{K})\,(\mathfrak{F}) = \varphi(\mathfrak{K})\,(V\mathfrak{F})$ is a zero-filter on $H_{\mathsf{u}}{}^n(X, Y)$. Hence φ is a continuous mapping of $L_{\mathsf{u}}{}^{n+1}(X, Y)$ onto $L_{\mathsf{q}}\big(X, H_{\mathsf{u}}{}^n(X, Y)\big)$ and therefore of $L_{\mathsf{u}}{}^{n+1}(X, Y)$ onto $L_{\mathsf{u}}\big(X, H_{\mathsf{u}}{}^n(X, Y)\big)$. Each equable zero-filter $\mathfrak{L}$ on $L_{\mathsf{u}}\big(X, H_{\mathsf{u}}{}^n(X, Y)\big)$ is also a zero-filter on $L_{\mathsf{q}}\big(X, H_{\mathsf{q}}{}^n(X, Y)\big)$. Thus, $\varphi^{-1}(\mathfrak{L})$ is a zero-filter on $L_{\mathsf{q}}{}^{n+1}(X, Y)$. In consequence, φ^{-1} is a continuous mapping of $L_{\mathsf{u}}\big(X, H_{\mathsf{u}}{}^n(X, Y)\big)$ onto $L_{\mathsf{u}}{}^{n+1}(X, Y)$. Therefore (6) holds with n replaced by $n + 1$. Under the corresponding assumptions, (5) and (7) can be proved analogously.

For an arbitrary family $(f_i)_{i \in I}$ of mappings of a set A into sets B_i by $\langle f_i \rangle_{i \in I}$ we denote the mapping of A into $\prod\limits_{i \in I} Y_i$ defined by $\langle f_i \rangle_{i \in I}\,(x) = \big(f_i(x)\big)_{i \in I}$ $(x \in A)$.

Theorem 3.4. *Let X be an arbitrary pseudo-topological* **K***-linear group and let Y be a finite product $Y_1 \times \cdots \times Y_m$ $(m > 0)$ of pseudo-topological* **K***-linear groups. Let a be equal to* s, c, q, h, u', u *or* d. *Then the mapping* $\langle f_1, \ldots, f_m \rangle \mapsto (f_1, \ldots, f_m)$ $\big(f_i \in H_a{}^n(X, Y_i)\big)$ *is a linear homeomorphism of* $H_a{}^n(Y, X)$ *onto* $H_a{}^n(X, Y_1) \times \cdots \times H_a{}^n(X, Y_m)$.

Proof. Concerning the cases $a = \mathsf{s}, \mathsf{c}, \mathsf{q}$ and h we refer to [6], vol. II (see especially 7.4.23, 7.14.25 and (7.14.19) of this book). If $n = 1$, by means of Theorem 2.4 and the assertion in the case $a = \mathsf{h}$ the assertion follows in the cases $a = \mathsf{u}'$, u and d. Now we suppose that the assertion is true for any n and for $a = \mathsf{u}'$, u and d. We use the abbreviations $Y_1' = H_a{}^n(X, Y_1), \ldots, Y_m' = H_a{}^n(X, Y_m)$, $Y' = H_a{}^n(X, Y)$. By $\pi_1, \ldots, \pi_m$ we denote the projections of $Y_1' \times \cdots \times Y_m'$ onto $Y_1', \ldots, Y_m'$, respectively. Obviously $l \mapsto \langle \pi_1 \circ l, \ldots, \pi_m \circ l \rangle$ $\big(l \in L_a(X, Y')\big)$ is a linear homeomorphism of $L_a(X, Y')$ onto $L_a(X, Y_1' \times \cdots \times Y_m')$ and $\langle \varphi_1, \ldots, \varphi_m \rangle \mapsto (\varphi_1, \ldots, \varphi_m)$ $\big(\varphi_i \in L_a(X, Y_1')\big)$ is a linear homeomorphism of $L_a(X, Y_1' \times \cdots \times Y_m')$ onto $L_a(X, Y_1') \times \cdots \times L_a(X, Y_m')$. Composing both mappings, the assertion follows in the cases $a = \mathsf{u}'$, u and d for $n + 1$.

4. Differentiabilities

Let $\mathfrak{X}$ be the class of all pseudo-topological linear groups. By a *kind of differentiability* we understand a pair $(\mathfrak{R}, \sigma)$ of mappings with the following properties:

(1) $\mathfrak{R}$ and σ have $\mathfrak{X} \times \mathfrak{X}$ as domain.

(2) For arbitrary X and Y of $\mathfrak{X}$, $\mathfrak{R}(X, Y)$ consists of mappings of subsets of X into Y and $\sigma(X, Y)$ is a pseudo-topology of $L(X, Y)$ such that $L_\sigma(X, Y)$, i.e. $L(X, Y)$ equipped wir $\sigma(X, Y)$, is a pseudo-topological linear group.

(3) For arbitrary X, Y, $Z \in \mathfrak{X}$ and $l \in L(Y, Z)$, from $r \in \mathfrak{R}(X, Y)$ it follows $l \circ r \in \mathfrak{R}(X, Z)$ and $m \mapsto l \circ m$ $\big(m \in L(X, Y)\big)$ is a continuous linear mapping of $L_\sigma(X, Y)$ into $L_\sigma(X, Z)$.

Let $(\mathfrak{R}, \sigma)$ be a kind of differentiability and let X and Y be pseudo-topological linear groups. We define recursively $H_\sigma{}^1(X, Y) = L_\sigma(X, Y)$ and $H_\sigma{}^n(X, Y) = L_\sigma\big(X, H_\sigma{}^{n-1}(X, Y)\big)$ $(n > 1)$ and identify each mapping $l \in H_\sigma{}^n(X, Y)$ with the multilinear mapping $(h_1, \ldots, h_n) \mapsto l(h_1) \cdots (h_n)$. Let f be a mapping of a subset of X into Y. We say that f is $\mathfrak{R}(X, Y)$-*differentiable* at a point $x \in \mathrm{db} f$ if a mapping $\mathrm{D}f(x) \in L(X, Y)$, and $\mathfrak{R}(X, Y)$-*derivative* of f *at* x, exists such that $h \mapsto f(x + h) - f(x) - \mathrm{D}f(x)(h)$ $(h \in \mathrm{db} f - x)$ belongs to $\mathfrak{R}(X, Y)$. Let U_1 be the set of all points of X at which $\mathfrak{R}(X, Y)$-derivatives of f exist. We call a mapping $\mathrm{D}f \colon U_1 \to L_\sigma(X, Y)$ an $\mathfrak{R}(X, Y)$-*derivative* of f if its values $\mathrm{D}f(x)$ $(x \in U_1)$ are $\mathfrak{R}(X, Y)$-derivatives of f at x. $\mathfrak{R}(X, Y)$-derivatives of f are also called $\mathfrak{R},\sigma$-*derivatives* of f *of order* 1 and will be denoted by $\mathrm{D}^1 f$. $\mathfrak{R},\sigma$-derivatives of higher order we define recursively. Let $\mathrm{D}^{n-1} f \colon U_{n-1} \to H_\sigma{}^{n-1}(X, Y)$ $(U_{n-1} \subseteq X,\ n > 1)$ be an $\mathfrak{R},\sigma$-derivative of f of order $n - 1$, by an $\mathfrak{R},\sigma$-*derivative* of f *of order* n we understand an $\mathfrak{R}\big(X, H_\sigma{}^{n-1}(X, Y)\big)$-derivative $\mathrm{D}^n f(x) = \mathrm{D}(\mathrm{D}^{n-1} f)(x)$ of $\mathrm{D}^{n-1} f$ at x provided that this multilinear mapping is symmetric.

There are two interesting special cases of kinds of differentiability, in both cases two mappings τ and σ are given with the following properties:

(1) σ is given as in the general case.

(2) τ has $\mathfrak{X} \times \mathfrak{X}$ as domain and for arbitrary X and Y of $\mathfrak{X}$, $\tau(X, Y)$ is a convergence structure of the set $\mathfrak{A}(X, Y)$ of all mappings of subsets of X into Y.

(3) For arbitrary X, Y, $Z \in \mathfrak{X}$ and $l \in L(Y, Z)$ and each filter $\mathfrak{K} \in \tau(X, Y)$ (0) the filter $[l] \dot{\circ} \mathfrak{K}$ on $\mathfrak{A}(X, Z)$, having as base $\{l \dot{\circ} K \mid K \in \mathfrak{K}\}$ with $l \dot{\circ} K = \{l \circ f \mid f \in K\}$, belongs to $\tau(X, Z)$ (0).

In the first case for arbitrary X and Y of $\mathfrak{X}$, $\mathfrak{R}(X, Y)$ consists of all mappings $r: U \to Y$ ($U \subseteq X$) such that the filters r_V with the base $\{r_V \mid V \in \boldsymbol{V}\}$ ($r_V = \{r_\alpha \mid \alpha \in V\}$) belongs to $\tau(X, Y)$ (0) where for each real number α, r_α is the mapping defined by $r_\alpha(h) = \frac{1}{\alpha} r(\alpha h) \times \left(\alpha \neq 0, h \in \frac{1}{\alpha} U\right)$ and $r_0(h) = 0$ ($h \in X$). In this case the $\mathfrak{R}$,σ-differentiation is also called *τ,σ-differentiation.* If for arbitrary X and Y of $\mathfrak{X}$, $\sigma(X, Y)$ is induced by $\tau(X, Y)$, we only speak of *τ-differentiation.*

In the second case for arbitrary X and Y of $\mathfrak{X}$, $\mathfrak{R}(X, Y)$ consists of all mappings $r: U \to Y$ of zero-neighbourhoods of X into Y, having the following property: There exists a mapping $\mu: U \to \mathfrak{A}(X, Y)$ such that (1) $\mu(h)$ ($h \in U$) are homogeneous mappings of X into Y and, especially, $\mu(0)$ is the zero-mapping 0, (2) μ is continuous at 0 with respect to $\tau(X, Y)$, and (3) $r(h) = \mu(h)$ (h) for each $h \in U$. In this case the $\mathfrak{R}$,σ-differentiation is also called *τ',σ-differentiation.* If for arbitrary X and Y of $\mathfrak{X}$, $\sigma(X, Y)$ is induced by $\tau(X, Y)$ we only speak of *τ'-differentiation.*

In an obvious manner as special cases of the τ,σ- and τ',σ-differentiation we define the s-, c-, q-, h- and *h'-differentiation* and the h,a- und *h',a-differentiation* with $a =$ u', u and d. By means of 8.4.6 and 8.4.7 of [6], vol. II, and of (2) we get following connections between these special kinds of differentiation

$$\begin{array}{ccccccccccccc}
\mathsf{h,d} & \to & \mathsf{h,u} & \to & \mathsf{h,u'} & \to & \mathsf{h} & & & & & & \\
\downarrow * & & \downarrow * & & \downarrow * & & \downarrow * & & & & & & \\
\mathsf{h'd} & \to & \mathsf{h',u} & \to & \mathsf{h'u'} & \to & \mathsf{h'} & \to & \mathsf{q} & \overset{**}{\to} & \mathsf{c} & \to & \mathsf{s}
\end{array}$$

where the stars mean that the corresponding implications are true if both spaces X and Y are equable and the double star means that the corresponding implication is true if X is a pseudo-topological vector space.

For each positive integer n by P_n we denote the set of all complexes $(a_1, \ldots, a_m)$ where $a_1, \ldots, a_m$ are finite sequences $(k_{11}, \ldots, k_{1j}), \ldots, (k_{1m}, \ldots, k_{mj})$ of numbers $k_{ij} \in \{1, \ldots, n\}$ such that each number $1, \ldots, n$ belongs to one and only one a_i ($i = 1, \ldots, m$) and we have $k_{i1} < \cdots < k_{ij_i}$ ($i = 1, \ldots, m$) and $k_{1j_1} < \cdots < k_{mj_m}$. For each finite sequence $a = (k_1, \ldots, k_j)$ of positive integers by $|a|$ we denote the number j of members of the sequence. If n is a positive integer, $a = (k_1, \ldots, k_j)$ is a finite sequence of numbers $k_i \in \{1, \ldots, n\}$ with $k_1 < \cdots < k_j$ and E is a vector space, by π_a we denote the mapping $(h_1, \ldots, h_n) \mapsto (h_{k_1}, \ldots, h_{k_j})$ of E^n into E^j. We can prove (see [6], vol. II):

Theorem 4.1 (*Chain rule of higher order*). *Let X, Y and Z be pseudo-topological linear groups. Especially let the $\mathfrak{R}$,σ-differentiation be the* c-, q-, h-, h,a-, h'- *or* h',a-*differentiation with $a =$* u', u *or* d. *In the case of* c-*differentiation we suppose that X and Y are pseudo-topological vector spaces. Then for each $n = 1, 2, \ldots$ we have: If a mapping f of a subset of X into Y has an $\mathfrak{R}$,σ-derivative $\mathrm{D}^n f(x)$ of order n at a point x and a mapping g of a subset of Y into Z has an $\mathfrak{R}$,σ-derivative $\mathrm{D}^n g(y)$ of order n at $y = f(x)$, then $g \circ f$ is n-times*

$\mathfrak{R},\sigma$-differentiable at x and

$$D^n(g \circ f)(x) = \sum_{(a_1,\ldots,a_m)\in P_n} D^m g(y) \circ \langle D^{|a_1|}f(x) \circ \pi_{a_1}, \ldots, D^{|a_m|}f(x) \circ \pi_{a_m}\rangle \tag{8}$$

is an $\mathfrak{R},\sigma$-derivative of order n of $g \circ f$ at x. In (8) $Df, \ldots, D^{n-1}f$ *and* $Dg, \ldots, D^{n-1}g$ *are the $\mathfrak{R},\sigma$-derivatives of f and g of order $< n$, appearing in the definition* of $D^n f(x)$ *and* $D^n g(y)$.

Let $(\mathfrak{R}, \sigma)$ be a differentiability kind and let X and Y be pseudo-topological linear groups. We say that a continuous mapping f of an open subset U of X into Y belongs to the *differentiability class* $C^n_{\mathfrak{R},\sigma}$ if f has continuous $\mathfrak{R},\sigma$-derivatives $D^k f \colon U \to H_\sigma^k(X, Y)$ with $k = 1, \ldots, n$. A mapping belongs to the *differentiability class* $C^\infty_{\mathfrak{R},\sigma}$ if it belongs to $C^n_{\mathfrak{R},\sigma}$ for all positive integers n. In the mentioned special cases we use obvious denotations.

We can prove (see [6], vol. II):

Theorem 4.2. (*Chain rule for differentiability classes*). *Let X, Y and Z be pseudo-topological linear groups. We suppose that X and Y are equable. Let the $\mathfrak{R},\sigma$-differentation be chosen as in Theorem* 4.1. *In the case of c-differentiation we suppose that X and Y are pseudo-topological vector spaces. Then for each $n = 1, 2 \ldots$ we have: If f is a mapping of an open subset U of X into Y and g is a mapping of an open subset V of Y into Z with $f[U] \subseteqq V$ and if f and g belong to $C^n_{\mathfrak{R},\sigma}$, then $g \circ f$ belongs to $C^n_{\mathfrak{R},\sigma}$, too.*

5. Strong continuity

By a *uniform convergence space* (see [6], vol. I) we understand a set M equipped with an $\wedge$-ideal $\mathfrak{u}$ of filters on $M \times M$, the *uniform convergence structure* of this space, such that

(1) $\mathfrak{F} \in \mathfrak{u} \Rightarrow \mathfrak{F}^{-1} = \{F^{-1} \mid F \in \mathfrak{F}\} \in \mathfrak{u}$,

(2) $\mathfrak{F}, \mathfrak{G} \in \mathfrak{u} \Rightarrow$ The filter $\mathfrak{G} \circ \mathfrak{F}$ with the base $\{G \circ F\} \mid F \in \mathfrak{F}, G \in \mathfrak{G}\}$ belongs to $\mathfrak{u}$, too.

A uniform convergence space $X = (M, \mathfrak{u})$ is called a *pseudo-uniform space* (see [3]) and $\mathfrak{u}$ a *pseudo-uniform structure* if the principal filter $[\Delta] = \{A \subseteqq M \times M \mid \Delta \subseteqq A\}$, generated by the diagonal Δ of M, belongs to $\mathfrak{u}$, too. It is well-known that the uniform spaces are the pseudo-uniform spaces $X = (M, \mathfrak{u})$ with $\bigcap_{\mathfrak{F}\in\mathfrak{u}} \mathfrak{F} \in \mathfrak{u}$. Each uniform convergence space $X = (M, \mathfrak{u})$ is a convergence space $\lambda(X) = (M, \lambda(\mathfrak{u}))$ where the convergence structure $\lambda(\mathfrak{u})$ is defined by $\lambda(\mathfrak{u})(x) = \{\mathfrak{F} \mid [x] \times \mathfrak{F} \in \mathfrak{u}\}$ $(x \in M)$.

Now let $X = (M, \mathfrak{u})$ and $Y = (N, \mathfrak{v})$ be uniform convergence spaces. A mapping $f \colon X \to Y$ is called *uniformly continuous* if for each $\mathfrak{F} \in \mathfrak{u}$ the filter $(f * f)(\mathfrak{F})$ belongs to $\mathfrak{v}$ where $f * f$ is the mapping $(x, y) \mapsto (f(x), f(y))$ of $M \times M$ into $N \times N$. Each uniformly continuous mapping is, of course, continuous (see [6], vol. I). We say that a mapping f of a subset U of X into Y is *strongly continuous* if for each family $(\mathfrak{F}_u)_{u\in U}$ of filters on X from $\bigcap_{u\in U} ([u] \times \mathfrak{F}_u) \in \mathfrak{u}$ it follows $\bigcap_{u\in U} \big([f(u)] \times f(\mathfrak{F}_u)\big) \in \mathfrak{v}$.

Theorem 5.1. *Each uniformly continuous mapping of X into Y is strongly continuous. Each strongly continuous mapping of a subset U of X into Y is continuous (with respect to $\lambda(\mathfrak{u})$ and $\lambda(\mathfrak{v})$), and if Y is a uniform space, a mapping f of a subset U of X into Y is strongly continuous if and only if f is continuous.*

Proof. 1 Let f be a uniformly continuous mapping of X into Y and let $(\mathfrak{F}_u)_{u\in X}$ be a family of filters on $M \times M$ with $\bigcap_{u\in X} ([u] \times \mathfrak{F}_u) \in \mathfrak{u}$. Then $(f * f)\left(\bigcap_{u\in X} ([u] \times \mathfrak{F}_u)\right) = \bigcap_{u\in X}\left([f(u)] \times f(\mathfrak{F}_u)\right)$ belongs to $\mathfrak{v}$, i.e., f is strongly continuous.

2. Let f be a strongly continuous mapping of a subset U of X into Y. Let x be a point of U and let $\mathfrak{F}_x$ be an arbitrary filter of $\lambda(\mathfrak{u})(x)$. For each $u \in U\setminus\{x\}$ let $\mathfrak{F}_u$ be the improper filter on $M \times M$. Then $\bigcap_{u\in U} ([u] \times \mathfrak{F}_u) = [x] \times \mathfrak{F}_x$ belongs to $\mathfrak{u}$, thus $\bigcap_{u\in U}\left([f(u)] \times f(\mathfrak{F}_u)\right) = [f(x)] \times f(\mathfrak{F}_x)$ belongs to $\mathfrak{v}$, i.e., f is continuous at x.

3. If f is a continuous mapping of a subset U of X into Y and Y is a uniform space, then obviously f is strongly continuous.

We can easily prove

Theorem 5.2. *Let f be a strongly continuous mapping of a subset of X into Y and let g be a strongly continuous mapping of a subset V of Y into a uniform convergence space Z with $f[U] \subseteqq V$. Then $g \circ f$ is strongly continuous, too.*

For the following theorem we suppose that Y is a product $\prod_{i\in I} Y_i$ of uniform convergence spaces $Y_i = (N_i, \mathfrak{v}_i)$ (see [6], vol. I).

Theorem 5.3. *Let U be a subset of X, and for arbitrary $i \in I$ let f_i be a strongly continuous mapping of U into Y_i. Then $f = \langle f_i\rangle_{i\in I}: U \to Y$ is strongly continuous, too.*

Proof. Let $(\mathfrak{F}_u)_{u\in U}$ be a family of filters on X with $\bigcap_{u\in U} ([u] \times \mathfrak{F}_u) \in \mathfrak{u}$. For each i by π_i we denote the i-th projection of Y onto Y_i. Because of

$$(\pi_i * \pi_i)\left(\bigcap_{u\in U}\left([f(u)] \times f(\mathfrak{F}_u)\right)\right) = \bigcap_{u\in U}\left([f_i(u)] \times f_i(\mathfrak{F}_u)\right) \in \mathfrak{v}_i$$

for arbitrary $i \in I$, by 4.6.10 of [6], vol. I, we see that $\bigcap_{u\in U}\left([f(u)] \times f(\mathfrak{F}_u)\right) \in \mathfrak{v}$, i.e. that f is strongly continuous.

Each pseudo-topological linear group $X = (E, \tau)$ has a natural pseudo-uniform structure $\mathfrak{u}$, the *canonical pseudo-uniform structure* of X (see [5] and [6], vol. II), which consists of all filters $\mathfrak{F}$ on $E \times E$ such that $w(\mathfrak{F})$ belongs to $\tau(0)$ where w is the mapping $(x, y) \mapsto x - y$ of $E \times E$ into E. A mapping f of a subset U of a pseudo-topological linear group $X = (E, \tau)$ into a pseudo-topological linear group $Y = (F, \sigma)$ is strongly continuous (with respect to the canonical pseudo-uniform structures of X and Y) if for each family $(\mathfrak{F}_u)_{u\in U}$ of filters on X from $\bigcap_{u\in U} (\mathfrak{F}_u - [u]) \in \tau(0)$ it follows $\bigcap_{u\in U}\left(f(\mathfrak{F}_u) - [f(u)]\right) \in \sigma(0)$.

Theorem 5.4. *Let $X = (M, \mathfrak{u})$ be a uniform convergence space and $Y = (F, \sigma)$ be a pseudo-topological linear group. Let $f_1, \ldots, f_m$ be mappings of a subset U of X into Y which are strongly continuous with respect to the canonical pseudo-uniform structure $\mathfrak{v}$ of Y. Then $f = f_1 + \cdots + f_m: x \mapsto f_1(x) + \cdots + f_m(x)$ ($x \in U$) is strongly continuous, too.*

Proof. Let $(\mathfrak{F}_u)_{u\in U}$ be a family of filters on X with $\bigcap_{u\in U} ([u] \times \mathfrak{F}_u) \in \mathfrak{u}$. Then $\bigcap_{u\in U}\left([f_i(u)] \times f_i(\mathfrak{F}_u)\right) \in \mathfrak{v}$ $(i = 1, \ldots, n)$ and by 5.6.20 of [6], vol. II, we get

$$\bigcap_{u\in U}\left([f(u)] \times f(\mathfrak{F}_u)\right) \supseteqq \bigcap_{u\in U}\left([f_1(u)] \times f_1(\mathfrak{F}_u)\right) + \cdots + \bigcap_{u\in U}\left([f_m(u)] \times f_m(\mathfrak{F}_u)\right) \in \mathfrak{u}.$$

Hence f is strongly continuous.

Under certain restrictions there exists an important characterization theorem of the strong continuity.

Theorem 5.5 (*Characterization of strong continuity*). *Let X be a topological linear group and let Y be a pseudo-topological union in the wide sense $\bigcup_{i \in I} Y_i$ of topological linear groups Y_i. Let f be a mapping of an open connected subset U of X into Y. Then f is strongly continuous (with respect to the canonical pseudo-uniform structures of X and Y) if and only if there exists an $i \in I$ such that $f[U]$ is contained in Y_i and f is continuous as mapping of U into Y_i.*

Proof. 1. We suppose that there exists an $i \in I$ with $f[U] \subseteqq Y_i$ such that $f\colon U \to Y_i$ is continuous. The canonical pseudo-uniform structure of Y_i is a uniform structure. Thus by Theorem 5.1, $f\colon U \to Y_i$ is strongly continuous. The identical mapping ι of Y_i into Y is continuous and linear and therefore uniformly continuous. By Theorem 5.1, ι is strongly continuous. Hence by Theorem 5.2, $\iota \circ f$, i.e. $f\colon U \to Y$ is strongly continuous.

2. Now we suppose that $f\colon U \to Y$ is strongly continuous. Let x be a point of U and let i_0 be an element of I with $f(x) \in Y_{i_0}$. For each $u \in X$ by $\mathfrak{N}_u$ we denote the neighbourhood filter of u. Of course, $\mathfrak{F} = \bigcap_{u \in U} \big(f(\mathfrak{N}_u) - [f(u)]\big)$ is a zero-filter on Y. Hence there exists an $i \in I$ with $i_0 \leqq i$ and a zero-filter $\mathfrak{F}_i$ on Y_i such that the filter on Y, generated by $\mathfrak{F}_i$, is contained in $\mathfrak{F}$. For each $u \in U$ there is an $N_u \in \mathfrak{N}_u$ with $f[N_u] - f(u) \subseteqq Y_i$. Therefore $f[N_x] \subseteqq Y_i$ and $f[N_u] - f[N_u] \subseteqq Y_i$ $(u \in U)$. Taking into account that U is connected we get $f[U] \subseteqq Y_i$. Therefore for each $u \in U$ the trace of $f(\mathfrak{N}_u) - [f(u)]$ on Y_i contains $\mathfrak{F}_i$. Hence $f\colon U \to Y_i$ is continuous.

We say that a subset B of a pseudo-topological linear group X is *evenly pointwise bounded* if $\bigcap_{X \in B} Vx$ is a zero-filter on X. Of course, each subset of a topological vector space is evenly pointwise bounded.

Theorem 5.6. *Let X be a finite product $X_1 \times \cdots \times X_n$ $(n > 0)$ of equable pseudo-topological linear groups and let Y be an arbitrary pseudo-topological linear group. Let U be an evenly pointwise bounded subset of X and let l be a continuous multilinear mapping of X into Y. Then the restriction of l to U is a strongly continuous mapping of U into Y.*

Proof. Let $(\mathfrak{F}_u)_{u \in U}$ be a family of filters on X such that $\bigcap_{u \in U} (\mathfrak{F}_u - [u])$ is a zero-filter. By π_i $(i = 1, \ldots, n)$ we denote the projections of X onto X_i. Of course for each i, $\bigcap_{u \in U} \big(\pi_i(\mathfrak{F}_u) - [\pi_i(u)]\big)$ and $\bigcap_{u \in U} V\pi_i(u)$ are zero-filters on X_i. $\bigcap_{u \in U} \big(l(\mathfrak{F}_u) - [l(u)]\big)$ contains a finite sum of intersections $\bigcap_{u \in U} l(\mathfrak{G}_{1u} \times \cdots \times \mathfrak{G}_{nu})$ where for each i and u, $\mathfrak{G}_{iu}$ is the filter $\pi_i(\mathfrak{F}_u) - [\pi_i(u)]$ or the filter $[\pi_i(u)]$, and for no one of u each $\mathfrak{G}_{iu}$ is equal to $[\pi_i(u)]$. We can replace in each such intersection every filter $\mathfrak{G}_{iu}$ which is equal to $\pi_i(\mathfrak{F}_u) - [\pi_i(u)]$ by an equable zero-filter $\mathfrak{H}_i \subseteqq \bigcap_{u \in U} \big(\pi_i(\mathfrak{F}_u) - [\pi_i(u)]\big)$ and therefore every filter $\mathfrak{G}_{iu}$ which is equal to $[\pi_i(u)]$ by the equable zero-filter $\mathfrak{H}_i = \bigcap_{u \in U} Vu_i$. Thus $\bigcap_{u \in U} \big(l(\mathfrak{F}_u) - [l(u)]\big)$ contains a finite sum of filters $l(\mathfrak{H}_1 \times \cdots \times \mathfrak{H}_n)$ where to each i, $\mathfrak{H}_i$ is a zero-filter on X_i. Taking into account that l is continuous we get that $\bigcap_{u \in U} \big(l(\mathfrak{F}_u) - [l(u)]\big)$ is a zero-filter on Y, i.e. l is strongly continuous.

The following statement is a consequence of Theorems 5.2, 5.5 and 5.6.

Theorem 5.7. *Let X be a topological linear group, let Y be a finite product $Y_1 \times \cdots \times Y_n$ of pseudo-topological vector spaces which is a pseudo-topological union $\bigcup_{i \in I} Y_i$ of topological vector spaces, and let Z be an arbitrary pseudotopological linear group. Let f be a strongly continuous mapping of an open connected subset U of X into Y and let g be a continuous multilinear mapping of Y into Z. Then $g \circ f$ is strongly continuous.*

Proof. Because of Theorem 5.5 there exists an $i \in I$ such that $f[U]$ is contained in Y_i and f is continuous as mapping of U into Y_i. Taking into account that Y_i is a topological vector space we see that $f[U]$ is evenly pointwise bounded in Y_i and therefore in Y. By Theorem 5.6 the restriction of g to $f[U]$ is strongly continuous. Hence by Theorem 5.2 $g \circ f$ is strongly continuous, too.

6. Differentiability classes with respect to strong continuity, the differentiability class $C_\Gamma{}^n$

If we replace in the definition of the differentiability class $C^n_{\mathfrak{R},\sigma}$ $(n = 1, 2, \ldots, \infty)$ the continuity by the strong continuity we get the notion of the (modified) *differentiability class* $\bar{C}^n_{\mathfrak{R},\sigma}$. In special cases of kinds of differentiability for the classes $\bar{C}^n_{\mathfrak{R},\sigma}$ we use the analogous denotations as those for the differentiability classes $C^n_{\mathfrak{R},\sigma}$.

Theorem 6.1 (*Chain rule for modified differentiability classes*). *Let f be a mapping of an open connected subset U of a topological vector space X into a pseudo-topological union Y and let g be a mapping of an open subset V of Y into a pseudo-topological linear group Z with $f[U] \subseteqq V$. Let a be equal to u or to d. In the case $a = \mathsf{d}$ let X be locally bounded and let Y be a pseudo-topological union of locally bounded topological vector spaces. Then for each $n = 1, 2, \ldots$, we have: If f and g belong to $\bar{C}^n_{\mathsf{h},a}$, then $g \circ f$ belongs to $\bar{C}^n_{\mathsf{h},a}$, too, and if f and g belong to $\bar{C}^n_{\mathsf{h}',a}$, then $g \circ f$ belongs to $\bar{C}^n_{\mathsf{h}',a}$, too.*

Proof. It is sufficient to restrict us to finite n. In all cases of considered differentiability classes we get that f and g are n-times q-differentiable (see 8.4.6 of [6], vol. II) and therefore $g \circ f$ is n-times q-differentiable, too. For the corresponding q-derivatives to arbitrary $u \in U$ we have

$$\mathrm{D}^n(g \circ f)(u) = \sum_{(a,\ldots,a_m) \in P_n} (\psi_{a_1,\ldots,a_m} \circ \langle \mathrm{D}^m g \circ f, \mathrm{D}^{|a_1|}f, \ldots, \mathrm{D}^{|a_m|}f\rangle)(u)$$

where $\psi_{a_1,\ldots,a_m}$ are the continuous multilinear mappings $(l, l_1, \ldots, l_m) \mapsto l \circ \langle l_1 \circ \pi_{a_1}, \ldots, l_m \circ \pi_{a_m}\rangle$ of $\overset{s}{H}_{\mathsf{q}}{}^m(Y, Z) \times H_{\mathsf{q}}{}^{|a_1|}(X, Y) \times \cdots \times H_{\mathsf{q}}{}^{|a_m|}(X, Y)$ into $H_a{}^n(X, Z)$ and $\overset{s}{H}_{\mathsf{q}}{}^m(Y, Z)$ is the subspace of $H_{\mathsf{q}}{}^m(Y, Z)$ consisting of all symmetric multilinear mappings $l \in H_{\mathsf{q}}{}^m(Y, Z)$ (see the proof of 8.14.8 in [6], vol. II). Let $(a_1, \ldots, a_m)$ be an element of P_n. By Theorems 5.2 and 5.3 we see that $\langle \mathrm{D}^m g \circ f, \mathrm{D}^{|a_1|}f, \ldots, \mathrm{D}^{|a_m|}f\rangle$ is strongly continuous. By Theorems 2.3, 2.5 and 3.3 it follows that $\psi_{a_1,\ldots,a_m}$ is a continuous multilinear mapping of $\overset{s}{H}_a{}^m(Y, Z) \times H_a{}^{|a_1|}(X, Y) \times \cdots \times H_a{}^{|a_m|}(Y, X)$ into $H_a{}^n(X, Z)$ where $\overset{s}{H}_a{}^m(Y, Z)$ is the subspace of $H_a{}^m(Y, Z)$ consisting of all symmetric multilinear mapping $l \in H_a{}^m(Y, Z)$. Therefore by Theorem 5.7, $\psi_{a_1,\ldots,a_m} \circ \langle \mathrm{D}^m g \circ f, \mathrm{D}_{|a_1|}f, \ldots, \mathrm{D}_{|a_m|}f\rangle$ is strongly continuous. In consequence of Theorem 5.4, $\mathrm{D}^n(g \circ f)$ is strongly continuous. Analogously we prove that $D^k(g \circ f)$ with $k < n$ is strongly continuous. By Theorem 5.2, $g \circ f$ is strongly continuous, too. Using Theorem 4.2 the proof can be finished.

In the following both theorems, under certain restrictions we characterize modified differentiability classes. For these theorems we suppose, that a is equal to u', u or d, that X is a topological vector space, and that in the first case of a, Y is a pseudo-topological union in the wide sense $\bigcup_{i\in I_0} Y_i$ and a pseudo-topological vector space, that in the second case of a, Y is a pseudo-topological union $\bigcup_{i\in I_0} Y_i$, and that in the third case of a, X is locally bounded and Y is a pseudo-topological union $\bigcup_{i\in I_0} Y_i$ of locally bounded topological vector spaces. Of course, in the first case of a, $L_a{}^k(X, Y)$ $(k = 1,2,\ldots)$ are pseudo-topological unions in the wide sense $\bigcup_{i\in I_k} L_{ai}^k(X, Y)$, in the second case of a, $L_a{}^k(X, Y)$ $(k = 1, 2, \ldots)$ are pseudo-topological unions $\bigcup_{i\in I_k} L_{ai}^k(X, Y)$ and in the third case of a, $L_a{}^k(X, Y)$ $(k = 1, 2, \ldots)$ are pseudo-topological unions $\bigcup_{i\in I_k} L_{ai}^k(X, Y)$ of locally bounded topological vector spaces. By Theorems 3.3 and 5.5 we get

Theorem 6.2. *Let f be a mapping of an open subset U of X into Y and let n be a positive integer.*

f belongs to $\bar{C}^n_{\mathsf{h},a}$ if and only if f has an h,a-derivative $\mathrm{D}^n f\colon U \to H_a{}^n(X, Y)$ of order n and for f, $\mathrm{D}^n f$ and the corresponding h,a-derivatives $\mathrm{D}^k f$ of order $k < n$ the following condition C *is fullfilled:*

For each $k = 0, \ldots, n$ there exists an $i \in I_k$ such that $\mathrm{D}^k f[U]$ is contained in $L_{ai}^k(X, Y)$ and $\mathrm{D}^k f\colon U \to L_{ai}^k(X, Y)$ is continuous where $\mathrm{D}^0 f = f$ and $L_{ai}^0(X, Y) = Y_i$.

f belongs to $\bar{C}^n_{\mathsf{h}',a}$ if and only if f has an h',a-derivative $\mathrm{D}^n f\colon U \to H_a{}^n(X, Y)$ of order n and for f, $\mathrm{D}^n f$ and the corresponding h',a-derivatives $D^k f$ of order $k < n$ the condition C *is fulfilled.*

Under certain restrictions on Y we can characterize the classes $\bar{C}^n_{\mathsf{h}',a}$ *by* weaker conditions. Especially we suppose that Y is *$t^\wedge$-regular*, i.e. that for each zero-filter $\mathfrak{F}$ on Y the filter which has as a base the set $\{\bar{F}^{t^\wedge} \mid F \in \mathfrak{F}\}$ of all $t^\wedge$-closed hulls of the sets $F \in \mathfrak{F}$ is a zero-filter, too; the *$t^\wedge$-closed hulls* are the closed hulls with respect to the finest locally convex vector topology coarser than the pseudo-topology of Y. We can prove (see 8.15.6'' in [6], vol. II):

Theorem 6.3. *Let f be a mapping of an open subset U of X into Y. Let Y be locally convex and $t^\wedge$-regular. Then for each positive integer n, f belongs to $\bar{C}^n_{\mathsf{h}',a}$ if and only if f has an* s-*derivative $\mathrm{D}^n f$ of order n with U as domain and for f, $\mathrm{D}^n f$ and the corresponding* s-*derivatives $\mathrm{D}^k f$ of order $k < n$ the condition C of Theorem* 6.2 *is fulfilled.*

If especially X and Y are locally convex topological vector spaces and a is equal to u or u', we can take as I_k the set Φ, defined in § 3, and can take as $L_{ai}^k(X, Y)$ $(i = \varphi \in \Phi)$ the spaces $L_\varphi{}^k(X, Y)$ and $L'^k_\varphi(X, Y)$, respectively. If X and Y are separated locally convex topological vector spaces and if Φ and $L_\varphi{}^k(X, Y)$ are taken as I_k and $L_{\mathsf{u}i}^k(X, Y)$ $(i = \varphi)$, respectively, then the condition in Theorem 6.3 that f belongs to $\bar{C}^n_{\mathsf{h}',\mathsf{u}}$ is nothing else but the definition that f is of differentiability class $C_\Gamma{}^n$ of GUTKNECHT [7]. Thus, up to the non-essential restriction that the domains of the mappings are connected, the differentiability class $\bar{C}^n_{\mathsf{h}',\mathsf{u}}$ is an extension of the class $C_\Gamma{}^n$. Therefore, up to this restriction, $\bar{C}^\infty_{\mathsf{h}',\mathsf{u}}$ is an extension of $C_\Gamma{}^\infty$ (see [7] and also the note of R. SCHMID of these Proceedings).

References

[1] В. И. Авербух и О. Г. Смолянов, Дифференцированние и псевдотопологии, Вестник МГУ, № 1 (1972), 3—7.

[2] C. H. Cook and H. R. Fischer, Uniform convergence structures, Math. Ann. **173** (1967), 290—306.

[3] H. R. Fischer, Limesräume, Math. Ann. **137** (1959), 269—303.

[4] A. Frölicher and W. Bucher, Calculus in Vector Spaces without Norm, Lecture Notes in Math. **30**, Springer-Verlag, Berlin—Heidelberg—New York 1966.

[5] W. Gähler, Über Limes- und Pseudouniformisierbarkeit, Demonstratio Math. **6** (1973), 613—632.

[6] W. Gähler, Grundstrukturen der Analysis. I, II, Akademie-Verlag, Berlin/Birkhäuser Verlag, Basel und Stuttgart 1977, 1978.

[7] J. Gutknecht, Die C_Γ^∞-Struktur auf der Diffeomorphismengruppe einer kompakten Mannigfaltigkeit, Diss., ETH Zürich 1977.

[8] H. H. Keller, Räume stetiger multilinearer Abbildungen als Limesräume, Math. Ann. **159** (1965), 259—270.

[9] H.-J. Kowalsky, Limesräume und Komplettierung, Math. Nachr. **12** (1954), 301—340.

[10] K. Kutzler, Zur Theorie der limitierten Vektorräume, Diss., FU Berlin (West) 1969.

Anschrift des Verfassers:

Akademie der Wissenschaften der DDR
Zentralinstitut für Mathematik und Mechanik
108 Berlin
Mohrenstraße 39
DDR

Generalized Bornologies and Projective Tensor Products

WERNER GÄHLER, Berlin

2. Hypo-bornologies

The first part of this note deals with generalizations of bornological vector spaces. We use definitions given in the preceding note [6] of these Proceedings without repetition.

Let $\boldsymbol{K}$ be equal to $\boldsymbol{R}$ or $\boldsymbol{C}$. By $\boldsymbol{E}$ we denote the set of all $\alpha \in \boldsymbol{K}$ with $|\alpha| \leqq 1$. Of course, a *bornological $\boldsymbol{K}$-vector space* (see [7]) is a $\boldsymbol{K}$-vector space E equipped with a dual filter $\mathfrak{B}$ on E, a *bornology* of E, such that the following conditions are fulfilled:

(1) $A, B \in \mathfrak{B} \Rightarrow A + B \in \mathfrak{B}$,

(2) $\alpha \in \boldsymbol{K}$, $B \in \mathfrak{B} \Rightarrow \alpha B \in \mathfrak{B}$,

(3) $\boldsymbol{E}B \in \mathfrak{B}$ for all $B \in \mathfrak{B}$,

(4) $\{x\} \in \mathfrak{B}$ for all $x \in E$.

Bornologies with the properties (1) to (4) are called *admissible* or *vector bornologies*. If we replace in the definition of a bornological vector space the condition (4) by the weaker condition

(4′) $\{0\} \in \mathfrak{B}$,

we get the more general notion of a bornological linear group. Bornologies with the properties (1) to (3) and (4′) are said to be *weakly admissible*. An example of a weakly admissible bornology is the set $\mathfrak{B}_b$ of all bounded subsets of a pseudo-topological linear group X. $\mathfrak{B}_b$ is admissible iff X is a pseudo-topological vector space.

There are useful generalizations of the notions of a bornological vector space and a bornological linear group defined by means of quasi-bornologies, semi-bornologies and hypo-bornologies.

Let M be a set. A *quasi-bornology* of M is an $\wedge$-ideal of filters on M, a *semi-bornology* of M is a $\vee$-ideal of dual filters on M, and a *hypo-bornology* of M is an $\wedge$-ideal of $\wedge$-ideals of dual filters on M. In general, the notion of an $\wedge$-ideal and the dual notion of an $\vee$-ideal are defined in lattices with a minimal and a maximal element (see [4], vol.I); in the cases of a quasi-bornology, a semi-bornology and a hypo-bornology of M, these lattices are the filter lattice $\mathscr{F}M$ of M, the dual filter lattice $\mathscr{D}M$ of M and the lattice $\mathscr{I}M$ of all $\wedge$-ideals of dual filters on M, respectively (the order in these lattices is $\subseteq$).

We identify each bornology $\mathfrak{B}$ of M with the quasi-bornology having as base the set of all principal filters $[B] = \{A \mid B \subseteqq A \subseteqq M\}$ $(B \in \mathfrak{B})$, and with the semi-bornology having as base the set of all principal dual filters $(B) = \{A \mid A \subseteqq B\}$ $(B \in \mathfrak{B})$. For each filter $\mathfrak{F}$ on M by $\mathfrak{u}_{\mathfrak{F}}$ we denote the $\wedge$-ideal of dual filters on M with base $\{(F) \mid F \in \mathfrak{F}\}$. We identify each quasi-bornology $\mathfrak{S}$ of M with the hypo-bornology of M having as base $\{\mathfrak{u}_{\mathfrak{F}} \mid \mathfrak{F} \in \mathfrak{S}\}$. Furthermore, we identify each semi-bornology $\mathfrak{L}$ of M with the hypo-bornology of M having as base the set of all principal $\wedge$-ideals $[\mathfrak{B}] = \{\mathfrak{C} \in \mathscr{D}M \mid \mathfrak{B} \subseteqq \mathfrak{C}\}$ $(\mathfrak{B} \in \mathfrak{L})$. Taking into account these identifications, we see that bornologies are special quasi- und semibornologies and quasi- and semi-bornologies are special hypo-bornologies.

We define *quasi-bornological, semi-bornological* and *hypo-bornological vector spaces* and *quasi-bornological, semi-bornological* and *hypo-bornological linear groups* analogously as bornological vector spaces and bornological linear groups, respectively. Thus, e.g., a hypo-bornological $\boldsymbol{K}$-vector space is a $\boldsymbol{K}$-vector space E equipped with a hypo-bornology $\mathfrak{B}$ of E with the properties:

(1) $\mathfrak{u}, \mathfrak{v} \in \mathfrak{B} \Rightarrow \mathfrak{u} + \mathfrak{v} \in \mathfrak{B}$,

(2) $\alpha \in \boldsymbol{K}, \mathfrak{v} \in \mathfrak{B} \Rightarrow \alpha\mathfrak{v} \in \mathfrak{B}$,

(3) $\boldsymbol{E}\mathfrak{v} \in \mathfrak{B}$ for all $\mathfrak{v} \in \mathfrak{B}$,

(4) $[(x)] \in \mathfrak{B}$ for all $x \in E$,

where the sum $\mathfrak{u} + \mathfrak{v}$ and the products $\alpha\mathfrak{v}$ and $\boldsymbol{E}\mathfrak{v}$ are defined obviously and $[(x)]$ $(x \in E)$ are the principal $\wedge$-ideals generated by the dual filters $(x) = \{A \mid x \in A \subseteqq E\}$. If we replace (4) by

(4′) $[(0)] \in \mathfrak{B}$,

of course, we get the notion of a hypo-bornological linear group. In an obvious manner we define the *admissibility* and the *weak admissibility* of quasi-bornologies, semi-bornologies and hypo-bornologies.

Let $X = (E, \tau)$ be a pseudo-topological linear group. An example of a weakly admissible quasi-bornology is the set $\mathfrak{S}_q$ of all quasi-bounded filters on X, an example of a weakly admissible semi-bornology is the set $\mathfrak{L}_e$ of all evenly bounded dual filters $\mathfrak{F}$ on X, i.e. of all dual filters $\mathfrak{F}$ with $\bigcap_{F \in \mathfrak{F}} VF \in \tau(0)$, and an example of a weakly admissible hypo-bornology is the set $\mathfrak{B}_h$ of all hypo-bounded $\wedge$-ideals of dual filters on X. $\mathfrak{S}_q$, $\mathfrak{L}_e$ and $\mathfrak{B}_h$ are admissible iff X is a pseudo-topological vector space.

By HLING we denote the category of all hypo-bornological linear groups, the morphisms of HLING are those linear mappings $l\colon (E, \mathfrak{B}) \to (F, \mathfrak{W})$ of hypo-bornological linear groups into hypo-bornological linear groups which are $\mathfrak{B}, \mathfrak{W}$-*faithful*, i.e. which have the property that for each $\mathfrak{v} \in \mathfrak{B}$ the $\wedge$-ideal $f(\mathfrak{v})$ (with the base $\{f(\mathfrak{F}) \mid \mathfrak{F} \in \mathfrak{v}\}$ belongs to $\mathfrak{W}$. In the case that $\mathfrak{B}$ and $\mathfrak{W}$ are identified with bornologies $\mathfrak{B}$ and $\mathfrak{C}$, with quasi-bornologies $\mathfrak{S}$ and $\mathfrak{T}$ or with semi-bornologies $\mathfrak{L}$ and $\mathfrak{M}$, instead of $\mathfrak{B},\mathfrak{W}$-faithful we also speak of $\mathfrak{B},\mathfrak{C}$-*faithful* (= *bounded*), $\mathfrak{S},\mathfrak{T}$-*faithful* or $\mathfrak{L},\mathfrak{M}$-*faithful*, respectively.

By BLING, QLING and SLING we denote the full subcategories of HLING of all bornological, all quasi-bornological and all semi-bornological linear groups, respectively, and by BVEC, QVEC, SVEC and HVEC we denote the full subcategories of HLING of all bornological, all quasi-bornological, all semi-bornological and all hypo-bornological vector spaces, respectively.

Let $\mathfrak{B}$ be a hypo-bornology of a vector space E. Then by $b(\mathfrak{B})$ we denote the bornology $\left\{\bigcap_{\mathfrak{F}\in\mathfrak{b}} \bigcup_{F\in\mathfrak{F}} F \mid \mathfrak{b} \in \mathfrak{B}\right\}$, by $q(\mathfrak{B})$ we denote the quasibornology $\left\{\bigvee_{\mathfrak{F}\in\mathfrak{b}} \bigcap_{F\in\mathfrak{F}} [F] \mid \mathfrak{b} \in \mathfrak{B}\right\}$ where the union is to be understood in the filter lattice $\mathscr{F}E$, and by $s(\mathfrak{B})$ we denote the semi-bornology $\left\{\bigcap_{\mathfrak{F}\in\mathfrak{b}} \mathfrak{F} \mid \mathfrak{b} \in \mathfrak{B}\right\}$. If $\mathfrak{B}$ is admissible or weakly admissible, then also $b(\mathfrak{B})$, $q(\mathfrak{B})$ and $s(\mathfrak{B})$ are admissible or weakly admissible, respectively. There eare (covariant) functors $\boldsymbol{b}^l$: HLING $\to$ BLING, $\boldsymbol{q}^l$: HLING $\to$ QLING and $\boldsymbol{s}^l$: HLING $\to$ SLING such that for each object $X = (E, \mathfrak{B})$ of HLING, $\boldsymbol{b}^l(X) = (E, b(\mathfrak{B}))$, $\boldsymbol{q}^l(X) = (E, q(\mathfrak{B}))$ and $\boldsymbol{s}^l(X) = (E, \boldsymbol{s}(\mathfrak{B}))$ and for each morphism $f: X \to Y$ of HLING, $\boldsymbol{b}^l(f)$ is the mapping $f: \boldsymbol{b}^l(X) \to \boldsymbol{b}^l(Y)$, $\boldsymbol{q}^l(f)$ is the mapping $f: \boldsymbol{q}^l(X) \to \boldsymbol{q}^l(Y)$ and $\boldsymbol{s}^l(f)$ is the mapping $f: \boldsymbol{s}^l(X) \to \boldsymbol{s}^l(Y)$. By restriction of $\boldsymbol{b}^l$, $\boldsymbol{q}^l$ and $\boldsymbol{s}^l$ we get functors $\boldsymbol{b}^v$: HVEC $\to$ BVEC, $\boldsymbol{q}^v$: HVEC $\to$ QVEC and $\boldsymbol{s}^v$: HVEC $\to$ SVEC, respectively.

We can prove that $\boldsymbol{q}^l$ is a bireflector but $\boldsymbol{s}^l$ is a bicoreflector and that the restriction of $\boldsymbol{b}^l$ to QLING is a bireflector $\boldsymbol{b}^{ql}$: QLING $\to$ BLING but the restriction of $\boldsymbol{b}^l$ to SLING is a bicoreflector $\boldsymbol{b}^{sl}$: SLING $\to$ BLING. Furthermore, we can prove that this assertion remains true if we replace l by v, BLING by BVEC, QLING by QVEC, QLING by QVEC and SLING by SVEC.

Let E be a $\boldsymbol{K}$-vector space. If F is a subset of E, by $F^\wedge$ we denote the convex hull of F. For each filter (dual filter) $\mathfrak{F}$ on E by $\mathfrak{F}^\wedge$ we denote the filter (dual filter) with base $\{F^\wedge \mid F \in \mathfrak{F}\}$ and for each $\wedge$-ideal $\mathfrak{u}$ of dual filters on E by $\mathfrak{u}^\wedge$ we denote the $\wedge$-ideal with base $\{\mathfrak{F}^\wedge \mid \mathfrak{F} \in \mathfrak{u}\}$. A filter (dual filter) $\mathfrak{F}$ on E is said to be *convex* if $\mathfrak{F} = \mathfrak{F}^\wedge$, and an $\wedge$-ideal $\mathfrak{u}$ of dual filters on E is said to be *convex* if $\mathfrak{u} = \mathfrak{u}^\wedge$. Furthermore, a hypo-bornological linear group $X = (E, \mathfrak{B})$ and its hypo-bornology are said to be *convex* if $\mathfrak{B}$ has a base of convex $\wedge$-ideals, i.e. $\mathfrak{B}$ is equal to the hypo-bornology $\mathfrak{B}^\wedge$ with the base $\{\mathfrak{b}^\wedge \mid \mathfrak{b} \in \mathfrak{B}\}$. If $\mathfrak{B}$ is identified with a bornology $\mathfrak{B}$, a quasi-bornology $\mathfrak{S}$ or a semi-bornology $\mathfrak{L}$, $\mathfrak{B}$ is convex iff the dual filter $\mathfrak{B}$ is convex, the quasi-bornology $\mathfrak{S}$ has a base of convex filters, or the semi-bornology $\mathfrak{L}$ has a base of convex dual filters, respectively.

For each subcategory $\mathfrak{C}$ of HLING by $\mathfrak{C}^\wedge$ we denote the full subcategory of $\mathfrak{C}$ of all convex objects of $\mathfrak{C}$.

There exists a functor $\boldsymbol{h}^{\curlywedge}$: HLING $\to$ HLING$^\wedge$ such that for each object $X = (E, \mathfrak{B})$ of HLING, $\boldsymbol{h}^{\curlywedge}(X) = (E, \mathfrak{B}^\wedge)$ and for each morphism $f: X \to Y$ of HLING, $\boldsymbol{h}^{\curlywedge}(f)$ is the mapping $f: \boldsymbol{h}^{\curlywedge}(X) \to \boldsymbol{h}^{\curlywedge}(Y)$. By restriction of $\boldsymbol{h}^{\curlywedge}$ to HVEC we get a functor $\boldsymbol{h}^\wedge$: HVEC $\to$ HVEC$^\wedge$. Composing $\boldsymbol{b}^l$, $\boldsymbol{q}^l$ and $\boldsymbol{s}^l$ with restrictions of $\boldsymbol{h}^{\curlywedge}$ we get functors $\boldsymbol{b}^{\curlywedge}$: HLING $\to$ BLING$^\wedge$, $\boldsymbol{q}^{\curlywedge}$: HLING $\to$ QLING$^\wedge$ and $\boldsymbol{s}^{\curlywedge}$: HLING $\to$ SLING$^\wedge$, respectively. Analogously by means of $\boldsymbol{b}^v$, $\boldsymbol{q}^v$, $\boldsymbol{s}^v$ and $\boldsymbol{h}^\wedge$ we define functors $\boldsymbol{b}^\wedge$: HVEC $\to$ BVEC$^\wedge$, $\boldsymbol{q}^\wedge$: HVEC $\to$ QVEC$^\wedge$ and $\boldsymbol{s}^\wedge$: HVEC $\to$ SVEC$^\wedge$. $\boldsymbol{h}^{\curlywedge}$, $\boldsymbol{h}^\wedge$, $\boldsymbol{q}^{\curlywedge}$, $\boldsymbol{q}^\wedge$ and the restriction $\boldsymbol{b}^{q\curlywedge}$: QLING $\to$ BLING$^\wedge$ of $\boldsymbol{b}^{\curlywedge}$ to QLING and the restriction $\boldsymbol{b}^{q\wedge}$: QVEC $\to$ BVEC$^\wedge$ of $\boldsymbol{b}^\wedge$ to QVEC are bireflectors.

By PLING, TLING, PVEC and TVEC we denote the category of all pseudo-topological linear groups, of all topological linear groups, of all pseudo-topological vector spaces and of all topological vector spaces, respectively. A pseudo-topological linear group $X = (E, \tau)$ is said to be *locally convex* if $\tau(0)$ has a base of convex filters. For each subcategory $\mathfrak{C}$ of PLING by $\mathfrak{C}^\wedge$ we denote the full subcategory of $\mathfrak{C}$ of all locally convex objects of $\mathfrak{C}$.

A pseudo-topology τ of a vector space E is said to be *compatible* or *weakly compatible* if (E, τ) is a pseudo-topological linear group or a pseudo-topological vector space, respectively. If $X = (E, \tau)$ is a pseudo-topological linear group, by $p^{\curlywedge}(\tau)$ we understand the finest locally convex weakly compatible pseudo-topology of E coarser than τ, by $t^l(\tau)$ we understand the finest weakly compatible topology of E coarser than τ, and by $t^{\curlywedge}(\tau)$ we understand the finest locally convex weakly compatible topology of E coarser than τ. If $X = (E, \tau)$ is a pseudo-topological vector space we define analogously a locally convex compatible pseudo-topology $p^{\wedge}(\tau)$ of E, a compatible topology $t^v(\tau)$ of E, and a locally convex compatible topology $t^{\wedge}(\tau)$ of E. There exists a bireflector $\boldsymbol{p}^{\curlywedge}$: PLING $\to$ PLING$^{\wedge}$ such that for each object $X = (E, \tau)$ of PLING, $\boldsymbol{p}^{\curlywedge}(X) = (E, p^{\curlywedge}(\tau))$ and for each morphism $f: X \to Y$ of PLING, $p^{\curlywedge}(f)$ is the mapping $f: \boldsymbol{p}^{\curlywedge}(X) \to \boldsymbol{p}^{\curlywedge}(Y)$. Analogously we can introduce bireflector $\boldsymbol{t}^l$: PLING $\to$ TLING, $\boldsymbol{t}^{\curlywedge}$: PLING $\to$ TLING$^{\wedge}$, $\boldsymbol{p}^{\wedge}$: PVEC $\to$ PVEC$^{\wedge}$, $\boldsymbol{t}^v$: PVEC $\to$ TVEC and $\boldsymbol{t}^{\wedge}$: PVEC $\to$ TVEC$^{\wedge}$.

Let $\mathfrak{B}$ be a hypo-bornology of a vector space E. By $\Phi(\mathfrak{B})$ (0) we denote the $\wedge$-ideal with the base $\left\{ \bigvee_{\mathfrak{F} \in \mathfrak{v}} \bigcap_{F \in \mathfrak{F}} VF \mid \mathfrak{v} \in \mathfrak{B} \right\}$. We get that if $\mathfrak{B}$ is admissible or weakly admissible, then $\Phi(\mathfrak{B}): x \mapsto [(x)] + \Phi(\mathfrak{B})\,(0)$ $(x \in E)$ is the finest compatible or weakly compatible pseudo-topology of E, respectively, such that $\mathfrak{B}$ consists of all $\wedge$-ideals of dual filters on E which are hypo-bounded with respect to $\Phi(\mathfrak{B})$ (see 5.24.14 of [4], vol. II). There exists a functor $\boldsymbol{\Phi}^{hl}$: HLING $\to$ PLING such that for each object $X = (E, \mathfrak{B})$ of HLING, $\boldsymbol{\Phi}^{hl}(X) = (E, \Phi(\mathfrak{B}))$ and for each morphism $f: X \to Y$ of HLING, $\boldsymbol{\Phi}^{hl}(f)$ is the mapping $f: \boldsymbol{\Phi}^{hl}(X) \to \boldsymbol{\Phi}^{hl}(Y)$. By restrictions of $\boldsymbol{\Phi}^{hl}$ we get functors $\boldsymbol{\Phi}^{bl}$: BLING $\to$ PLING, $\boldsymbol{\Phi}^{ql}$: QLING $\to$ PLING, $\boldsymbol{\Phi}^{sl}$: SLING $\to$ PLING, $\boldsymbol{\Phi}^{bv}$: PVEC $\to$ PVEC, $\boldsymbol{\Phi}^{qv}$: QVEC $\to$ PVEC, $\boldsymbol{\Phi}^{sv}$: SVEC $\to$ PVEC and $\boldsymbol{\Phi}^{hv}$: HVEC $\to$ PVEC. Composing the restriction of $\boldsymbol{\Phi}^{bl}$ to BLING$^{\wedge}$ with the functor $\boldsymbol{p}^{\curlywedge}$ we get a functor $\boldsymbol{\Phi}^{b\curlywedge}$: BLING$^{\wedge}$ $\to$ PLING$^{\wedge}$. Analogously we can define functors $\boldsymbol{\Phi}^{q\curlywedge}$: QLING$^{\wedge}$ $\to$ PLING$^{\wedge}$, $\boldsymbol{\Phi}^{s\curlywedge}$: SLING$^{\wedge}$ $\to$ PLING$^{\wedge}$, $\boldsymbol{\Phi}^{h\curlywedge}$: HLING$^{\wedge}$ $\to$ PLING$^{\wedge}$, $\boldsymbol{\Phi}^{b\wedge}$: BVEC$^{\wedge}$ $\to$ PVEC$^{\wedge}$, $\boldsymbol{\Phi}^{q\wedge}$: QVEC$^{\wedge}$ $\to$ PVEC$^{\wedge}$, $\boldsymbol{\Phi}^{s\wedge}$: SVEC$^{\wedge}$ $\to$ PVEC$^{\wedge}$ and $\boldsymbol{\Phi}^{h\wedge}$: HVEC$^{\wedge}$ $\to$ PVEC$^{\wedge}$; in the last four cases these functors are compositions with $\boldsymbol{p}^{\wedge}$.

For each weakly compatible pseudo-topology τ of a vector space E let $\psi^b(\tau)$ be the bornology $\mathfrak{B}_b$, let $\psi^q(\tau)$ be the quasi-bornology $\mathfrak{S}_q$, let $\psi^s(\tau)$ be the semi-bornology $\mathfrak{L}_s$, and let $\psi^h(\tau)$ be the hypo-bornology $\mathfrak{B}_h$ defined with respect to τ, respectively. There exists a functor $\boldsymbol{\Psi}^{bl}$: PLING $\to$ BLING such that for each object $X = (E, \tau)$, $\boldsymbol{\Psi}^{bl}(X) = (E, \psi^b(\tau))$ and for each morphism $f: X \to Y$ of PLING, $\boldsymbol{\Psi}^{bl}(f)$ is the mapping f: $\boldsymbol{\Psi}^{bl}(X) \to \boldsymbol{\Psi}^{bl}(Y)$. Analogously we can introduce functors $\boldsymbol{\Psi}^{ql}$: PLING $\to$ QLING, $\boldsymbol{\Psi}^{sl}$: PLING $\to$ SLING and $\boldsymbol{\Psi}^{hl}$: PLING $\to$ HLING. By restrictions of $\boldsymbol{\Psi}^{bl}$ we get functors $\boldsymbol{\Psi}^{b\curlywedge}$: PLING$^{\wedge}$ $\to$ BLING$^{\wedge}$, $\boldsymbol{\Psi}^{bv}$: PVEC $\to$ BVEC and $\boldsymbol{\Psi}^{b\wedge}$: PVEC$^{\wedge}$ $\to$ BVEC$^{\wedge}$. Analogously we can introduce functors $\boldsymbol{\Psi}^{a\curlywedge}$, $\boldsymbol{\Psi}^{av}$ and $\psi^{a\wedge}$ with $a = q$, s and h.

In the following theorem we give some connections between these functors.

Theorem 1.1. *We have*

$$\boldsymbol{\Psi}^{sl} = \boldsymbol{s}^l \circ \boldsymbol{\Psi}^{hl}, \quad \boldsymbol{\Psi}^{s\curlywedge} = \boldsymbol{s}^{\curlywedge\prime} \circ \boldsymbol{\Psi}^{h\curlywedge}, \quad \boldsymbol{\Psi}^{sv} = \boldsymbol{s}^v \circ \boldsymbol{\Psi}^{hv}, \quad \boldsymbol{\Psi}^{s\wedge} = \boldsymbol{s}^{\wedge\prime} \circ \boldsymbol{\Psi}^{h\wedge}$$

and

$$\boldsymbol{\Psi}^{bl} = \boldsymbol{b}^{ql} \circ \boldsymbol{\Psi}^{ql}, \quad \boldsymbol{\Psi}^{b\curlywedge} = \boldsymbol{b}^{q\curlywedge\prime} \circ \boldsymbol{\Psi}^{q\curlywedge}, \quad \boldsymbol{\Psi}^{bv} = \boldsymbol{b}^{qv} \circ \boldsymbol{\Psi}^{qv}, \quad \boldsymbol{\Psi}^{b\wedge} = \boldsymbol{b}^{q\wedge\prime} \circ \boldsymbol{\Psi}^{q\wedge}$$

where $\boldsymbol{s}^{\curlywedge\prime}$, $\boldsymbol{s}^{\wedge\prime}$, $\boldsymbol{b}^{q\curlywedge\prime}$ *and* $\boldsymbol{b}^{q\wedge\prime}$ *are the restrictions of* $\boldsymbol{s}^{\curlywedge}$ *to* HLING$^{\wedge}$, *of* $\boldsymbol{s}^{\wedge}$ *to* HVEC$^{\wedge}$, *of* $\boldsymbol{b}^{q\curlywedge}$ *to* QLING$^{\wedge}$ *and of* $\boldsymbol{b}^{q\wedge}$ *to* QVEC$^{\wedge}$, *respectively.*

Theorem 1.2. *If a is equal to b, q, s or h, $\Phi^{a!}$ is a left adjoint of $\Psi^{\cdot a!}$, $\Phi^{a\curlywedge}$ is a left adjoint of $\Psi^{\cdot a\curlywedge}$, Φ^{av} is a left adjoint of $\Psi^{\cdot av}$, and $\Phi^{a\wedge}$ is a left adjoint of $\Psi^{\cdot a\wedge}$.*

Remarks to the proof: Concerning the first and third case we refer to [4], vol. II, Theorem 5.24.16 and Corollary (partially in this book we use easily changed denotations). The assertion in the other two cases can be proved analogously.

2. On adjoint functors

If $\mathfrak{C}$ is a category, by ob $\mathfrak{C}$ we denote the class of all objects of $\mathfrak{C}$, by mor $\mathfrak{C}$ we denote the class of all morphisms of $\mathfrak{C}$, and for each $X \in \text{ob}\,\mathfrak{C}$ by 1_X we denote the identical morphism of X. A functor T of a category $\mathfrak{C}$ into a subcategory $\mathfrak{D}$ is called a *retraction* if the composition of the identical embedding $\text{emb}_{\mathfrak{D}}$ of $\mathfrak{D}$ into $\mathfrak{C}$ with T is the identical functor $\text{id}_{\mathfrak{D}}$ of $\mathfrak{D}$: $T\,\text{emb}_{\mathfrak{D}} = \text{id}_{\mathfrak{D}}$. A functor $S: \mathfrak{C} \to \mathfrak{C}$ is called *idempotent* if $SS = S$. For each retraction $T: \mathfrak{C} \to \mathfrak{D}$ the composition $\text{emb}_{\mathfrak{D}}\,T$ is idempotent. On the other hand if $S: \mathfrak{C} \to \mathfrak{C}$ is an idempotent functor, there exists a subcategory $\mathfrak{D}$ of $\mathfrak{C}$ with $\text{ob}\,\mathfrak{D} = \{S(X) \mid X \in \text{ob}\,\mathfrak{C}\}$ and $\text{mor}\,\mathfrak{D} = \{S(f) \mid f \in \text{mor}\,\mathfrak{C}\}$ and from S by restriction of the range of S we get a retraction $T: \mathfrak{C} \to \mathfrak{D}$; T is called the *retraction corresponding to* S, and $\mathfrak{D}$ is called the *subcategory of* $\mathfrak{C}$ *corresponding to* S.

For each adjunction $(S: \mathfrak{D} \to \mathfrak{C}, T: \mathfrak{C} \to \mathfrak{D}, \mu)$, from 5.21.4 and 5.21.5 of [4], vol. II, it follows: If $TS: \mathfrak{D} \to \mathfrak{D}$ is idempotent, then the retraction corresponding to TS is a reflector, and if $ST: \mathfrak{C} \to \mathfrak{C}$ is idempotent, then the retraction corresponding to ST is a coreflector. By means of the following two theorems under certain assumptions we can characterize the objects of the subcategories corresponding to TS and to ST.

Theorem 2.1. *Let $S: \mathfrak{D} \to \mathfrak{C}$ and $T: \mathfrak{C} \to \mathfrak{D}$ be functors. We assume that $STS = S$ (then ST and TS are idempotent), that for each $X \in \text{ob}\,\mathfrak{D}$ there exists a morphism u_X: $X \to TS(X)$ of $\mathfrak{D}$ with $S(u_X) = 1_{S(X)}$, and that for each $X \in \text{ob}\,\mathfrak{D}$ and each morphism $g: TS(X) \to X$ of $\mathfrak{D}$ only $S(g) = 1_{S(X)}$ if $g = 1_X$. Moreover we assume that the subcategory $\mathfrak{C}'$ of $\mathfrak{C}$ corresponding to ST is full. Let $\mathfrak{D}'$ be the subcategory of $\mathfrak{D}$ corresponding to TS. Then for each $X \in \text{ob}\,\mathfrak{D}$ we have*

$X \in \text{ob}\,\mathfrak{D}' \Leftrightarrow$ *For each $Y \in \text{ob}\,\mathfrak{D}$ and each morphism $f: S(Y) \to S(X)$ of $\mathfrak{C}$ there exists a morphism $g: Y \to X$ of $\mathfrak{D}$ with $S(g) = f$.*

Proof 1. We assume that $X \in \text{ob}\,\mathfrak{D}'$. Let Y be an object of $\mathfrak{D}$ and let $f: S(Y) \to S(X)$ be an morphism of $\mathfrak{C}$. Since $\mathfrak{D}'$ is a full subcategory of $\mathfrak{C}$ and $STS = S$, we have $f \in \text{mor}\,\mathfrak{C}'$ and therefore $ST(f) = f$. Because of $X \in \text{ob}\,\mathfrak{D}'$ it follows that $g = T(f)_{u_Y}$ is a morphism of Y to X with $S(g) = f$.

2. We assume that the condition of the right-hand side of the equivalence is fulfilled. $f = 1_{S(X)}$ is a morphism of $S(Y)$ to $S(X)$ with $Y = TS(X)$. Thus, there exists a morphism $g: Y \to X$ with $S(g) = 1_{S(X)}$ and therefore with $g = 1_X$. It follows $TS(X) = X$, i.e. $X \in \text{ob}\,\mathfrak{D}'$.

By dualization of this theorem we obtain the

Theorem 2.2. *Let $S: \mathfrak{D} \to \mathfrak{C}$ and $T: \mathfrak{C} \to \mathfrak{D}$ be functors. We assume that $TST = T$, that for each $X \in \text{ob}\,\mathfrak{D}$ there exists a morphism $v_X: ST(X) \to X$ of $\mathfrak{C}$ with $T(v_X) = 1_{T(X)}$, and that for each $X \in \text{ob}\,\mathfrak{C}$ and each morphism $g: X \to ST(X)$ of $\mathfrak{C}$ only $T(g) = 1_{T(X)}$*

if $g = 1_X$. *Moreover we assume that the subcategory* $\mathfrak{D}'$ *of* $\mathfrak{D}$ *corresponding to* ST *is full. Let* $\mathfrak{C}'$ *be the subcategory of* $\mathfrak{C}$ *corresponding to* ST. *Then for each* $X \in \operatorname{ob} \mathfrak{C}$ *we have*

$X \in \operatorname{ob} \mathfrak{C}' \Leftrightarrow$ *For each* $Y \in \operatorname{ob} \mathfrak{C}$ *and each morphism* $f\colon T(X) \to T(Y)$ *of* $\mathfrak{D}$ *there exists a morphism* $g\colon X \to Y$ *of* $\mathfrak{C}$ *with* $T(g) = f$.

In the following part of this section we shall apply Theorem 2.1 and 2.2.

1. We suppose that $\mathfrak{C}$ is a full bireflective subcategory of PLING. Let $R\colon \mathrm{PLING} \to \mathfrak{C}$ be the uniquely determined reflector such that for each $X \in \operatorname{ob} \mathrm{PLING}$, X and $R(X)$ have the same underlying vector space and for each morphism $f\colon X \to Y$ of PLING, the corresponding morphism of $\mathfrak{C}$ by R is $f\colon R(X) \to R(Y)$. Examples of R are the identical functor $\mathrm{id}_{\mathrm{PLING}}$ of PLING, $\boldsymbol{p}^{\curlywedge}$, $\boldsymbol{t}^{l}$ and $\boldsymbol{t}^{\curlywedge}$. Let $\mathrm{emb}_{\mathfrak{C}}\colon \mathfrak{C} \to \mathrm{PLING}$ be the identical embedding of $\mathfrak{C}$ into PLING. Of course, R is a left adjoint of $\mathrm{emb}_{\mathfrak{C}}$. Thus, by a well-known theorem on products of adjoint functors (see e.g. 5.21.1 of [4], vol. II) and by Theorem 1.2 we have hat $R\boldsymbol{\Phi}^{al}$ is a left adjoint of $\boldsymbol{\Psi}^{al}\,\mathrm{emb}_{\mathfrak{C}}$ for $a = b$, q, s and h.

For each weakly compatible pseudo-topology τ we have $\Phi(\psi^a(\tau)) \leqq \tau$ ($a = b$, q, s and h), and if $\mathfrak{B}$ is a weakly admissible bornology, $\mathfrak{S}$ is a weakly admissible quasi-bornology, $\mathfrak{L}$ is a weakly admissible semi-bornology and $\mathfrak{B}$ is a weakly admissible hypo-bornology, we have $\mathfrak{B} \subseteqq \Psi^b(\Phi(\mathfrak{B}))$, $\mathfrak{S} \subseteqq \Psi^q(\Phi(\mathfrak{S}))$, $\mathfrak{L} \subseteqq \Psi^l(\Phi(\mathfrak{L}))$ and $\mathfrak{B} \subseteqq \Psi^h(\Phi(\mathfrak{B}))$. Thus, if $S = \boldsymbol{\Phi}^{al}$ and $T = \boldsymbol{\Psi}^{al}\,\mathrm{emb}_{\mathfrak{C}}$ ($a = b$, q, s or h), we can easily prove that the assumptions of the Theorems 2.1 and 2.2 are fulfilled. By $\mathrm{BLING}^{\mathfrak{C}}$ we denote the full reflective subcategory of BLING corresponding to $TS = \boldsymbol{\Psi}^{bl}\,\mathrm{emb}_{\mathfrak{C}}\,R\boldsymbol{\Phi}^{bl}$, and by $\mathfrak{C}_{\mathrm{BLING}}$ we denote the full coreflective subcategory of $\mathfrak{C}$ corresponding to $ST = R\boldsymbol{\Phi}^{bl}\boldsymbol{\Psi}^{bl}\,\mathrm{emb}_{\mathfrak{C}}$. Analogously we introduce full reflective subcategories $\mathrm{QLING}^{\mathfrak{C}}$ of QLING, $\mathrm{SLING}^{\mathfrak{C}}$ of SLING and $\mathrm{HLING}^{\mathfrak{C}}$ of HLING and full coreflective subcategories $\mathfrak{C}_{\mathrm{QLING}}$, $\mathfrak{C}_{\mathrm{SLING}}$ and $\mathfrak{C}_{\mathrm{HLING}}$ of $\mathfrak{C}$.

From Theorem 2.1 it follows

Theorem 2.3. *A bornological linear group* $X = (E, \mathfrak{B})$ *is an object of* $\mathrm{BLING}^{\mathfrak{C}}$ *iff each continuous linear mapping* $f\colon R\boldsymbol{\Phi}^{bl}(Y) \to R\boldsymbol{\Phi}^{bl}(X)$ ($Y = (E', \mathfrak{B}') \in \operatorname{ob} \mathrm{BLING}$) *is a* $\mathfrak{B},\mathfrak{B}'$*-faithful (= bounded) mapping of* Y *into* X.

Of course, by Theorem 2.1 we can characterize analogously the objects of $\mathrm{QLING}^{\mathfrak{C}}$, $\mathrm{SLING}^{\mathfrak{C}}$ and $\mathrm{HLING}^{\mathfrak{C}}$.

In the cases $\mathfrak{C} = \mathrm{PLING}$ and TLING (i.e. $R = \mathrm{id}_{\mathrm{PLING}}$ and $\boldsymbol{t}^{l}$) we introduce special denotations. A bornological linear group X is said to be *pseudo-topological* or *topological with respect to* BLING if X is an object of $\mathrm{BLING}^{\mathrm{PLING}}$ or $\mathrm{BLING}^{\mathrm{TLING}}$, respectively. Analogously for the objects of QLING, SLING and HLING we define the properties to be *pseudo-topological with respect to* QLING, SLING and HLING, respectively, and the properties to be *topological with respect to* QLING, SLING and HLING, respectively.

We can easily prove

Theorem 2.4. *Each bornological linear group is pseudo-topological with respect to* BLING. *A quasi-bornological linear group* $X = (E, \mathfrak{S})$ *is pseudo-topological with respect to* QLING *iff* $\bigvee_{M \in V\mathfrak{F}} [\alpha_M M] \in \mathfrak{S}$ *for all* $\mathfrak{F} \in \mathfrak{S}$ *and all families* $(\alpha_M)_{M \in V\mathfrak{F}}$ *of positive real numbers. A semi-bornological linear group* $X = (E, \mathfrak{L})$ *is pseudo-topological with respect to* SLING *iff*

$\bigcap_{M \in \mathfrak{M}_{\mathfrak{F}}} \bigvee_{\alpha>0} (\alpha M) \in \mathfrak{L}$ *for each* $\mathfrak{F} \in \mathfrak{S}$ *and* $\mathfrak{M}_{\mathfrak{F}} = \bigcap_{F \in \mathfrak{F}} VF$. *A hypo-bornological linear group* $X = (E, \mathfrak{B})$ *is pseudo-topological with respect to* HLING *iff* $\bigvee_{M \in \mathfrak{M}_{\mathfrak{v}}} \bigcap_{\alpha>0} [(\alpha M)] \in \mathfrak{B}$ *for each* $\mathfrak{v} \in \mathfrak{B}$ *and* $\mathfrak{M}_{\mathfrak{v}} = \bigvee_{\mathfrak{F} \in \mathfrak{v}} \bigcap_{F \in \mathfrak{F}} VF$. *In each case the unions are to be understood in the corresponding lattice.*

A pseudo-topological linear group X is said to be *bornological, quasi-bornological, semi-bornological* or *hypo-bornological with respect to* $\mathfrak{C}$ if X is an object of $\mathfrak{C}_{\mathrm{BLING}}$, $\mathfrak{C}_{\mathrm{QLING}}$, $\mathfrak{C}_{\mathrm{SLING}}$ or $\mathfrak{C}_{\mathrm{HLING}}$, respectively.

From Theorem 2.2 we obtain

Theorem 2.5. *A pseudo-topological linear group* X *is bornological with respect to* $\mathfrak{C}$ *iff each bounded linear mapping* $f\colon \Psi^{bl}(X) \to \Psi^{bl}(Y)$ $(Y \in \mathrm{ob}\,\mathfrak{C})$ *is a continuous mapping of* X *into* Y.

Of course, by Theorem 2.2 we can characterize analogously pseudo-topological linear groups which are quasi-bornological, which are semi-bornological, and which are hypo-bornological with respect to $\mathfrak{C}$, respectively.

A pseudo-topological linear group is bornological with respect to PLING iff it is equable and locally bounded. It is quasi-bornological and also hypo-bornological with respect to PLING iff it is equable.

2. Now we suppose that $\mathfrak{C}$ is a full bireflective subcategory of PLING. By R: PLING^ $\to \mathfrak{C}$ we denote the reflector defined analogously as above, and by $\mathrm{emb}_{\mathfrak{C}}$ we denote the identical embedding of $\mathfrak{C}$ into PLING^. In this case examples of R are the identical functor $\mathrm{id}_{\mathrm{PLING}}$, of course, and the restriction $t^{\wedge\prime}$: PLING^ $\to$ TLING^ of $t^{\wedge}$ to PLING^. If $S = R\Phi^{a\wedge}$ and $T = \Psi^{a\wedge}\,\mathrm{emb}_{\mathfrak{C}}$ $(a = b, q, s$ or $h)$, the assumptions of Theorems 2.1 and 2.2 are fulfilled. By BLING$^{\wedge\mathfrak{C}}$ we denote the full reflective subcategory of BLING^ corresponding to $TS = \Psi^{b\wedge}\,\mathrm{emb}_{\mathfrak{C}}\,R\Phi^{b\wedge}$ and by $\mathfrak{C}_{\mathrm{BLING}^\wedge}$ we denote the full coreflective subcategory of $\mathfrak{C}$ corresponding to $ST = R\Phi^{b\wedge}\Psi^{b\wedge}\,\mathrm{emb}_{\mathfrak{C}}$. Analogously we introduce full reflective subcategories QLING$^{\wedge\mathfrak{C}}$ of QLING^, SLING$^{\wedge\mathfrak{C}}$ of SLING^ and HLING$^{\wedge\mathfrak{C}}$ of HLING^ and full coreflective subcategories $\mathfrak{C}_{\mathrm{QLING}^\wedge}$, $\mathfrak{C}_{\mathrm{SLING}^\wedge}$ and $\mathfrak{C}_{\mathrm{HLING}^\wedge}$ of $\mathfrak{C}$. We have

$$\mathrm{ob}\,\mathrm{BLING}^{\wedge\mathfrak{C}} \subseteqq \mathrm{ob}\,\mathrm{BLING}^{\mathfrak{C}} \quad \text{and} \quad \mathrm{ob}\,\mathfrak{C}_{\mathrm{BLING}} \subseteqq \mathrm{ob}\,\mathfrak{C}_{\mathrm{BLING}^\wedge},$$

and in the other cases analogous inclusions hold.

By means of Theorem 2.1 we can easily characterize the objects of BLING$^{\wedge\mathfrak{C}}$, QLING$^{\wedge\mathfrak{C}}$, SLING$^{\wedge\mathfrak{C}}$ and HLING$^{\wedge\mathfrak{C}}$. E.g., we obtain the characterization of the objects of BLING$^{\wedge\mathfrak{C}}$ if in Theorem 2.3 we replace bornological by convex bornological, BLING by BLING^ and Φ^{bl} by $\Phi^{b\wedge}$.

A convex bornological linear group X is said to be *locally convex pseudo-topological* or *locally convex topological with respect to* BLING^ if X is an object of BLING$^{\wedge\mathrm{PLING}^\wedge}$ or BLING$^{\wedge\mathrm{TLING}^\wedge}$, respectively. Analogously for the objects of QLING^, SLING^ and HLING^ we define the property to be *locally convex pseudo-topological with respect to* QLING^, SLING^ and HLING^, respectively, and the property to be *locally convex topological with respect to* QLING^, SLING^ and HLING^, respectively.

We can easily prove

Theorem 2.6. *Each convex bornological linear group is locally convex pseudo-topological with respect to* BLING^. *A convex quasi-bornological linear group is locally convex pseudo-topological with respect to* QLING^ *iff it is locally convex pseudo-topological with respect to* QLING. *A convex semi-bornological linear group* $X = (E, \mathfrak{L})$ *is locally convex pseudo-topological with respect to* SLING^ *iff* $\bigcap_{N \in \mathfrak{N}_{\mathfrak{F}}} \bigvee_{\alpha > 0} (\alpha N) \in \mathfrak{L}$ *for each* $\mathfrak{F} \in \mathfrak{L}$ *and* $\mathfrak{N}_{\mathfrak{F}} = \left(\bigcap_{F \in \mathfrak{F}} VF\right)^{\wedge}$. *A convex hypo-bornological linear group* $X = (E, \mathfrak{B})$ *is locally convex pseudo-topological with respect to* HLING^ *iff* $\bigvee_{N \in \mathfrak{N}_{\mathfrak{v}}} \bigcap_{\alpha > 0} [(\alpha N)] \in \mathfrak{B}$ *for each* $\mathfrak{v} \in \mathfrak{B}$ *and* $\mathfrak{N}_{\mathfrak{v}} = \left(\bigvee_{\mathfrak{F} \in \mathfrak{v}} \bigcap_{F \in \mathfrak{F}} VF\right)^{\wedge}$.

A locally convex pseudo-topological linear group X is said to be *convex bornological, convex quasi-bornological, convex semi-bornological* or *convex hypo-bornological with respect to* $\mathfrak{C}$ if X is an object of $\mathfrak{C}_{\mathrm{BLING}^\wedge}$, $\mathfrak{C}_{\mathrm{QLING}^\wedge}$, $\mathfrak{C}_{\mathrm{SLING}^\wedge}$ or $\mathfrak{C}_{\mathrm{HLING}^\wedge}$, respectively. By means of Theorem 2.2 we can easily characterize those locally convex pseudo-topological linear groups which are convex bornological, which are convex quasi-bornological, which are convex semi-bornological, and which are convex hypo-bornological with respect to $\mathfrak{C}$, respectively. E.g., we obtain the characterization of the locally convex pseudo-topological linear groups which are convex bornological with respect to $\mathfrak{C}$ if in Theorem 2.5 we replace pseudo-topological by locally convex pseudo-topological, bornological by convex bornological and $\boldsymbol{\Psi}^{bl}$ by $\boldsymbol{\Psi}^{b\wedge}$.

3. Let $\mathfrak{C}$ be a full bireflective subcategory of PVEC (case 3.1) or of PVEC^ (cas 3.2). Then we can introduce analogously as in the cases 1 and 2 reflectors R: PVEC $\to \mathfrak{C}$ and R: PVEC^ $\to \mathfrak{C}$, respectively, and embedding functors $\mathrm{emb}_{\mathfrak{C}}: \mathfrak{C} \to$ PVEC and $\mathrm{emb}_{\mathfrak{C}}: \mathfrak{C} \to$ PVEC^, respectively. In case 3.1 examples of R are the identical functor $\mathrm{id}_{\mathrm{PVEC}}$, of course, and $\boldsymbol{p}^\wedge$, $\boldsymbol{t}^v$ and $\boldsymbol{t}^\wedge$, and in case 3.2 examples of R are the identical functor $\mathrm{id}_{\mathrm{PVEC}^\wedge}$ and the restriction of $\boldsymbol{t}^\wedge$ to PVEC^. In the case 3.1 for $S = R\boldsymbol{\Phi}^{av}$ and $T = \boldsymbol{\Psi}^{av}\,\mathrm{emb}_{\mathfrak{C}}$ ($a = b$, q, s and h) and in the case 3.2 for $S = R\boldsymbol{\Phi}^{a\wedge}$ and $T = \boldsymbol{\Psi}^{a\wedge}\,\mathrm{emb}_{\mathfrak{C}}$ ($a = b$, q, s and h) the assumptions of Theorems 2.1 and 2.2 are fulfilled. Thus, analogously as in the first two cases, in the case 3.1 we can define full reflective subcategories $\mathrm{BVEC}^{\mathfrak{C}}$ of BVEC, $\mathrm{QVEC}^{\mathfrak{C}}$ of QVEC, $\mathrm{SVEC}^{\mathfrak{C}}$ of SVEC and $\mathrm{HVEC}^{\mathfrak{C}}$ of HVEC and full coreflective subcategories $\mathfrak{C}_{\mathrm{BVEC}}$, $\mathfrak{C}_{\mathrm{QVEC}}$, $\mathfrak{C}_{\mathrm{SVEC}}$ and $\mathfrak{C}_{\mathrm{HVEC}}$ of $\mathfrak{C}$, and in the case 3.2 we can define full reflective subcategories $\mathrm{BVEC}^{\wedge\mathfrak{C}}$ of BVEC^, $\mathrm{QVEV}^{\wedge\mathfrak{C}}$ of QVEC^, $\mathrm{SVEC}^{\wedge\mathfrak{C}}$ of SVEC^ and $\mathrm{HVEC}^{\wedge\mathfrak{C}}$ of HVEC^ and full coreflective subcategories $\mathfrak{C}_{\mathrm{BVEC}^\wedge}$, $\mathfrak{C}_{\mathrm{QVEC}^\wedge}$, $\mathfrak{C}_{\mathrm{SVEC}^\wedge}$ and $\mathfrak{C}_{\mathrm{HVEC}^\wedge}$ of $\mathfrak{C}$. For these categories analogous results can be proved as those mentioned in the cases 1 and 2. But these results are not far from results of the cases 1 and 2 as can be shown by the following theorem (if $\mathfrak{D} = \mathfrak{C}$).

Theorem 2.7. *In the case* 3.1 *we have*

$$\mathrm{ob\ BVEC}^{\mathfrak{C}} = \mathrm{ob\ BLING}^{\mathfrak{D}} \cap \mathrm{ob\ BVEC} \quad \textit{and} \quad \mathfrak{C}_{\mathrm{BVEC}} = \mathfrak{D}_{\mathrm{BLING}}$$

where $\mathfrak{D}$ *is an arbitrary full bireflective subcategory of* PLING *such that* $\mathfrak{C}$ *is the full subcategory of* $\mathfrak{D}$ *of all objects of* $\mathfrak{D}$ *which are pseudo-topological vector spaces. In the case* 3.2 *we have*

$$\mathrm{ob\ BVEC}^{\wedge\mathfrak{C}} = \mathrm{ob\ BLING}^{\wedge\mathfrak{D}} \cap \mathrm{ob\ BVEC} \quad \textit{and} \quad \mathfrak{C}_{\mathrm{BVEC}^\wedge} = \mathfrak{D}_{\mathrm{BLING}^\wedge}$$

where $\mathfrak{D}$ *is an arbitrary full bireflective subcategory of* PLING^ *such that* $\mathfrak{C}$ *is the full subcategory of* $\mathfrak{D}$ *of all objects of* $\mathfrak{D}$ *which are pseudo-topological vector spaces. In both cases* $\mathfrak{D}$ *may be equal to* $\mathfrak{C}$. *These assertions remain true if we replace* BVEC *and* BLING *by* QVEC *and* QLING, *by* SVEC *and* SLING *and by* HVEC *and* HLING, *respectively.*

If a bornological vector space is pseudo-topological with respect to BLING, it is also said to be *pseudo-topological with respect to* BVEC, and if it is topological with respect to BLING, it is also said to be *topological with respect to* BVEC. Analogously we define *pseudo-topological with respect to* QVEC, SVEC and HVEC, *topological with respect to* QVEC, SVEC and HVEC, *locally convex pseudo-topological with respect to* BVEC^, QVEC^, SVEC^ and HVEC^ and *locally convex topological with respect to* BVEC^, QVEC^, SVEC and HVEC^.

3. Projective tensor products

In the following we introduce projective tensor products in PLING and in some subcategories of PLING (see [4], vol. II, and [5]). Let $X = (E, \tau)$ and $Y = (F, \sigma)$ be pseudo-topological linear groups. By $\tau \otimes \sigma$ we denote the finest weakly compatible pseudo-topology of the tensor product $E \otimes F$ such that the canonical mapping $\varkappa: E \times F \to E \otimes F$ is continuous with respect to $\tau \times \sigma$ and $\tau \otimes \sigma$ where $\tau \times \sigma$ is the product pseudo-topology of τ and σ. $X \otimes Y$, i.e. E equipped with $\tau \otimes \sigma$ is called the *projective tensor product* of X and Y in PLING.

Theorem 3.1 (see [4], vol. II). *If X and Y are pseudo-topological vector spaces, then also $X \otimes Y$ is a pseudo-topological vector space. If X and Y are equable or locally bounded, then also $X \otimes Y$ is equable or locally bounded, respectively. If X and Y are topological vector spaces of infinite dimensions, then $X \otimes Y$ is not a topological vector space.*

Let $\mathfrak{P}$ be a category and let $\mathfrak{C}$ and $\mathfrak{D}$ be subcategories of $\mathfrak{P}$. A functor $R: \mathfrak{P} \to \mathfrak{C}$ is called a *reflector* on $\mathfrak{D}$ if for each $X \in \text{ob}\, \mathfrak{D}$ there exists a morphism $R_X: X \to R(X)$ of $\mathfrak{P}$, a corresponding *reflexion* of X, such that for each $Y \in \text{ob}\, \mathfrak{C}$ and each morphism $f: X \to Y$ of $\mathfrak{P}$ there is one and only one morphism $g: R(X) \to Y$ of $\mathfrak{D}$ with $f = gR_X$. Of course, if $R: \mathfrak{P} \to \mathfrak{C}$ is a reflector, R is a reflector on $\mathfrak{D}$. Under the assumption that $\mathfrak{D}$ is a subcategory of $\mathfrak{C}$, every retraction $R: \mathfrak{P} \to \mathfrak{C}$ is a reflector on $\mathfrak{D}$. If $\mathfrak{E}$ and $\mathfrak{F}$ are subcategories of $\mathfrak{C}$, the composition of a reflector $R: \mathfrak{P} \to \mathfrak{C}$ on $\mathfrak{D}$ with a reflector $S: \mathfrak{C} \to \mathfrak{E}$ on $\mathfrak{F}$ is a reflector $SR: \mathfrak{P} \to \mathfrak{E}$ on $\mathfrak{F}$.

Now let $\mathfrak{C}$ be a full subcategory of PLING. By $\mathfrak{C} \otimes \mathfrak{C}$ we denote the full subcategory of PLING of all projective tensor products $X \otimes Y$ ($X, Y \in \text{ob}\, \mathfrak{C}$). Let R: PLING $\to \mathfrak{C}$ be a reflector on $\mathfrak{C} \times \mathfrak{C}$. Then for each pair of objects X and Y of $\mathfrak{C}$ we call $R(X \otimes Y)$ the *projective tensor product* of X and Y *corresponding to R*. In many cases of $\mathfrak{C}$ there exists a "canonical" functor R: PLING $\to \mathfrak{C}$ being a reflector on $\mathfrak{C} \times \mathfrak{C}$. Projective tensor products corresponding to a canonical functor R: PLING $\to \mathfrak{C}$ are also called *projective tensor products in* $\mathfrak{C}$. The bireflectors id_{PLING}, $\boldsymbol{p}^{\wedge}$, $\boldsymbol{t}^{l}$ and $\boldsymbol{t}^{\wedge}$ are examples of such canonical functors. The corresponding projective tensor products are the projective tensor products in PLING, in PLING^, in TLING and in TLING^, respectively. There exists a functor v: PLING $\to$ PVEC such that for each object $X = (E, \tau)$ of PLING, $\boldsymbol{v}(X)$ is the vector space E, equipped with the finest compatible pseudo-topology coarser than τ, and for each morphism $f: X \to Y$ of PLING, $\boldsymbol{v}(f)$ is the mapping $f: \boldsymbol{v}(X) \to \boldsymbol{v}(Y)$. The bireflectors $\boldsymbol{v}$, $\boldsymbol{p}^{\wedge}\boldsymbol{v}$, $\boldsymbol{t}^{v}\boldsymbol{v}$ and $\boldsymbol{t}^{\wedge}\boldsymbol{v}$ are further examples of canonical functors. The corresponding projective tensor products are the projective tensor products in PVEC, PVEC^, TVEC and TVEC^, respectively. By Theorem 3.1 we see that also $\boldsymbol{\Phi}^{bl}\boldsymbol{\Psi}^{\cdot bl}$, $\boldsymbol{\Phi}^{ql}\boldsymbol{\Psi}^{\cdot ql}$ ($= \boldsymbol{\Phi}^{hl}\boldsymbol{\Psi}^{\cdot hl}$), $\boldsymbol{\Phi}^{bv}\boldsymbol{\Psi}^{\cdot bv}\boldsymbol{v}$ and $\boldsymbol{\Phi}^{qv}\boldsymbol{\Psi}^{\cdot qv}\boldsymbol{v}$ ($= \boldsymbol{\Phi}^{hv}\boldsymbol{\Psi}^{\cdot hv}\boldsymbol{v}$) are examples of such canonical functors. The corresponding projective tensor products are the projective tensor products in

$\mathrm{PLING}_{\mathrm{BLING}}$, in $\mathrm{PLING}_{\mathrm{QLING}}$ ($= \mathrm{PLING}_{\mathrm{HLING}}$), in $\mathrm{PVEC}_{\mathrm{BVEC}}$ and in $\mathrm{PVEC}_{\mathrm{QVEC}}$ ($= \mathrm{PVEC}_{\mathrm{HVEC}}$), respectively. We use the shorter denotations $\boldsymbol{\delta} = \boldsymbol{\Phi}^{bl}\boldsymbol{\Psi}^{bl}$, $\# = \boldsymbol{\Phi}^{ql}\boldsymbol{\Psi}^{ql}$, $\boldsymbol{v}^{\delta} = \boldsymbol{\Phi}^{bv}\boldsymbol{\Psi}^{bv}\boldsymbol{v}$, $\boldsymbol{v}^{\#} = \boldsymbol{\Phi}^{qv}\boldsymbol{\Psi}^{qv}\boldsymbol{v}$, $\mathrm{PLING}^{\delta} = \mathrm{PLING}_{\mathrm{BLING}}$, $\mathrm{PLING}^{\#} = \mathrm{PLING}_{\mathrm{QLING}}$, $\mathrm{PVEC}^{\delta} = \mathrm{PVEC}_{\mathrm{BVEC}}$ and $\mathrm{PVEC}^{\#} = \mathrm{PVEC}_{\mathrm{QVEC}}$. $(\boldsymbol{p}^{\wedge} \mid \mathrm{PLING}^{\delta})\, \boldsymbol{\delta}$, $(\boldsymbol{p}^{\wedge} \mid \mathrm{PLING}^{\#})\, \#$, $(\boldsymbol{p}^{\wedge} \mid \mathrm{PVEC}^{\delta})\, \boldsymbol{v}^{\delta}$ and $(\boldsymbol{p}^{\wedge} \mid \mathrm{PVEC}^{\#})\, \boldsymbol{v}^{\#}$ are examples of canonical functors, too. The corresponding projective tensor products are the projective tensor products in $\mathrm{PLING}^{\delta\wedge}$, in $\mathrm{PLING}^{\#\wedge}$, in $\mathrm{PVEC}^{\delta\wedge}$ and in $\mathrm{PVEC}^{\#\wedge}$, respectively. It is easy to write down more canonical functors.

We suppose now that $\mathfrak{C}$ is a subcategory of PLING which is closed with respect to products, that $\boldsymbol{K}$ ($= \boldsymbol{R}$ or $\boldsymbol{C}$) belongs to $\mathfrak{C}$, and that R: PLING $\to \mathfrak{C}$ is a reflector on $\mathfrak{C} \otimes \mathfrak{C}$. By $\mathfrak{C} \times \mathfrak{C}$ we denote the category of all pairs of objects of $\mathfrak{C}$; morphisms are all pairs of morphisms of $\mathfrak{C}$. There exists a functor $\otimes_T: \mathfrak{C} \times \mathfrak{C} \to \mathfrak{C}$ such that for each pair of objects X and Y of $\mathfrak{C}$, $\otimes_T(X, Y) = T(X \otimes Y)$ and for each pair of morphisms $f: X \to Y$ and $g: Z \to U$ of $\mathfrak{C}$, $\otimes_T(f, g)$ is the mapping $f \otimes g: T(X \otimes Z) \to T(Y \otimes U)$. We can prove (see 5.23.18 of [4], vol. II)

Theorem 3.2. $\mathfrak{C}$, *together with* $\otimes_T$, *with the object* $\boldsymbol{K}$ *of* $\mathfrak{C}$ *and with certain natural isomorphisms is a symmetric monoidal category.*

If such a symmetric monoidal category is closed (in the sense of [9]) for simplicity we say that $\mathfrak{C}$ is closed. We can prove (see [4], vol. II)

Theorem 3.3. PLING, $\mathrm{PLING}^{\wedge}$, PLING^{δ}, $\mathrm{PLING}^{\#}$, $\mathrm{PLING}^{\delta\wedge}$, $\mathrm{PLING}^{\#\wedge}$, PVEC, $\mathrm{PVEC}^{\wedge}$, PVEC^{δ}, $\mathrm{PVEC}^{\#}$, $\mathrm{PVEC}^{\delta\wedge}$ *and* $\mathrm{PVEC}^{\#\wedge}$ *are closed, but* TLING, $\mathrm{TLING}^{\wedge}$, TVEC *and* $\mathrm{TVEC}^{\wedge}$ *are not closed.*

References

[1] F. BERQUIER, Calcul différentiel entre espaces quasibornologiques, Esquisses Math., no 20, Paris 1973.

[2] S. BJON, Die projektive absolutkonvexe Tensorproduktlimitierung, Ann. Acad. Sci. Fennicae, Ser. AI, Nr. 593 (1975).

[3] S. BJON, Vervollständigung streng ausgeglichener Limesvektorräume I, Math. Scand. 41 (1977), 249–264.

[4] W. GÄHLER, Grundstrukturen der Analysis. I, II, Akademie-Verlag, Berlin/Birkhäuser Verlag, Basel und Stuttgart 1977, 1978.

[5] W. GÄHLER, Categorical aspects in the theory of convergence spaces, Proc. Third Int. Sympos. in Yugoslavia on Top. and its Applications (to appear).

[6] W. GÄHLER, Generalized Marinescu convergence, these Proceedings, 89–103.

[7] H. HOGBE-NLEND, Théorie des Bornologies et Applications, Lecture Notes in Math. 213, Springer-Verlag, Berlin–Heidelberg–New York 1971.

[8] H. JARCHOW, Marinescu-Räume, Commentarii math. Helvet. 44 (1969), 138–163.

[9] S. MAC LANE, Kategorien, Begriffssprache und mathematische Theorie, Springer-Verlag, Berlin–Heidelberg–New York 1972.

Anschrift des Verfassers:

Akademie der Wissenschaften der DDR
Zentralinstitut für Mathematik und Mechanik
108 Berlin
Mohrenstraße 39
DDR

Характеризация некоторых классов псевдотопологических линейных пространств

Марк П. Кац, Ленинград

Characterization of some classes of pseudo-topological linear spaces

Summary

We consider some classes of linear pseudo-topological spaces (l.p.s.) which are important both in theory itself and in its applications. Studying the duality between l.p.s. and locally convex linear topological spaces (l.c.t.s.) $X \mapsto L^{\#}(X, R^1)$ we describe the class of all polynormable admissible spaces (as the class of all subspaces in $L^{\#}(X, R^1)$ where X is a l.c.t.s.) and its sublcasses (for example: the class of all regular admissible countably-polynormable spaces consists of all subspaces in $L^{\#}(X, R^1)$ where X is a quasinormable l.c.t.s.). We define the quasiregular spaces to be the spaces X with $Y^{\#} \leqq X \leqq Y$ for a suitable regular space Y and we show that X is quasiregular iff every compact filter in X is bounded.

We obtain a necessary and sufficient condition for an admissible space to be equipped with the convergence on a system of filters. We apply it to the problem of description of the extension of the class of all l.c.t.s. being minimal and closed under the operations of inductive and projective limits.

Введение

Хотя теория линейных псевдотопологических пространств еще очень молода, из класса всех изучаемых ею объектов уже выделилось некоторое число подклассов, используемых как для приложений, так и для получения содержательных результатов внутри самой теории. Это, в первую очередь, класс пространств отображений с псевдотопологиями сходимости на системах фильтров, классы луковичных, правильных, пригодных и полинормированных пространств. Способы определения новых классов можно разбить на две группы: дескриптивные — когда класс определяется как совокупность всех объектов ранее определенного класса, обладающих заданными свойствами, и конструктивные — когда класс определяется как совокупность результатов применения некоторой конструкции к объектам ранее определенного класса. Конструктивно определенными являются, например, класс пространств со сходимостью на системе фильтров и класс всех пространств, которые можно получить исходя из локально выпуклых топологических линейных пространств (ЛВП) и применяя любое число раз операции перехода к индуктивному и проективному пределу. Примеры дескриптивно определенных классов: луковичные, правильные, пригодные, полинормированные пространства. Естественно возникает задача конструктивной характеризации дескриптивно определенных и дескриптивной характеризации конструктивно определенных классов, которой мы и занимаемся в этой работе.

В § 1 мы даем конструктивную характеризацию класса пригодных полинормированных пространств (как всевозможных подпространств в $L^{\#}(X, R^1)$ где X — ЛВП и двух его подклассов: класса „пространств с борнологией" (как всевозможных подпространств в $L^{\#}(X, R^1)$ где X — бочечное ЛВП) и класса правильных счетно-полинормированных пригодных пространств (как всевозможных подпространств в $L^{\#}(X, R^1)$ где X — квазинормируемое пространство Фреше). Для этого изучается естественная двойственность $Y \mapsto L^{\#}(Y, R^1)$ между классом пригодных полинормированных пространств и классом отделимых ЛВП. В § 2 дается дескриптивная характеризация пространств Y таких, что $X^{\#} \leqq Y \leqq X$ для некоторого правильного пространства X (такие пространства мы называем квазиправильными): пространство квазиправильно тогда и только тогда, когда в нем каждый компактный фильтр ограничен. Приведен пример луковичного пригодного не квазиправильного пространства. В § 3 дана дескриптивная характеризация пространств со сходимостью на системе фильтров. Наконец, в § 4 при помощи некоторой конструкции показывается, что каждое квазиправильное луковичное локально выпуклое пространство изоморфно фактор-пространству луковичного пространства со сходимостью на системе фильтров. Этот результат применяется к задаче описания класса пространств, которые можно получить исходя из ЛВП и применяя любое число раз операции индуктивного и проективного пределов.

В этой работе мы придерживаемся терминологии, принятой в [1] и [2]. Все используемые без пояснений понятия теории псевдотопологий определены в книге [2].

Некоторые определения, обозначения и сокращения:

ЛП — линейное пространство,

ЛПП — линейное псевдотопологическое пространство,

ЛВП — локально выпуклое линейное топологическое пространство (даже если это не оговорено явно, мы рассматриваем только отделимые ЛВП и ЛПП),

$X(x) = \bigcap_{\mathfrak{F}\downarrow_x X} \mathfrak{F}$ (X — ЛПП),

$\Gamma(E)$ — абсолютно выпуклая оболочка множества E в ЛП,

$\Gamma_Y(E)$ — замкнутая в Y^0 абсолютно выпуклая оболочка множества E(Y — ЛПП),

$\Gamma_Y(\mathfrak{F})$ — фильтр, порожденный множествами $\Gamma_Y(A)$, $A \in \mathfrak{F}$ ($\mathfrak{F}$ — фильтр в ЛПП Y),

Y^0 — ЛВП, ассоциированное с ЛПП Y (см. [2]),

$Y^{\#}$ — луковичное ЛПП, ассоциированное с ЛПП Y (см. [2], где вместо слова „луковичное" употребляется слово „уравновешенное"),

$\mathrm{UF}(\mathfrak{F})$ — совокупность всех ультрафильтров, мажорирующих фильтр $\mathfrak{F}$ ($\mathfrak{F}$ — фильтр во множестве X),

$b(X)$ — совокупность всех ограниченных множеств ЛПП X,

$\Gamma b(X)$ — совокупность всех абсолютно выпуклых ограниченных в ЛПП X множеств,

$L_{\mathrm{MB}}(X, Y)$, $L_{\mathrm{FB}}(X, Y)$ — пространство всех линейных непрерывных отображений ЛПП X в ЛПП Y наделенное псевдотопологиями сходимости на системе всех сходящихся и всех ограниченных в X фильтров соответственно,

$L^{\#}(X, Y) = L_{\mathrm{MB}}(X, Y)^{\#} = L_{\mathrm{FB}}(X, Y)^{\#}$,

$Y' = \{f\colon Y \to R^1 \mid f$ линейно и непрерывно$\}$ (Y — ЛПП),
$Y^* = \{f\colon Y \to R^1 \mid f$ линейно$\}$ (Y — ЛП),
$X(E)$ — ЛВП (X — ЛП, $E = \Gamma(E) \subset X$), $\underline{X(E)} = R^1 E$, $X(E)\,(0) = R^1(0)\,E$,
$j_Y\colon Y \to (Y')^*$, $j_Y(y)\,(\varphi) = \varphi(y)\ \forall\, y \in Y\ \forall\, \varphi \in Y'$ (Y — ЛПП)
R^A — ЛВП всех числовых функций на множестве A с топологией произведения,
$R^\infty = R^{\mathbf{N}}$ ($\mathbf{N}$ — совокупность всех натуральных чисел).

Фильтр $\mathfrak{F}$ в ЛПП X называется компактным, если $\Psi \in \mathrm{UF}(\mathfrak{F}) \Rightarrow \exists\, x \in X\colon \Psi{\downarrow_x}X$.

Фильтр $\mathfrak{F}$ в ЛПП X называется ограниченным (в [2] — квазиограниченным) если $R^1(0)\,\mathfrak{F}{\downarrow_0}X$.

ЛПП X называется полинормированным, если $\mathfrak{F}{\downarrow_0}X \Leftrightarrow \exists\, A \in \Gamma b(X)\colon \mathfrak{F} \supset R^1(0)\,A$.

Все индуктивные и проективные пределы мы рассматриваем в категории отделимых ЛПП.

§ 1. Пригодные полинормированные пространства

1.1. Теорема. *Класс пригодных полинормированных ЛПП совпадает с классом подпространств в ЛПП вида $L^\#(X, R^1)$ где X отделимое ЛВП.*

Поскольку $L^\#(X, R^1)$ и любое его подпространство пригодно и полинормировано ($\forall$ ЛВП X), то 1.1 вытекает из следующих двух теорем.

1.2. Теорема. *Если Y пригодное полинормированное ЛПП, то $L^\#(Y, R^1)$ (полное) отделимое ЛВП.*

1.3. Теорема. *Если Y отделимое ЛВП или пригодное полинормированное ЛПП, то отображение $j_Y\colon Y \to L^\#\big(L^\#(Y, R^1), R^1\big)$ изоморфизм Y с подпространством $j_Y(Y)$.*

Мы опускаем простые доказательства этих теорем. Отметим только, что 1.2 проверяется непосредственно, а доказательство 1.3 сводится к применению теоремы о биноляре ([3], гл. II, т-ма 4). Теоремы 1.1—1.3 устанавливают двойственность между классом отделимых ЛВП и классом пригодных полинормированных ЛПП. Укажем необходимые и достаточные условия рефлексивности относительно этой двойственности.

1.4. Определение. Пусть Y — отделимое ЛВП или пригодное полинормированное ЛПП. Скажем, что $Y\#$-рефлексивно, есил $j_Y\colon Y \to L^\#\big(L^\#(Y, R^1), R^1\big)$ изоморфизм.

В силу 1.3 $\#$-рефлексивность Y эквивалентна сюръективности j_Y.

1.5. Теорема. *Пусть Y полинормированное пригодное ЛПП. Следующие условия эквивалентны:*

а) *Y $\#$-рефлексивно;*

б) *каждое ограниченное в Y множество $\sigma(Y, Y')$ относительно компактно.*

Доказательство. Пусть выполнено а) и A — ограниченное в Y и замкнутое в Y^0 абсолютно выпуклое множество. Тогда $j_Y(A)$ есть поляра в $L^\#(Y, R^1)'$ окрестности A^0 нуля в ЛВП $L^{\#}(Y, R^1)$. По теореме Алаоглу-Бурбаки $j_Y(A)$ — $\sigma(L^{\#}(Y, R^1)', L^{\#}(Y, R^1))$-компактно, а так как $j_Y(Y) = L^{\#}(Y, R^1)'$, то отсюда следует компактность A в топологии $\sigma(Y, Y')$.

Пусть выполнено б). Тогда $L^{\#}(Y, R^1)' = \bigcup A^{00} = \bigcup \overline{j_Y(A)} = \bigcup j_Y(A) = A$ — абсолютно випукло, $\sigma(Y, Y')$ замкнуто и ограничено в

$$Y = j_Y(\bigcup A) = j_Y(Y)$$

(черта обозначает здесь замыкание в топологии $\sigma(L^{\#}(Y, R^1)', L^{\#}(Y, R^1))$; мы воспользовались теоремой о биноляре и замкнутностью $j_Y(A)$ как непрерывного образа компакта). Теорема доказана.

1.6. Теорема. *Пусть X отделимое ЛВП. Следующие условия эквивалентны:*

а) *X $\#$-рефлексивно;*

б) *X полурефлексивно и $L^{\#}(X, R^1)' = L_b(X, R^1)'$;*

в) *X полурефлексивно и $L^{\#}(X, R^1)^0 = L_b(X, R^1)$.*

Доказательство. 1. а) $\Rightarrow$ б). Если выполнено а), то $L^{\#}(X, R^1)' = j_X(X) \subset L_b(X, R^1)' \subset L^{\#}(X, R^1)'$ (так как $L^{\#}(X, R^1) \leqq L_b(X, R^1)$). Отсюда $L^{\#}(X, R^1)' = L_b(X, R^1)'$.

2. б) $\Rightarrow$ в). Пусть $\mathfrak{A}_1$ и $\mathfrak{A}_2$ совокупности всех $\sigma(X', X)$ замкнутых абсолютно выпуклых окрестностей в $L^{\#}(X, R^1)^0$ и $L_b(X, R^1)$. В силу б) $\sigma(X', X)$-замкнутость, $\sigma(L^{\#}(X, R^1)^0, L^{\#}(X, R^1)')$-замкнутость и $\sigma(L_b(X, R^1), L_b(X, R^1)')$-замкнутость для абсолютно выпуклых множеств совпадают, поэтому $\mathfrak{A}_1$ и $\mathfrak{A}_2$ — базисы окрестностей нуля в $L^{\#}(X, R^1)^0$ и в $L_b(X, R^1)$. Так как $L^{\#}(X, R^1)^0 \leqq L_b(X, R^1)$ то $\mathfrak{A}_1 \supset \mathfrak{A}_2$. Пусть $U \in \mathfrak{A}_1$. По теореме о биноляре $U = U^{00}$ (поляры берутся в двойственности (X, X')). Поэтому достаточно доказать ограниченность U^0 в X. Но $(V \in X(0), V = \Gamma_X(V)) \Rightarrow V^0$ ограничено в $L^{\#}(X, R^1) \Rightarrow V^0$ ограничено в $L^{\#}(X, R^1)^0 \Rightarrow U$ поглощает $V^0 \Rightarrow V^{00} = V$ поглощает U^0. Таким образом $\mathfrak{A}_1 = \mathfrak{A}_2$ и б) $\Rightarrow$ в) доказано.

Наконец, импликации в) $\Rightarrow$ б) и б) $\Rightarrow$ а) очевидны. Теорема доказана.

1.7. Теорема. *Пусть X рефлексивное ЛВП. Тогда X — $\#$-рефлексивно $\Leftrightarrow$ $L_b(X, R^1)$ борнологично.*

Доказательство. Если X $\#$-рефлексивно, то в силу 1.6 $L_b(X, R^1) = L^{\#}(X, R^1)^0$ и его борнологичность вытекает из леммы.

1.7.1. Лемма. *Пусть Y пригодное полинормированное ЛПП. Тогда Y^0 борнологическое ЛВП.*

Доказательство. Действительно, Y^0 в этом случае является индуктивным пределом борнологических пространств, а, следовательно, ([4], гл. 2, 8.2) Y^0 борнологично.

Пусть $L_b(X, R^1)$ борнологично, $\varphi \in L^{\#}(X, R^1)'$. Тогда φ ограничен на всяком равностепенно непрерывном $=$ ограниченном в $L_b(X, R^1)$ (так как X бочечно) множестве. Так как $L_b(X, R^1)$ борнологично, $\varphi \in L_b(X, R^1)'$. Осталось применить импликацию б) $\Rightarrow$ а) теоремы 1.6.

1.7.2. Следствие. *Для пространств Фреше $\#$-рефлексивность совпадает с рефлексивностью.*

Доказательство. Так как сильное сопряженное к рефлексивному пространству Фреше борнологично ([4], гл. IV, стр. 196, следствие 1), то 1.7.2 вытекает из 1.7.

1.8. Теорема. *Пусть Y отделимое ЛВП или пригодное полинормированное ЛПП. Тогда Y $\#$-рефлексивно $\Leftrightarrow$ $L^{\#}(Y, R^1)$ $\#$-рефлексивно.*

Доказательство очевидно.

Таким образом классы $\#$-рефлексивных ЛВП и $\#$-рефлексивных пригодных полинормированных ЛПП двойственны друг другу совершенно аналогично классам компактных и конечных групп.

1.9. Определение. Полинормированное ЛПП X назовем счетнополинормированным, если $\exists\ \mathfrak{B} \subset \Gamma b(X)$, $\mathfrak{B}$ счетно и $\forall\ B_1 \in \Gamma b(X)\ \exists\ B_2 \in \mathfrak{B}: B_2 \supset B_1$.

Легко видеть, что пригодное ЛПП Y является счетнополинормированным тогда и только тогда, когда $L^{\#}(Y, R^1)$ пространство Фреше.

1.10. Предложение. *Пусть Y пригодное счетнополинормированное ЛПП. Тогда Y $\#$-рефлексивно $\Leftrightarrow$ $L^{\#}(Y, R^1)$ рефлексивное пространство Фреше.*

Доказательство. Это вытекает из 1.7.2, 1.8 и предыдущего замечания.

Следующая конструкция, в отличие от рассмотрения всевозможных подпространств в ЛПП вида $L^{\#}(X, R^1)$ (X — ЛВП) дает, как будет показано ниже, только часть всего класса пригодных полинормированных пространств.

1.11. Определение. Пусть X отделимое ЛВП. ЛПП $X_{\text{БОРН}}$, определенное условиями $\underline{X_{\text{БОРН}}} = \underline{X}$, $\mathfrak{F} \downarrow_0 X_{\text{БОРН}} \Leftrightarrow \exists\ A$ — ограниченное в X множество, $\mathfrak{F} \supset R^1(0)\, A$ назовем ассоциированным с X пространством с борнологией.

Легко видеть, что $X_{\text{БОРН}}$ пригодное полинормированное ЛПП. Если Y подпространство в $X_{\text{БОРН}}$, X_1 подпространство в X такое, что $\underline{Y} = \underline{X_1}$, то $Y = (X_1)_{\text{БОРН}}$, так что класс ЛПП вида $X_{\text{БОРН}}$ замкнут относительно перехода к подпространствам.

1.12. Предложение. *Пусть Y пригодное полинормированное ЛПП. Следующие условия эквивалентны:*

а) *Y пространство с борнологией (ассоциированное с некоторым отделимым ЛВП Z);*

б) $b(Y) = b(Y^0)$;

в) $Y = (Y^0)_{\text{БОРН}}$.

Доказательство. а) $\Rightarrow$ б). Включение $b(Y) \subset b(Y^0)$ выполнено для любого ЛПП Y. Пусть $Y = Z_{\text{БОРН}}$, $A \in b(Y^0)$. Тогда $\varphi(A)$ ограничено в R^1 $\forall\ \varphi \in (Y^0)' = Y'$. Но $Z' \subset Y'$, поскольку $Y \leqq Z$. Следовательно $\varphi \in Z' \Rightarrow \varphi(A)$ ограничено в R^1. Поэтому $A \in b(Z)$. Но по определению $b(Z) = b(Y)$. Таким образом $A \in b(Y^0)$ $\Rightarrow A \in b(Y)$ и $b(Y^0) = b(Y)$.

Импликации б) $\Rightarrow$ в) и в) $\Rightarrow$ а) очевидны.

1.13. Теорема. *Пусть Y пригодное полинормированное ЛПП. Каждое из следующих условий достаточно для того, чтобы Y было пространством с борнологией:*

а) *$Y = L^{\#}(X, R^1)$, X квазибочечное ЛВП;*

б) *$L^{\#}(Y, R^1)$ квазибочечное ЛВП;*

в) *Y счетнополинормированное ЛПП.*

Доказательство. Пусть $Y = L^{\#}(X, R^1)$, X квазибочечное ЛВП. Тогда $b(Y) = b(L_b(X, R^1))$ поэтому Y пространство с борнологией. Если выполнено б) то из теоремы 1.3 следует, что Y изоморфно подпространству в $L^{\#}(X, R^1)$, где $X = L^{\#}(Y, R^1)$ квазибочечное ЛВП. Наконец, очевидно, в) $\Rightarrow$ б). Теорема доказана.

Следующие два условия, очевидно, необходимы для того, чтобы пригодное полинормированное ЛПП Y было пространством с борнологией:

(α) $B \notin b(Y) \Rightarrow \exists\, B_1 \subset B$: $(B_2 \subset B_1$, B_2 бесконечно$) \Rightarrow B_2 \notin b(Y)$.

(β) Если $B = \{b_n \mid n \in \mathbf{N}\}$ и $B \notin b(Y)$, то $\exists\, \{\lambda_n\}_{n=1}^{\infty}$, $\lambda_n > 0$, $\lambda_n \to 0$ и $\{\lambda_n b_n \mid n \in \mathbf{N}\} \notin b(Y)$.

Приведем примеры пригодных полинормированных ЛПП Y не являющихся пространствами с борнологией.

1. $Y = L^{\#}(X, R^1)$, X σ-квазибочечное, но не квазибочечное ЛВП (ЛВП X называется σ-квазибочечным, если каждое счетное множество, ограниченное в $L_b(X, R^1)$ равностепенно непрерывно. Пример требуемого ЛВП X см., например, [5]).

В этом случае Y не удовлетворяет условию (α).

2. Пусть Z метризуемое ЛВП, в котором существует ограниченное множество, не являющееся относительно компактным, $\mathfrak{A}$ — совокупность всех абсолютно выпуклых компактов Z, $\underline{Y} = \underline{Z}$, $\mathfrak{F} \downarrow_0 Y \Leftrightarrow \exists\, A \in \mathfrak{A} : \mathfrak{F} \supset R^1(0)\, A$.

В этом случае (α) выполнено, но (β) не выполнено.

1.14. Задача. *Всякое ли пригодное полинормированное ЛПП, удовлетворяющее условиям* (α) *и* (β), *является пространством с борнологией?*

Для пространств с борнологией имеет место следующее описание, аналогичное теореме 1.1.

1.15. Теорема. *Следующие классы ЛПП совпадают:*

1. *Класс пространств с борнологией (обозначим его $\mathcal{K}_1$);*
2. *Класс всевозможных подпространств в ЛПП вида $L^{\#}(X', R^1)$ где X*
   а) *квазибочечное ЛВП (обозначим этот класс $\mathcal{K}_2$);*
   б) *борнологическое ЛВП (обозначим $\mathcal{K}_3$);*
   в) *бочечное ЛВП (обозначим $\mathcal{K}_4$).*

Доказательство. Во-первых, в силу 1.13 $\mathcal{K}_2 \subset \mathcal{K}_1$. Далее, очевидно, $\mathcal{K}_3 \subset \mathcal{K}_2$ и $\mathcal{K}_4 \subset \mathcal{K}_2$. Так как полное квазибочечное ЛВП бочечно, то из следующей очевидной леммы вытекает теперь равенство $\mathcal{K}_2 = \mathcal{K}_4$.

1.15.1. Лемма. *Пусть Z — ЛВП и $\tilde{Z}$ его пополнение. Тогда $L^{\#}(Z, R^1) = L^{\#}(\tilde{Z}, R^1)$.*

Итак, $\mathcal{K}_3 \subset \mathcal{K}_2 = \mathcal{K}_4 \subset \mathcal{K}_1$. Осталось доказать, что $\mathcal{K}_1 \subset \mathcal{K}_3$. Пусть Y пространство с борнологией. В силу 1.12 $Y = (Y^0)_{\text{БОРН}}$. Пусть $X = L^{\#}(Y^0, R^1)^0$. Из леммы 1.7.1 следует, что X и Y^0 борнологические пространства. Покажем, что отображение j_Y осуществляет изоморфизм Y с подпространством в $L^{\#}(X, R^1)$.

Для этого достаточно проверить, что $b(Y) = \{A \subset Y \mid \exists\, U \in X(0) : A \subset U^0\}$. Так как X борнологично, то $\{A \subset Y \mid \exists\, U \in X(0) : A \subset U^0\} = b(Z)$, где Z — ЛВП, получающееся при наделении $\underline{Y}$ топологией равномерной сходимости на ограниченных в X множествах.

Имеем $b(L^{\#}(Y^0, R^1)) \subset b(X) \subset b(L_b(Y^0, R^1))$. Так как ЛВП Y^0 борнологично, то $b(L^{\#}(Y^0, R^1)) = b(L_b(Y^0, R^1))$, а, значит, $b(L^{\#}(Y^0, R^1)) = b(X)$. Поэтому $Z = Y^0$ и $b(Z) = b(Y^0) = b(Y)$ в силу 1.12.

Теорема доказана.

1.16. Замечание. Предыдущая теорема наводит на мысль, что $L^{\#}(Y, R^1)$ квазибочечно для всякого пространства с борнологией Y. Однако, это не так. Пусть Z такое ЛВП, что $L_b(Z, R^1)$ не квазибочечно (такие ЛВП существуют: см. [4], гл. IV, упр. 13). Положим $Y = Z_{\text{БОРН}}$. Легко видеть, что $L^{\#}(Y, R^1) = L_b(Z, R^1)$ не квазибочечно.

В заключение параграфа рассмотрим вопрос о правильности полинормированных пространств.

1.17. Теорема. *Пусть полинормированное ЛПП X удовлетворяет следующему условию*

$$(*)\quad \forall B_1 \in \Gamma b(X)\ \exists B_2 \in \Gamma b(X),\quad B_2 \supset B_1 \colon \forall B_3 \in \Gamma b(X)\ \forall a > 0$$
$$\exists \varepsilon > 0 \colon B_1 \cap \varepsilon B_3 \subset a B_2$$

Тогда ЛПП X правильно.

Доказательство. Легко видеть, что условие $(*)$ для полинормированных X эквивалентно следующему условию

$$(**)\quad \forall B_1 \in \Gamma b(X)\ \exists B_2 \in \Gamma b(X)\ B_2 \supset B_1 \colon$$
$$(\mathfrak{F} \downarrow_0 X, \mathfrak{F} \ni B_1) \Leftrightarrow (\mathfrak{F} \supset R^1(0)\, B_2, \mathfrak{F} \ni B_1)$$

Поэтому утверждение теоремы вытекает из следующей общей леммы.

1.17.1. Лемма. *Пусть ЛПП X и система множеств $\mathfrak{E}$ в X таковы, что*

1. $\forall E_1 \in \mathfrak{E}\ \forall E_2 \in \mathfrak{E}\ \exists E_3 \in \mathfrak{E} \colon E_3 \supset E_1 \cup E_2$
2. $\mathfrak{F} \downarrow_0 X \Rightarrow \mathfrak{F} \cap \mathfrak{E} \neq \emptyset$
3. $\forall E \in \mathfrak{E}\ \exists \mathfrak{F}_E \downarrow_0 X \colon (\mathfrak{F} \downarrow_0 X, \mathfrak{F} \ni E) \Leftrightarrow (\mathfrak{F} \supset \mathfrak{F}_E, \mathfrak{F} \ni E)$.

Тогда ЛПП X правильно.

Доказательство леммы. Пусть $\Psi \in \mathrm{UF}(\mathfrak{F}) \Rightarrow \Psi \downarrow_0 X$. Тогда $\mathfrak{F} \cap \mathfrak{E} \neq \emptyset$. Действительно, если $\mathfrak{F} \cap \mathfrak{E} = \emptyset$ то существует фильтр $\mathfrak{F}_1 \supset \mathfrak{F} \cup \{X \setminus E \mid E \in \mathfrak{E}\}$ и $\forall \Psi_1 \in \mathrm{UF}(\mathfrak{F}_1) \subset \mathrm{UF}(\mathfrak{F})$ имеем $\Psi_1 \cap \mathfrak{E} = \emptyset$ а, значит, $\Psi_1 \not\downarrow_0 X$. Пусть $E \in \mathfrak{F} \cap \mathfrak{E}$. В силу условия 3, $\Psi \in \mathrm{UF}(\mathfrak{F}) \Rightarrow \Psi \supset \mathfrak{F}_E$. Имеем $\mathfrak{F} = \bigcap_{\Psi \in \mathrm{UF}(\mathfrak{F})} \Psi \supset \mathfrak{F}_E \downarrow_0 X$.

Лемма, а тем самым и теорема 1.17. доказаны.

1.18. Теорема. *Пусть X счетнополинормированное ЛПП. Тогда X правильно $\Leftrightarrow$ X удовлетворяет условию* $(*)$.

Доказательство. В силу 1.17 достаточно показать, что всякое правильное счетнополинормированное ЛПП удовлетворяет условию $(*)$.

Пусть X не удовлетворяет (*). По индукции легко построить такую последовательность $\{B_n\} \subset \Gamma b(X)$ что

1. $B \in \Gamma b(X) \Rightarrow \exists\, n \in \mathbf{N}: B \subset B_n$;
2. $\forall\, n \in \mathbf{N}\ \exists\, a_n > 0\ \forall\, \varepsilon > 0\ \ B_1 \cap \varepsilon B_{n+1} \not\subset a_n B_n$;
3. $B_{n+1} \supset B_n \quad \forall\, n \in \mathbf{N}$.

Пусть $R_+^\infty = \{\alpha: \mathbf{N} \to R^1 \mid \alpha(n) > 0\ \forall\, n \in \mathbf{N}\}$. Определим $\beta: R_+^\infty \to \mathbf{N}$ и $\alpha_0 \in R_+^\infty$ равенствами $\beta(\alpha) = \left[\frac{1}{\alpha(1)}\right] + 1$ ([.] — целая часть), $\alpha_0(n) = n^{-1}a_n\ \forall\, n \in \mathbf{N}$. Для $\alpha \in R_+^\infty$ положим $E_\alpha = \bigcup_{i=1}^{\beta(\alpha)} \alpha(i)\, B_i$. Тогда $\forall\, \alpha_1, \ldots, \alpha_k \in R_+^\infty\ \forall\, n \in \mathbf{N}\ \bigcap_{j=1}^{k} E_{\alpha_j} \not\subset \alpha_0(n)\, B_n$. Действительно, можно считать, что $\exists\, 1 < j_0 < k: 1 \leqq j \leqq j_0 \Rightarrow \alpha_j(1) \leqq \frac{1}{n}$, $k_0 < j \leqq k \Rightarrow \alpha_j(1) > \frac{1}{n}$. Тогда $\forall\ 1 \leqq j \leqq j_0\ \beta(\alpha_j) \geqq n + 1$. Имеем $\bigcap_{j=1}^{k} E_{\alpha_j} \supset \frac{1}{n} B_1 \cap \varepsilon\, B_{n+1}$ дла некоторого $\varepsilon > 0$. Применяя 2., получаем

$$\alpha_0(n)\, B_n = n^{-1}a_n B_n \not\supset n^{-1}(B_1 \cap n\varepsilon B_{n+1}) = \frac{1}{n} B_1 \cap \varepsilon B_{n+1} \subset \bigcap_{j=1}^{k} E_{\alpha_j}.$$

Пусть $\mathfrak{F}$ фильтр, базисом которого является множество $\left\{\bigcap_{j=1}^{k} E_{\alpha_j} \mid \alpha_1, .., \alpha_k \in R_+^\infty\right\}$. В силу доказанно выше соотношения $\mathfrak{F} \not\downarrow_0 X$. Пусть $\Psi \in \mathrm{UF}(\mathfrak{F})$. Если $\Psi \not\downarrow_0 X$ то $\exists\, \alpha_1 \in R_+^\infty: \alpha_1(n)\, B_n \notin \Psi\ \forall\, n \in \mathbf{N}$. Тогда $\forall\ N \in \mathbf{N}\ \bigcup_{n=1}^{N} \alpha_1(n)\, B_n \notin \Psi$. С другой стороны $\Psi \ni E_{\alpha_1} = \bigcup_{n=1}^{\beta(\alpha_1)} \alpha_1(n)\, B_n$. Полученное противоречие показывает, что $\Psi \in \mathrm{UF}(\mathfrak{F}) \Rightarrow \Psi \downarrow_0 X$.

Теорема доказана.

1.19. Следствие. *Пусть ЛВП X удовлетворяет условию*

$$(***)\quad \forall\ U \in X(0)\ \exists\ V \in X(0)\ V \subset U: \forall\ W \in X(0)\ \forall\ \varepsilon > 0\ \exists\ K > 0\quad V \subset \varepsilon\, U + KW.$$

Тогда $L^\#(X, R^1)$ правильно.

1.20. Следствие. *Пусть X метризуемое ЛВП. Тогда $L^\#(X, R^1)$ правильно $\Leftrightarrow X$ удовлетворяет условию* (***).

Так как X удовлетворяет (***) $\Leftrightarrow L^\#(X, R^1)$ удовлетворяет (*), то 1.19 следует из 1.17, а 1.20 — из 1.18.

Легко видеть, что каждое квазинормируемое ЛВП удовлетворяет условию (***). Следующий пример показывает, что обратное неверно.

1.21. Пример не квазинормируемого ЛВП X, удовлетворяющего условию (***). Пусть

$$\underline{X} = \left\{x \in R^\infty \,\middle|\, \sum_{n=1}^{\infty} 2^n\, |x(n)| < \infty\right\},$$

$$\underline{Y} = \left\{y \in R^\infty \,\middle|\, \sup_n 2^{-n}\, |y(n)| < \infty\right\},$$

$$(x, y) = \sum_{n=1}^{\infty} x(n)\, y(n)$$

$\forall\, x \in \underline{X}$, $\forall\, y \in \underline{Y}$ (тем самым определена дуальная пара $(\underline{X}, \underline{Y})$). Фиксируем алгебраическое прямое дополнение Y_1 к $\bigcup_{p \geqq 1} l_p$ до $\underline{Y}$.

Положим

$$B_k = \left\{ y \in \underline{Y} \,\middle|\, \sum_{n=1}^{\infty} |y(n)|^k \leqq 1 \right\} \quad (\forall\, k \in \mathbf{N}),$$

$$B_{a,k,y_1,\ldots,y_m} = aB_k + \sum_{i=1}^{m} [-1, 1]\, y_i \quad (\forall\, a > 0\, \forall\, k \in \mathbf{N}, \forall\, y_1, \ldots, y_m \in Y_1),$$

$$\mathfrak{A} = \{B_{a,k,y_1,\ldots,y_m} \mid k > 1, a > 0, \quad y_1, \ldots, y_m \in Y_1\}.$$

Тогда $\mathfrak{A}$ — совокупность абсолютно выпуклых $\sigma(Y, X)$ ограниченных множеств, удовлетворяющая условиям B1—B3 п. 2 гл. III книги [3]. Пусть X — ЛВП, получающееся при наделении $\underline{X}$ топологией $\mathfrak{A}$ — сходимости, а Y — при наделении $\underline{Y}$ топологией, определяемой нормой $\|y\| = \sup_{n \in \mathbf{N}} 2^{-n} |y(n)|$. Так как тождественное вложение $i_k : l_k \to Y$ непрерывно, а, значит, и непрерывно при наделении l_k топологией $\sigma(l_k, l_k')$ а Y топологией $\sigma(Y, X)$ более слабой, чем $\sigma(Y, Y')$. Поскольку при $k > 1$ B_k — $\sigma(l_k, l_k')$-компактно, имеем $B_k = i_k(B_k)$ — $\sigma(Y, X)$-компактно. Отсюда следует, что $A \in \mathfrak{A} \Rightarrow A$ $\sigma(Y, X)$-компактно. По теореме Макки-Аренса (т-ма 7 гл. III [3]) топология в X согласуется с двойственностью (X, Y)). Пусть Z нормируемое пространство, получающееся при наделении $\underline{X}$ топологией, определяемой нормой $\|x_1\| = \sum_{n=1}^{\infty} 2^n |x(n)|$. Легко видеть, что топология в Z согласуется с той же двойственностью, а топология в Y есть $\beta(Y, Z) = \beta(Y, X)$.

Допустим, что X квазинормируемо. Тогда $\forall\, A \in \mathfrak{A}\, \exists\, \tilde{A} \in \mathfrak{A}\, \forall\, \varepsilon > 0\, \exists\, \delta > 0$: $(y \in A, \|y\| < \delta) \Rightarrow y \in \varepsilon \tilde{A}$. В частности это выполнено и для $A = B_2$. Пусть $\tilde{B}_2 = B_{a,k,y_1,\ldots,y_m}$. Тогда $\forall\, \varepsilon > 0\, \exists\, \delta > 0$: $(y \in B_2, \|y\| < \delta) \Rightarrow y \in \left(\varepsilon a B_k + \sum_{i=1}^{m} [-\varepsilon, \varepsilon]\, y_i\right) \cap l_2 \subset \varepsilon a B_k$ (так как $Y_1 \cap \left(\bigcup_{p \geqq 1} l_p\right) = \{0\}$). Но пусть $e_n \in Y$ $e_n(m) = \begin{cases} 0, & n \neq m \\ 1, & n = m \end{cases}$ и n_0 столь велико, что $2^{-n_0} < \delta$. Тогда $\|e_{n_0}\| = 2^{-n_0} < \delta$, $e_{n_0} \in B_2$ но $e_{n_0} \notin \frac{1}{2} B_k\, \forall\, k \in \mathbf{N}$. Полученное противоречие показывает, что X не квазинормируемо.

Так как выполнение для X условия (***) эквивалентно выполнению для $L^{\#}(X, R^1)$ условия (*) а

$$\Gamma b\big(L^{\#}(X, R^1)\big) = \{B \subset \underline{Y} \mid B \text{ — абсолютно выпукло и } \exists\, A \in \mathfrak{A} : B \subset A\}$$

то осталось доказать, что $\forall\, A_1 \in \mathfrak{A}\, \exists\, A_2 \in \mathfrak{A}\, \forall\, A_3 \in \mathfrak{A}\, \forall\, \varepsilon > 0\, \exists\, \delta > 0$: $A_1 \cap \delta A_3 \subset \varepsilon A_2$.

Пусть $\sum_{n=1}^{\infty} |y(n)|^k = a < \infty$, $\sum_{n=1}^{\infty} |y(n)|^{k+2} = b < \infty$. Тогда, по неравенству Коши, $\sum_{n=1}^{\infty} |y(n)|^{k+1} < \sqrt{ab}$. Таким образом $b < 1 \Rightarrow B_k \cap bB_{k+2} \subset b^{\frac{k+2}{2k+1}} B_{k+1} \subset \sqrt{b}\, B_{k+1}$ а при $r \geqq k + 2, b < 1$ имеем: $B_k \cap bB_r \subset B_k \cap (B_{r-2} \cap bB_r) \subset B_k \cap \sqrt{b}\, B_{r-1} \subset \cdots \subset B_k \cap b^{2^{k+1-r}} B_{k+1}$.

Пусть $A_1 = B_{a,k,y_1,\ldots,y_m}$. Положим $A_2 = B_{a,k+1,y_1,\ldots,y_m}$. Пусть $A_3 = B_{a_1,r,z_1,\ldots,z_p}$. Можно считать, что $a = a_1 = 1$, $r > k + 1$. Имеем

$$A_1 \cap \delta A_3 = B_k \cap \delta B_r + \left(\sum_{i=1}^{p} [-\delta, \delta]\, Z_i\right) \cap \left(\sum_{i=1}^{m} [-1, 1]\, y_i\right) \; (\forall\, \delta > 0).$$

Так как $K = \sum_{i=1}^{p} [-1, 1] Z_i \cap \sum_{i=1}^{m} R^1 y_i$ компакт в конечномерном пространстве $\sum_{i=1}^{m} R^1 y_i$, а $\sum_{i=1}^{m} [-1, 1] y_i$ окрестность нуля в этом пространстве, то $\forall \varepsilon > 0 \exists \delta_1 > 0: \delta_1 K \subset \varepsilon \sum_{i=1}^{m} [-1, 1] y_i$.

Имеем: $(\delta < \delta_1, \delta < 1, \delta < \varepsilon^{2^{r-k+1}}) \Rightarrow A_1 \cap \delta A_3 \subset B_k \cap \delta B_r + \varepsilon \sum_{i=1}^{m} [-1, 1] y_i \subset \varepsilon B_{k+1} + \varepsilon \sum_{i=1}^{m} [-1, 1] y_i = \varepsilon A_2$.

Проверка выполнения для X условия (***) завершена.

1.22. Определение. Пригодное полинормированное ЛПП Y назовем дуально-квазинормируемым, если $\forall A \in \Gamma b(Y) \exists B \in \Gamma b(Y)$ на A совпадают топологии, индуцированные из Y^0 и из $\underline{Y}(B)$.

Следующее предложение объясняет термин ,,дуально-квазинормируемое ЛПП''.

1.23. Теорема. а) *Если X квазинормируемое ЛВП, то $L^{\#}(X, R^1)$ дуально-квазинормируемо.*

б) *Если Y дуально-квазинормируемое ЛВП, то $L^{\#}(Y, R^1)$ квазинормируемо.*

Доказательство. Поскольку квазинормируемость ЛВП X эквивалентна тому, что $\forall U \in X(0) \exists V \in X(0) \forall \varepsilon > 0 \exists W_\varepsilon \in L_b(X, R^1)(0)$: $W_\varepsilon \cap U^0 \subset \varepsilon V^0$, а то, что $L^{\#}(X, R^1)$ дуально-квазинормируемо, эквивалентно условию: $\forall U \in X(0) \exists V \in X(0) \forall \varepsilon > 0 \exists W_\varepsilon \in (L^{\#}(X, R^1)^0)(0)$: $W_\varepsilon \cap U^0 \subset \varepsilon V^0$. Утверждение а) вытекает из неравенства $L^{\#}(X, R^1)^0 \leqq L_b(X, R^1)$.

Пусть Y дуально-квазинормируемое ЛПП. Так как Y пригодно, то $A \in \Gamma b(Y) \Rightarrow \Gamma_{Y^0}(A) \in \Gamma b(Y)$. Поэтому $\forall A \in \Gamma b(Y) \exists B \in \Gamma b(Y): B = \Gamma_{Y^0}(B)$, $\forall \varepsilon > 0 \exists E_\varepsilon \in Y^0(0)$, $E_\varepsilon = \Gamma_{Y^0}(E_\varepsilon)$: $\Gamma_{Y^0}(A) \cap E_\varepsilon \subset \varepsilon B$.

Переходя к полярам (поскольку $\Gamma_{Y^0}(A)$ и E_ε абсолютно выпуклы и $\sigma(Y, Y'$-замкнуты) получаем $\frac{1}{\varepsilon} B^0 \subset A^0 + E_\varepsilon{}^0$, или $B^0 \subset \varepsilon A^0 + \varepsilon E_\varepsilon{}^0$. Поскольку множества вида C^0, $C \in \Gamma b(Y)$ образуют базис окрестностей нуля в $L^{\#}(Y, R^1)$ то $E_\varepsilon{}^0$ ограничено в $L^{\#}(Y, R^1)$, а следовательно $L^{\#}(Y, R^1)$ квазинормируемо.

1.24. Теорема. *Класс дуально-квазинормируемых ЛПП совпадает с классом всевозможных подпространств в ЛПП вида $L^{\#}(X, R^1)$ где X — квазинормируемое ЛВП.*

Доказательство. Это немедленное следствие 1.23, 1.3 и того простого факта, что всякое подпространство дуально-квазинормируемого ЛПП является дуально-квазинормируемым.

1.25. Замечание. Можно показать, что для ЛВП X построенного в примере 1.21 ЛПП $L^{\#}(X, R^1)$ является дуально-квазинормируемым. Отсюда следует, что $L^{\#}(X, R^1)$ дуально-квазинормируемо $\nRightarrow X$ квазинормируемо.

Дадим теперь окончательное описание правильных счетнополинормированных пригодных ЛПП.

1.26. Теорема. *Класс правильных счетнополинормированных пригодных ЛПП совпадает с классами всевозможных подпространств в ЛПП вида $L^{\#}(X, R^1)$ где X квазинормируемое пространство Фреше.*

Это утверждение вытекает из I.18, I.3, I.23 и следующей теоремы.

1.27. Теорема. *Пусть Y счетнополинормированное пригодное ЛПП. Выполнение для Y условия $(*)$ эквивалентно тому, что Y дуально-квазинормируемо.*

Доказательство. Каждое дуально-квазинормируемое ЛПП удовлетворяет, очевидно, условию $(**)$, эквивалентному условию $(*)$.

Пусть Y счетнополинормированное пригодное ЛПП, удовлетворяющее условию $(*)$, $A \in \Gamma b(Y)$. По индукции легко построить последовательность $\{B_n\} \subset \Gamma b(Y)$ такую, что:

1. $B_0 = A$, $B_{n+1} \supset B_n \ \forall\, n \in \mathbf{N} \cup \{0\}$;
2. $\forall\, B \in b(Y) \ \exists\, n \in \mathbf{N}\colon B_n \supset B$;
3. $\forall\, n \in \mathbf{N} \ \forall\, \varepsilon > 0 \ \exists\, \delta > 0\colon B_{n+1} \cap \delta B_{n-1} \subset \varepsilon B_n$.

Положим для $n \in \mathbf{N}$, $\varepsilon > 0$:

$$a_n(\varepsilon) = \frac{1}{2} \sup \{\delta > 0 \mid B_{n-1} \cap \delta B_{n+1} \subset \varepsilon B_n\}.$$

В силу свойства 3. $a_n(\varepsilon) > 0 \ \forall\, n \in \mathbf{N} \ \forall\, \varepsilon > 0$.

Пусть выбрано $\varepsilon > 0$. Построим по индукции две последовательности $\{\lambda_n\}$ и $\{\mu_n\}$ положительных чисел. Положим $\mu_0 = \lambda_0 = \varepsilon$. Пусть определены $\mu_0, \ldots, \mu_n$, $\lambda_0, \ldots, \lambda_n$. Выберем

$$\mu_{n+1} = \begin{cases} a_n\left(\frac{1}{4}\lambda_n\right), & \text{если} \quad n \geqq 1, \\ \varepsilon \qquad\qquad , & \text{если} \quad n = 0, \end{cases}$$

а $\lambda_{n+1} > 0$ выберем столь малым, чтобы:

а) $\sum\limits_{i=1}^{n+1} \lambda_n < 1$;

б) $\lambda_{n+1} < \mu_{n+1}$;

в) $\lambda_{n+1} < a_{n+1}\left(\frac{1}{4}\mu_{n+1}\right)$.

Покажем, что $\forall\, n \in \mathbf{N}$ $B_0 \cap \left(\sum\limits_{i=1}^{n} \lambda_i B_i + \mu_{n+1} B_{n+1}\right) \subset B_0 \cap \left(\sum\limits_{i=1}^{n-1} \lambda_i B_i + \mu_n B_n\right)$.

Действительно, $B_0 \cap \left(\sum\limits_{i=1}^{n} \lambda_i B_i + \mu_{n+1} B_{n+1}\right) \subset B_0 \cap \left(\sum\limits_{i=1}^{n-1} \lambda_i B_i + (\lambda_n + \mu_{n+1})\, B_{n+1}\right)$. Но $\sum\limits_{i=1}^{n-1} \lambda_i B_i \subset \left(\sum\limits_{i=1}^{n-1} \lambda_i\right) B_{n+1} \subset B_{n-1}$ поскольку $\sum\limits_{i=1}^{k} \lambda_i < 1 \quad \forall\, k \in \mathbf{N}$. Поэтому

$$B_0 \cap \left(\sum_{i=1}^{n} \lambda_i B_i + \mu_{n+1} B_{n+1}\right) \subset B_0 \cap \left[\left(\sum_{i=1}^{n-1} \lambda_i B_i\right) + 2B_{n-1} \cap (\lambda_n + \mu_{n+1})\, B_{n+1}\right]$$

$$= B_0 \cap \left[\sum_{i=1}^{n-1} \lambda_i B_i + 2\left(B_{n-1} \cap \frac{1}{2}\lambda_n B_{n+1}\right) + 2\left(B_{n-1} \cap \frac{1}{2}\mu_{n+1} B_{n+1}\right)\right]$$

$$\subset B_0 \cap \left[\sum_{i=1}^{n-1} \lambda_i B_i + 2 \cdot \frac{1}{4}\mu_n B_n + 2 \cdot \frac{1}{4}\lambda_n B_n\right] \subset B_0 \cap \left(\sum_{i=1}^{n-1} \lambda_i B_i + \mu_n B_n\right).$$

Применяя полученное соотношение n раз, получим: $B_0 \cap \sum_{i=1}^{n+1} \lambda_i B_i \subset B_0 \cap \left(\sum_{i=1}^{n} \lambda_i B_i + \mu_{n+1} B_{n+1}\right) \subset B_0 \cap \mu_1 B_1 \subset \varepsilon B_1$.

Пусть $W_\varepsilon = \bigcup_{n=1}^{\infty} \left(\sum_{i=1}^{n} \lambda_i B_i\right)$. Тогда $W_\varepsilon \cap B_0 \subset \varepsilon B_1$, а из свойства 2. и абсолютной выпуклости W_ε следует, что $W_\varepsilon \in Y^0(0)$. В силу произвольности $B_0 = A \in \Gamma b(Y)$ и $\varepsilon > 0$ дуальная квазинормируемость Y доказана.

1.28. Замечание. Существует правильное пригодное полинормируемое ЛПП, не удовлетворяющее условию (*) (так что требование счетности в 1.18 существенно).

Пусть E — несчетное множество, $\underline{Y} = \left\{f\colon E \to R^1 \,\middle|\, \sum_{e \in E} |f(e)| < \infty\right\}$. Для каждого счетного множества $A \subset E$ определим норму p_A на $\underline{Y}$ равенством $p_A(y) = \sum_{e \notin A} |y(e)| + \left(\sum_{e \in A} y(e)^2\right)^{1/2}$. Положим $\mathfrak{F} \downarrow_0 Y \Leftrightarrow \exists\, A \subset E$, A счетно, $p_A(\mathfrak{F}) \downarrow_0 R^1$. Тогда Y пригодное полинормированное ЛПП, дающее требуемый пример (мы опускаем несложное доказательство этого факта).

1.29. Задача. *Всякое ли пригодное полинормируемое ЛПП, удовлетворяющее условию (*) дуально-квазинормируемо?*

§ 2. Квазиправильные линейные псевдотопологические пространства

2.1. Определение. ЛПП X называется квазиправильным, если существует такое правильное ЛПП Y, что $Y^\# \leqq X \leqq Y$.

2.2. Теорема. *ЛПП X квазиправильно тогда и только тогда, когда каждый компактный фильтр в X ограничен.*

Доказательство. Пусть $Y^\# \leqq X \leqq Y$, Y правильно, $\mathfrak{F}$ компактный фильтр в X. Тогда $\mathfrak{F}$ компактен и в Y. Так как Y правильно, $R^1(0)\,\mathfrak{F} \downarrow_0 Y$ ([1], т-ма 2.2.5.) Но тогда $R^1(0)\,\mathfrak{F} \downarrow_0 Y^\#$ и $R^1(0)\,\mathfrak{F} \downarrow_0 X$. Таким образом, фильтр $\mathfrak{F}$ ограничен в X.

Пусть каждый компактный фильтр ограничен в X. Положим $\underline{Y} = \underline{X}$, $\mathfrak{F} \downarrow_0 Y \Leftrightarrow (\Psi \in \mathrm{UF}(\mathfrak{F}) \Rightarrow \Psi \downarrow_0 X)$.
Тогда Y правильное ЛПП и $X \leqq Y$. Пусть $\Phi \downarrow_0 Y^\#$ то есть $\Phi \supset R^1(0)\,\mathfrak{F}$, $\mathfrak{F} \downarrow_0 Y$. Тогда фильтр $\mathfrak{F}$ компактен в X. По предположению, отсюда следует, что $R^1(0)\,\mathfrak{F} \downarrow_0 X$. Значит и $\Phi \downarrow_0 X$. Таким образом $Y^\# \leqq X$. Теорема доказана.

2.3. Замечание. Следствие 2.4.6 в [1] показывает, что квазиправильное луковичное пригодное пространство может не быть правильным.

2.4. Теорема. *Существует луковичное пригодное не квазиправильное ЛПП.*

Доказательство. Пусть $\underline{X} = \{x\colon \mathbf{N} \to R^1 \mid \exists\, n \in \mathbf{N}\ x(k) = 0\ \forall\, k \geqq n\}$, $\|x\| = \sum_{i=1}^{\infty} 2^{-i} |x(i)|$ для $x \in \underline{X}$, $K_n = \{x \in \underline{X} \mid x(i) = 0\ \forall\, 1 \leqq i \leqq n\}$ $(n \in \mathbf{N})$, $\varphi\colon \underline{X} \to R^1$, $\varphi(x) = \sum_i x(i)$, $K = \operatorname{Ker} \varphi$. Для $\alpha > 0$ и $n \in \mathbf{N}$ положим $K_{n,\alpha} = \{x \in K \cap K_n \mid \|x\| \leqq \alpha\}$.

Очевидно $K_{n,\alpha} \neq \emptyset \quad \forall \alpha > 0 \,\forall n \in \mathbf{N}$ и $(n_1 > n_2,\ \alpha_1 < \alpha_2) \Rightarrow K_{n_1,\alpha_1} \subset K_{n_2,\alpha_2}$. Поэтому существует фильтр $\mathfrak{K}$ в $\underline{X}$ с базисом $\{K_{n,\alpha} \mid \alpha > 0,\ n \in \mathbf{N}\}$. Положим $\mathfrak{F} \downarrow_0 X \Leftrightarrow \exists\, x_1, \ldots, x_n \in \underline{X}\ \mathfrak{F} \supset \sum_{i=1}^{n} R^1(0)\, x_i + \mathfrak{K}$. Легко видеть, что X луковичное пригодное пространство. Пусть $e_n \in X$, $e_n(i) = \begin{cases} 0, & i \neq n \\ 1, & i = n \end{cases}$, $x_n = e_{2n} - e_{2n-1}$. Представим $\mathbf{N} = \bigcup_{i=1}^{\infty} \mathbf{N}_i$, $i \neq j \Rightarrow \mathbf{N}_i \cap \mathbf{N}_j = \emptyset$, $\operatorname{card} \mathbf{N}_i = \operatorname{card} \mathbf{N} \quad \forall i \in \mathbf{N}$. Пусть для $j \in \mathbf{N}$ $i(j) \in \mathbf{N}$ таково, что $j \in \mathbf{N}_{i(j)}$ (это условие однозначно определяет отображение $\mathbf{N} \to \mathbf{N}$). Положим $y_j = 2^{-j} x_{i(j)} \quad \forall j \in \mathbf{N}$ и пусть Φ фильтр в $\underline{X}$ порожденный множествами $\{y_j \mid j \geqq s\}$ где s пробегает $\mathbf{N}$. Тогда $\mathfrak{F} = R^1(0)\, \Phi \not\downarrow_0 X$. Действительно, допустим, что $\mathfrak{F} \supset \sum_{r=1}^{n} R^1(0)\, z_r + \mathfrak{K}$, $z_r \in X$. Но $\exists\, N \in \mathbf{N}$: $j > N \Rightarrow z_r(j) = 0 \quad \forall\, 1 \leqq r \leqq n$. Поэтому $\mathfrak{F} \supset \sum_{r=1}^{N} R^1 e_r + \mathfrak{K}$ и $\exists\, \varepsilon > 0\ \exists\, F \in \mathfrak{F}$: $[-\varepsilon, \varepsilon] F \subset \sum_{r=1}^{N} R^1 e_r + K_{N+1,1}$ a, значит, $x \in F \Rightarrow x(N+1) = 0$. С другой стороны, пусть $p = \dfrac{N+1}{2}$ если N нечетно, и $p = \dfrac{N}{2} + 1$ если N четно. Так как $\operatorname{card} \mathbf{N}_p = \operatorname{card} \mathbf{N}$ имеем: $\forall\, j \in \mathbf{N}\ \exists\, j' > j$: $i(j') = p$, $y_{j'}(N+1) = 2^{-j'}(e_{2p} - e_{2p-1})\,(N+1) \neq 0$. Поэтому $\exists\, x \in F\ x(N+1) \neq 0$. Полученное противоречие показывает, что $R^1(0)\, \Phi \not\downarrow_0 X$. Покажем теперь, что фильтр Φ компактен. Пусть $\Psi \in \mathrm{UF}(\mathfrak{F})$. Допустим, что $\Psi \not\downarrow_0 X$. Так как $\|\Phi\| \downarrow_0 R^1$ то $\|\Psi\| \downarrow_0 R^1$. Поэтому $\Psi \not\supseteq \left[\sum_{i=1}^{n} R^1 e_i\right] + \tilde{\mathfrak{K}}$ $(\forall\, n \in \mathbf{N})$ где $\tilde{\mathfrak{K}}$ фильтр в X порожденный множествами $K_{m,1}$, $m \in \mathbf{N}$. Следовательно $\forall\, n \in \mathbf{N}\ \exists\, \bar{n} > n$: $\Psi \not\ni \sum_{i=1}^{n} R^1 e_i + K_{\bar{n},1}$. Так как $K \in \Psi$ то $\Psi \not\ni \{x \in \underline{X} \mid x(i) = 0\ \forall\, n < i \leqq \bar{n}\} \quad \forall\, n \in \mathbf{N}$. Пусть $A = \{x \in \underline{X} \mid \operatorname{card} \{n \in \mathbf{N} \mid x(n) \neq 0\} = 2\}$, $B_n = \{x \in \underline{X} \mid \exists\, i\colon n < i \leqq \bar{n},\ x(i) \neq 0\}$. Тогда $A \in \Phi \subset \Psi$ a так как Ψ ультрафильтр, $\Psi \not\ni X \setminus B_n$, то $\Psi \in B_n \quad \forall\, n \in \mathbf{N}$. Но $A \cap B_1 \cap B_{\bar{1}} \cap B_{(\bar{1})^-} = \emptyset$. Полученное противоречие показывает, что $\Psi \downarrow_0 X$. Теорема доказана.

2.5. Предложение. а) *Проективный предел квазиправильных ЛПП квазиправилен.* б) *Каждое пригодное полинормированное ЛПП квазиправильно.*

Доказательство. Утверждение а) вытекает из 2.2, утверждение б) из 1.1 и квазиправильности подпространств квазиправильного пространства.

В заключение этого параграфа приведем вариант теоремы 2.2, относящийся к пригодным пространствам.

2.6. Теорема. *Пусть X пригодное квазиправильное ЛПП. Тогда существует такое правильное пригодное ЛПП Y что $Y^{\#} \leqq X \leqq Y$.*

Доказательство. Определим ЛПП Y условиями $\underline{Y} = \underline{X}$, $\mathfrak{F} \downarrow_0 Y \Leftrightarrow R^1(0)\, \mathfrak{F} \downarrow_0 X$ и $\mathfrak{F} \downarrow_0 X^0$. Очевидно, что Y пригодно. Пусть $\Psi \in \mathrm{UF}(\mathfrak{F}) \Rightarrow \Psi \downarrow_0 Y$. Имеем $\Psi \in \mathrm{UF}\left(R^1(0)\, \mathfrak{F}\right) \Rightarrow \exists\, \Psi_1 \in \mathrm{UF}(\mathfrak{F})$: $\Psi \supset R^1(0)\, \Psi_1 \downarrow_0 X$. Таким образом фильтр $R^1(0)\, \mathfrak{F}$ компактен в X. По теореме 2.2 отсюда следует, что $R^1(0)\, \mathfrak{F} = R^1(0) \cdot R^1(0)\, \mathfrak{F} \downarrow_0 X$. Так как X^0 ЛВП, а потому правильно, $\mathfrak{F} \downarrow_0 X^0$ и, значит, $\mathfrak{F} \downarrow_0 Y$. Ясно, что $X \leqq Y$ а из определения Y немедленно следует: $Y^{\#} \leqq X$. Теорем доказана.

§ 3. Псевдотопологии сходимости на системе фильтров

3.1. Определение. Пусть (X, Y) дуальная пара линейных пространств, $\mathscr{A}$ — множество фильтров в X такое, что

$$(\mathscr{A}1) \quad \forall\, x \in X \ \exists\, \mathfrak{F} \in \mathscr{A} : [x] \supset \mathfrak{F};$$

$$(\mathscr{A}2) \quad R^1(0)\,\varphi(\Phi) \downarrow_0 R^1 \ \forall\, \Phi \in \mathscr{A} \ \forall\, \varphi \in Y.$$

Псевдотопология $\tau_{\mathscr{A}}$ в Y, определенная условиями $\mathfrak{F} \downarrow_0 \tau_{\mathscr{A}} \Leftrightarrow (\Phi, \mathfrak{F}) \downarrow_0 R^1 \ \forall\, \Phi \in \mathscr{A}$, $\mathfrak{F} \downarrow_y Y \Leftrightarrow \mathfrak{F} - y \downarrow_0 Y$ называется псевдотопологией $\mathscr{A}$-сходимости в Y, ЛПП, получающееся при наделении Y псевдотопологией $\tau_{\mathscr{A}}$ будем обозначать $Y_{\mathscr{A}}$.

3.2. Определение. ЛПП Z будем называть пространством со сходимостью на системе фильтров, если $Z \approx Y_{\mathscr{A}}$ для некоторой дуальной пары (X, Y) и множества $\mathscr{A}$ фильтров в X удовлетворяющего $(\mathscr{A}1)$ и $(\mathscr{A}2)$.

Пусть X пригодное ЛПП, $\mathfrak{B}$ множество всех ограниченных, а C_0 всех сходящихся (не обязательно к нулю) фильтров в X. Рассмотрим дуальную пару (X, X'). Очевидно, что $X_{\mathfrak{B}}' = L_{\mathrm{FB}}(X, R^1)$, $X_{C}' = L_{\mathrm{MB}}(X, R^1)$.

Отметим, что класс пространств со сходимостью на системе фильтров замкнут относительно перехода к подпространствам. Ниже мы покажем, что этот класс исчерпывается подпространствами ЛПП вида $L_{MB}(X, R^1)$ где X — пригодное ЛПП.

3.3. Теорема. *Следующие условия на ЛПП Z эквивалентны.*

а) *Z пространство со сходимостью на системе фильтров.*

б) *Z пригодно и выполнено следующее условие:*

$(\Gamma *)$ Если фильтр $\mathfrak{F} \not\downarrow_0 Z$ то можно так выбрать по множеству $A_\Psi \in \Psi$ из каждого сходящегося к нулю в Z фильтра Ψ, что $\Gamma_X \left(\bigcup_{j=1}^{s} A_{\Psi_j} \right) \notin \mathfrak{F} \ \forall\ \Psi_1 \downarrow_0 Z, \ldots, \Psi_s \downarrow_0 Z$.

в) *Z изоморфно подпространству в $L_{\mathrm{MB}}(L_{\mathrm{MB}}(Z, R^1), R^1)$.*

Доказательство. а) $\Rightarrow$ б). Пусть (X, Y) дуальная пара, $\mathscr{A}$ — множество фильтров в X, удовлетворяющее условиям $(\mathscr{A}1)$ и $(\mathscr{A}2)$, $Z = Y_{\mathscr{A}}$. Пригодность $Y_{\mathscr{A}}$ очевидна. Пусть $\mathfrak{F} \not\downarrow_0 Z$. Тогда $\exists\, \Phi \in \mathscr{A} : (\Phi, \mathfrak{F}) \not\downarrow_0 R^1$, то есть $(\Phi, \mathfrak{F}) \not\ni [-\alpha, \alpha]$ для некоторого $\alpha > 0$. Это значит, что $\mathfrak{F} \not\ni \alpha B^0 \ \forall\, B \in \Phi$. Пусть $\Psi \downarrow_0 Z$. Тогда $\exists\, A_\Psi \in \Psi$ $\exists\, B_\Psi \in \Phi : A_\Psi \subset \alpha B_\Psi{}^0$ (поскольку $(\Phi, \Psi) \downarrow_0 R^1$). Если $\Psi_1 \downarrow_0 Z, \ldots, \Psi_n \downarrow_0 Z$ то $\bigcup_{i=1}^{n} A_{\Psi_i} \subset \alpha \left(\bigcap_{i=1}^{n} B_{\Psi_i} \right)^0$, $\Gamma_Z \left(\bigcup_{i=1}^{n} A_{\Psi_i} \right) \subset \Gamma_Z \left(\alpha \left(\bigcap_{i=1}^{n} B_{\Psi_i} \right)^0 \right) = \alpha \left(\bigcap_{i=1}^{n} B_{\Psi_i} \right)^0 \notin \mathfrak{F}$ так как $\bigcap_{i=1}^{n} B_{\Psi_i} \in \Phi$.

б) $\Rightarrow$ в). Пусть $X = Z'$, $\mathscr{A} = \{\Phi \mid \Phi$ фильтр в $X, \mathfrak{F} \downarrow_0 Z \Rightarrow \Phi(\mathfrak{F}) \downarrow_0 R^1\}$. Рассмотрим дуальную пару (с обычной двойственностью) (Z', Z) и покажем, что $Z = \underline{Z}_{\mathscr{A}}$. Очевидно, $\mathfrak{F} \downarrow_0 Z \Rightarrow (\Phi, \mathfrak{F}) = \Phi(\mathfrak{F}) \downarrow_0 R^1 \ \forall\, \Phi \in \mathscr{A}$, то есть $\mathfrak{F} \downarrow_0 Z \Rightarrow \mathfrak{F} \downarrow_0 \underline{Z}_{\mathscr{A}}$. Пусть $\mathfrak{F} \downarrow_0 \underline{Z}_{\mathscr{A}}$. Допустим, что $\mathfrak{F} \not\downarrow_0 Z$. В силу $(\Gamma *)$ можно выбрать $A_\Psi \in \Psi$ $(\Psi \downarrow_0 Z)$. Так, что $\Gamma_Z(A_{\Psi_1} \cup \cdots \cup A_{\Psi_n}) \notin \mathfrak{F}$. Тогда существует фильтр $\mathfrak{F}_1 \supset \mathfrak{F} \cup \{Z \setminus \Gamma_Z(A_{\Psi_1} \cup \cdots \cup A_{\Psi_n}) \mid \Psi_i \downarrow_0 Z\}$. Пусть Φ фильтр в Z' порожденный множествами $(A_{\Psi_1} \cup \cdots \cup A_{\Psi_n})^0$, $\Psi_1 \downarrow_0 Z, \ldots, \Psi_n \downarrow_0 Z$. Покажем, что $\Phi \in \mathscr{A}$. Пусть $\Psi \downarrow_0 Z$, $\varepsilon > 0$. Тогда $\varepsilon^{-1}\Psi \downarrow_0 Z$, $(A_{\varepsilon^{-1}\Psi})^0 \in \Phi$, $\varepsilon A_{\varepsilon^{-1}\Psi} \in \Psi$ и $(A_{\varepsilon^{-1}\Psi})^0\, (\varepsilon A_{\varepsilon^{-1}\Psi}) \subset [-\varepsilon; \varepsilon]$. Таким образом $\Psi \downarrow_0 Z \Rightarrow \Phi(\Psi) \downarrow_0 R^1$ то есть $\Phi \in \mathscr{A}$. Но $(\Phi, \mathfrak{F}_1) = \Phi(\mathfrak{F}_1) \not\downarrow_0 R^1$. Действительно, $\forall\ \Psi_1 \downarrow_0 Z, \ldots, \Psi_n \downarrow_0 Z$ $\mathfrak{F}_1 \supset Z \setminus \Gamma_Z(A_{\Psi_1} \cup \cdots \cup A_{\Psi_n}) = Z \setminus \Gamma_{Z^0}(A_{\Psi_1} \cup \cdots \cup A_{\Psi_n}) = Z \setminus (A_{\Psi_1} \cup \cdots \cup A_{\Psi_n})^{00}$ (по теореме о биполяре ([3], гл. II, т-ма 4)). Поэтому $\forall\, B \in \Phi\ \mathfrak{F}_1 \not\ni B^0$ и, значит $\Phi(\mathfrak{F}_1) \not\ni [-1; 1]$. Пришли к противоречию с тем, что $\mathfrak{F}_1 \supset \mathfrak{F} \downarrow_0 \underline{Z}_{\mathscr{A}}$. Итак, $Z = \underline{Z}_{\mathscr{A}}$. Осталось показать, что отображение $\varphi = j_Z \colon Z \to L_{\mathrm{MB}}(L_{\mathrm{MB}}(Z, R^1), R^1)$ осуществляет изоморфизм $\varphi\colon Z = \underline{Z}_{\mathscr{A}} \xrightarrow{\sim} \varphi(Z)$, Z с подпространством в $L_{\mathrm{MB}}(L_{\mathrm{MB}}(Z, R^1), R^1)$. Так как Z^0 отделимо, $\operatorname{Ker} \varphi = \{0\}$. Так как $\Phi \downarrow_f L_{\mathrm{MB}}(Z, R^1) \Rightarrow \Phi \in \mathscr{A}$, то $\mathfrak{F} \downarrow_0 Z = \underline{Z}_{\mathscr{A}} \Rightarrow \varphi(\mathfrak{F}) \downarrow_0 L_{\mathrm{MB}}(L_{\mathrm{MB}}(Z, R^1), R^1)$. Пусть $\varphi(\mathfrak{F}) \downarrow_0 L_{\mathrm{MB}}(L_{\mathrm{MB}}(Z, R^1), R^1)$. Это значит, что $\Phi(\mathfrak{F}) \downarrow_0 R^1$ коль скоро $\exists\, f \in Z'\ \Phi \downarrow_f L_{\mathrm{MB}}(Z, R^1)$. Отсюда следует, что $\Phi(\mathfrak{F}) \downarrow_0 R^1$ для всякого компактного в $L_{\mathrm{MB}}(Z, R^1)$ фильтра Φ. Поэтому достаточно доказать, что каждый фильтр из $\mathscr{A}$ компактен в $L_{\mathrm{MB}}(Z, R^1)$. Пусть $\Phi \in \mathscr{A}$, $\Phi_1 \in \mathrm{UF}(\Phi)$. Тогда $\Phi_1 \in \mathscr{A}$. Имеем $R^1(0)\, z \downarrow_0 Z$ поэтому $\Phi_1(R^1(0)\, z) = R_1(0)\ \Phi_1(z) \downarrow_0 R^1\ \forall\, z \in Z$. Следовательно $\Phi_1(z)$ содержит ограниченное в R^1 множество. Так как $\Phi_1(z)$ ультрафильтр, а каждое ограниченное множество в R^1 относительно компактно, то $\exists\, f(z) \in R^1 \colon \Phi_1(z) \downarrow_{f(z)} R^1$. Легко видеть, что отображение $f\colon Z \to R^1$ линейно. Пусть $\Psi \downarrow_0 Z$. Тогда $\Phi_1(\Psi) \downarrow_0 R^1$ и $f(\Psi) \supset \Gamma_{R^1}(\Phi_1(\Psi)) \downarrow_0 R^1$ (поскольку $f(A) \subset \overline{F(A)}\ \forall\, A \subset Z\ \forall\, F \in \Phi_1$ — это вытекает из определения f). Таким образом $f \in Z'$. Наконец, поскольку $(F - f)\,(A) \subset F(A) - f(A) \subset F(A) - \overline{F(A)} \subset 2\Gamma_{R^1}(F(A))$ имеем $(\Phi_1 - f)\,(\Psi) \supset 2\Gamma_{R^1}(\Phi_1(\Psi)) \downarrow_0 R^1\ \forall\, \Psi \downarrow_0 Z$. Тем самым доказано, что $\Phi_1 \downarrow_f L_{\mathrm{MB}}(Z, R^1)\ \forall\, \Phi_1 \in \mathrm{UF}(\Phi)$, то есть фильтр Φ компактен. Поскольку импликация в) $\Rightarrow$ а) очевидна, теорема полностью доказана.

3.4. Следствие. *Пусть X — множество, Y — ЛВП, L — линейное пространство отображений $X \to Y$, $\mathscr{A}$ множество фильтров в X такое, что $R^1(0)\, f(\Phi) \downarrow_0 Y$ $\forall\, f \in L\ \forall\, \Phi \in \mathscr{A}$ и $\forall\, x \in X\ \exists\, \Phi \in \mathscr{A} \colon [x] \supset \Phi$. Тогда ЛПП $L_{\mathscr{A}}$ определенное условиями $L_{\mathscr{A}} = L$, $\mathfrak{F} \downarrow_f L_{\mathscr{A}} \Leftrightarrow (\mathfrak{F} - f)\,(\Phi) \downarrow_0 R^1\ \forall\, \Phi \in \mathscr{A}$ является пространством со сходимостью на системе фильтров.*

Доказательство. Легко проверить, что $L_{\mathscr{A}}$ пригодное ЛПП, обладающее свойством ($\Gamma *$). Осталось применить теорему 3.3.

3.5. Предложение. *Каждое ЛПП со сходимостью на системе фильтров правильно и пригодно.*

Доказательство. Пригодность, как уже отмечалось, очевидна, а правильность следует из теоремы 3.3 и замечания 2.4.2 статьи [1].

3.6. Задача. *Является ли всякое правильное пригодное ЛПП пространством со сходимостью на системе фильтров?* (То есть справедливо ли обращение предложения 3.5).

Для счетнополинормированных пригодных ЛПП утверждение задачи 3.6 выполняется.

3.7. Теорема. *Каждое правильное счетнополинормированное ЛПП есть пространство со сходимостью на системе фильтров.*

В силу 1.26 достаточно доказать следующую теорему.

3.8. Теорема. *Если X квазинормируемое ЛВП, то $L^{\#}(X, R^1)$ пространство со сходимостью на системе фильтров.*

Доказательство. Из теоремы 5.2.1 [1] следует, что $L^{\#}(X, R^1) = L_{\mathrm{FB}}(X, R^1)$.

3.9. Теорема. *Каждое дуально-квазинормируемое ЛПП есть пространство со сходимостью на системе фильтров.*

Доказательство. Это вытекает из 3.8 и 1.24.

3.10. Теорема. *Пусть X ЛВП. Тогда $\big(L^{\#}(X, R^1)$ — пространство со сходимостью на системе фильтров$\big)$ $\Leftrightarrow$ $L^{\#}(X, R^1) = L_{\mathrm{FB}}(X, R^1)$.*

Доказательство. Имликация $\Leftarrow$ очевидна. Пусть $L^{\#}(X, R^1)$ пространство со сходимостью на системе фильтров и $\mathfrak{F} \downarrow_0 L^{\#}(X, R^1)$. Согласно ($\Gamma*$) в этом случае $\forall\, U \in X(0)\ \exists\, \lambda_U > 0\ \forall\, U_1, \ldots, U_n \in X(0)\ \Gamma\left(\bigcup\limits_{i=1}^{n} \lambda_{U_i} U_i^0\right) \notin \mathfrak{F}$. Пусть Φ фильтр в X базисом которого являются всевозможные множества вида $\left(\Gamma\left(\bigcup\limits_{i=1}^{n} \lambda_{U_i} U_i^0\right)\right)^0$, $U_i \in X(0)$, $n \in \mathbf{N}$. Если U замкнутся абсолютно выпуклая окрестность нуля в X то $\Phi \ni \lambda_U^{-1} U$ так что Φ ограниченный фильтр в X. Но $\Gamma(\bigcup \lambda_{U_i} U_i^0)$ $\sigma(X', X)$-компактное множество (как абсолютно выпуклая оболочка конечного объединения абсолютно выпуклых $\sigma(X', X)$-компактных множеств). По теореме о биполяре $\Gamma\left(\bigcup\limits_{i=1}^{n} \lambda_{U_i} U_i^0\right) = \left(\Gamma\left(\bigcup\limits_{i=1}^{n} \lambda_{U_i} U_i^0\right)\right)^{00}$. Значит $A \in \Phi \Rightarrow A^0 \notin \mathfrak{F}$, то есть $\mathfrak{F}(\Phi) \not\ni [-1, 1]$. Таким образом $\mathfrak{F} \downarrow_0 L^{\#}(X, R^1) \Rightarrow \mathfrak{F} \downarrow_0 L_{\mathrm{FB}}(X, R^1)$ и $L^{\#}(X, R^1) = L_{\mathrm{FB}}(X, R^1)$. Теорема доказана.

3.11. Следствие. *Если X пространство Фреше, то $L^{\#}(X, R^1)$ правильно $\Leftrightarrow$ $L^{\#}(X, R^1) = L_{\mathrm{FB}}(X, R^1)$.*

Доказательство. Это следует из 3.7 и 3.10.

3.12. Задачи. а) *Пусть Y — полинормированное пространство со сходимостью на системе фильтров. Верно ли, что*

$$L^{\#}\big(L^{\#}(Y, R^1), R^1\big) = L_{\mathrm{FB}}\big(L^{\#}(Y, R^1), R^1\big)?$$

б) *Пусть X пространство Фреше и $L^{\#}(X, R^1) = L_{\mathrm{FB}}(X, R^1)$. Вытекает ли отсюда квазинормируемость X?*

3.13. Замечания. 1. Если ЛВП X не является пространством Фреше, то $L^{\#}(X, R^1) = L_{\mathrm{FB}}(X, R^1) \not\Rightarrow X$ — квазинормируемо. Действительно, пусть X ЛВП примера 1.21. Тогда $L^{\#}(X, R^1)$ дуально квазинормируемо. В силу 3.9 и 3.10 $L^{\#}(X, R^1) = L_{\mathrm{FB}}(X, R^1)$. Но X не квазинормируемо.

2. Существует полинормированное пространство со сходимостью на системе фильтров, не являющееся дуально квазинормируемым.

3.14. В связи с теоремой 3.3 возникает вопрос: не изоморфно ли каждое пространство со сходимостью на системе фильтров подпространству в ЛПП вида

а) $L_{\mathrm{MB}}(X, R^1)$, X ЛВП?

б) $L_{\mathrm{MB}}(X, Y)$, X, Y ЛВП?

в) $L_{\mathrm{FB}}(X, Y)$, X, Y ЛВП?

Отрицательный ответ на вопрос а) вытекает уже из следующей любопытной теоремы.

3.14.1. Теорема. *Никакое бесконечномерное ЛВП не изоморфно подпространству в ЛПП вида $L_{\text{MB}}(X, R^1)$, X ЛВП.*

Мы опускаем несложное доказательство.

Эта теорема показывает, что ЛПП вида $L_{\text{MB}}(X, R^1)$ (X — ЛВП) сугубо псевдотопологические: никакое их бесконечномерное подпространство не топологизируемо.

Каждое подпространство Z в ЛПП вида $L_{\text{MB}}(X, Y)$ (X, Y — ЛВП) обладает следующим свойством:

(****) Каждый ограниченный в Z и сходящийся к нулю в Z^0 фильтр сходится в Z.

3.14.2. Пусть $\underline{Y}$ линейное пространство всех ограниченных отображений $f: [-1, 1] \to R^1$. Положим $\mathfrak{F} \downarrow_0 Y \Leftrightarrow \left(\mathfrak{F}([x]) \downarrow_0 R^1 \ \forall\, x \in R^1 \text{ и } \mathfrak{F}(R^1(0)) \downarrow_0 R^1\right)$.

Тогда Y пространство со сходимостью на системе фильтров, не обладающее свойством (****). Таким образом, ответ на вопрос б) тоже отрицательный.

Наконец, приведем пример пространства со сходимостью на системе фильтров не изоморфного подпространству ни в ЛПП вида $L_{\text{MB}}(X, Y)$ (X, Y — ЛВП) ни в ЛПП вида $L_{\text{FB}}(X, Y)$ (X, Y — ЛВП). Заметим, что если Z подпространство в $L_{\text{MB}}(X, Y)$ или $L_{\text{FB}}(X, Y)$ то $Z^{\#}$ подпространство в $L^{\#}(X, Y)$ а потому является проективным пределом пригодных полинормированных пространств.

3.14.3. Пусть Z подпространство в $R^{[0;1]}$, $\underline{Z} = \{f: [0, 1] \to R^1 \mid f^{-1}(R^1 \setminus \{0\})$ не более, чем счетно$\}$, $\mathscr{E}$ совокупность всех сепарабельных подпространств в Z. Определим ЛПП Y так: $\underline{Y} = \underline{Z}$, $\mathfrak{F} \downarrow_0 Y \Leftrightarrow (\mathfrak{F} \downarrow_0 Z, \mathfrak{F} \cap \mathscr{E} \neq \emptyset)$.

Тогда Y пространство со сходимостью на системе фильтров. Нетрудно проверить, что если $\varphi \in L(Y, W)$, W полинормированное ЛПП, то $\varphi(Y)$ конечномерно. Поэтому $Y = Y^{\#}$ не является проективным пределом полинормированных пространств.

§ 4. Конструкция $\mathscr{S}(X)$ и представление луковичных локально выпуклых линейных псевдотопологических пространств

4.1. Пусть A множество с отмеченной точкой a_0. (Ниже, рассматривая в качестве A ЛПП, мы всегда будем выбирать $a_0 = 0$.) Для $f: A \to R^1$ определим носитель $\operatorname{supp} f = \{a_0\} \cup \{a \in A \mid f(a) \neq 0\}$. Если E множество функций $A \to R^1$, положим $\operatorname{supp} E = \bigcup_{f \in E} \operatorname{supp} f$. Наконец, если $\mathfrak{F}$ фильтр во множестве X отображений $A \to R^1$, то пусть $\operatorname{supp} \mathfrak{F}$ — фильтр в A базис которого образуют множества вида $\operatorname{supp} E$, $E \in \mathfrak{F}$. Пусть $\mathscr{S}$ совокупность фильтров в A а X ЛВП, элементами которого являются числовые функции на A, причем:

($\mathscr{S}X1$) $\quad \mathrm{UF}(\mathfrak{F}) \subset \mathscr{S} \Rightarrow \mathfrak{F} \in \mathscr{S}$ $\qquad$ ($\mathfrak{F}$ фильтр в A);

($\mathscr{S}X2$) $\quad f \in X \Rightarrow [\operatorname{supp} f] \in \mathscr{S}$;

($\mathscr{S}X3$) $\quad a \in A \Rightarrow \varepsilon_a \in X' \left(\varepsilon_a(x) = x(a) \quad \forall\, a \in A \ \forall\, x \in X\right)$.

Определим ЛПП $\mathscr{S}(X)$ условиями:

$$\underline{\mathscr{S}(X)} = \underline{X}, \ \mathfrak{F} \downarrow_0 \mathscr{S}(X) \Leftrightarrow \mathfrak{F} \downarrow_0 X \text{ и } \operatorname{supp} \mathfrak{F} \in \mathscr{S}.$$

4.2. Отметим следующие простые свойства носителей:

4.2.1. $\operatorname{supp} \left(\bigcup_\alpha E_\alpha\right) = \bigcup_\alpha \operatorname{supp} E_\alpha$ (для любого семейства $\{E_\alpha\}$ подмножеств R^A;

4.2.2. $\operatorname{supp}(E_1 + E_2) = \operatorname{supp} E_1 \cup \operatorname{supp} E_2$ ($\forall\, E_1, E_2 \subset R^A$);

4.2.3. $\operatorname{supp} \Gamma_X(E) = \operatorname{supp} E$ ($\forall\, E \subset X$);

4.2.4. $\operatorname{supp}(F \cdot E) = \operatorname{supp} E$ ($\{0\} \neq F \subset R^1$, $E \subset R^A$).

Для фильтров из 4.2.1—4.2.4 получаем:

4.2.5. $\operatorname{supp}(\bigcap \mathfrak{F}_\alpha) = \bigcap\limits_\alpha \operatorname{supp} \mathfrak{F}_\alpha$ (для всякого семейства $\{\mathfrak{F}_\alpha\}$ фильтров в X;

4.2.6. $\operatorname{supp}(\mathfrak{F}_1 + \mathfrak{F}_2) = \operatorname{supp} \mathfrak{F}_1 \cap \operatorname{supp} \mathfrak{F}_2$ ($\mathfrak{F}_1, \mathfrak{F}_2$ фильтры в X);

4.2.7. $\operatorname{supp}(F \cdot \mathfrak{F}) = \operatorname{supp} \mathfrak{F}$ ($\{0\} \neq F \subset R^1$, $\mathfrak{F}$ фильтр в X);

4.2.8. $\operatorname{supp}\big(R^1(0)\, \mathfrak{F}\big) = \operatorname{supp} \mathfrak{F}$ ($\mathfrak{F}$ фильтр в X);

4.2.9. $\operatorname{supp} \Gamma_X(\mathfrak{F}) = \operatorname{supp} \mathfrak{F}$ ($\mathfrak{F}$ фильтр в X).

4.3. Теорема. *$\mathscr{S}(X)$ луковичное ЛПП со сходимостью на системе фильтров.*

Доказательство. То, что $\mathscr{S}(X)$ пригодное ЛПП непосредственно следует из 4.2.5—4.2.9.

Пусть $\mathfrak{F} \downarrow_0 \mathscr{S}(X)$. Положим $\mathfrak{F}^\# = \bigcap\limits_{\alpha>0} \alpha[-1, 1]\, \mathfrak{F}$. В силу 4.2.5 и 4.2.7 $\operatorname{supp} \mathfrak{F}^\# = \operatorname{supp} \mathfrak{F}$, поэтому $\mathfrak{F}^\# \downarrow_0 \mathscr{S}(X)$. Но $\mathfrak{F}^\#$ луковичный фильтр и $\mathfrak{F} \supset \mathfrak{F}^\#$. Луковичность $\mathscr{S}(X)$ доказана.

В силу 3.3 осталось доказать, что $\mathscr{S}(X)$ удовлетворяет условию ($\Gamma*$). С учетом 4.2.1 и 4.2.3 условие ($\Gamma*$) эквивалентно следующему свойству системы $\mathscr{S}$:

Если $\mathfrak{F} \notin \mathscr{S}$, то можно так выбрать по множеству $A_\Phi \in \Phi$ из каждого фильтра $\Phi \in \mathscr{S}$, что $\bigcup\limits_{i=1}^n A_{\Phi_i} \notin \mathfrak{F}$ $\forall\, \Phi_1, \ldots, \Phi_n \in \mathscr{S}$.

Пусть $\mathfrak{F} \notin \mathscr{S}$. В силу ($\mathscr{S}X1$) $\exists\, \Psi \in \mathrm{UF}(\mathfrak{F})$: $\Psi \notin \mathscr{S}$. Тогда $\forall\, \Phi \in \mathscr{S}\ \exists\, A_\Phi \in \Phi$: $\Psi \ni A \setminus A_\Phi$. Имеем $\forall\, \Phi_1, \ldots, \Phi_n \in \mathscr{S}$, $\Psi \in \bigcap\limits_{i=1}^n (A \setminus A_{\Phi_i}) = A \setminus \bigcup\limits_{i=1}^n A_{\Phi_i}$ а, следовательно, $\Psi \not\ni \bigcup\limits_{i=1}^n A_{\Phi_i}$. Теорема доказана.

4.4. Пусть A — ЛПП (отмеченная точка — 0), $\mathscr{S}$ множество всех компактных фильтров в A, $\underline{X} = \big\{f \in R^A \mid f^{-1}(R^1 \setminus \{0\})\big\}$ конечное множество, топология в X задается нормой $\|f\|_1 = \sum\limits_{a \in A} |f(a)|$. Легко видеть, что условия ($\mathscr{S}X1$) — ($\mathscr{S}X3$) выполнены. Положим $\hat{A} = \mathscr{S}(X)$. Формула $\pi(f) = \sum\limits_{a \in A} f(a)\, a$ определяет лиейное отображение $\pi\colon \hat{A} \to A$. Пусть $\mathfrak{F}$ фильтр в A, $\hat{\mathfrak{F}}$ фильтр в $\hat{A}$ базис которого образуют множества вида $F_\varepsilon = \{\varphi_{a,\delta} \mid a \in F, 0 < \delta \leqq \varepsilon\}$ $\left(\varphi_{a,\delta}(b) = \begin{cases} \delta, & a = b \\ 0, & a \neq b \end{cases}\right)$, $F \in \mathfrak{F}$, $\varepsilon > 0$. Очевидно $\operatorname{supp} \hat{\mathfrak{F}} = \mathfrak{F} \cap [0]$, $\pi(\hat{\mathfrak{F}}) = R^1(0)\, \mathfrak{F}$.

4.5. Теорема. *Пусть A локально выпуклое ЛПП. Следующие условия эквивалентны:*

а) A квазиправильно;
б) $\pi \in L(\hat{A}, A)$.

Доказательство. 1. Пусть A квазиправильно, $\mathfrak{F} \downarrow_0 \hat{A}$. Тогда $\|\mathfrak{F}\|_1 \downarrow_0 R^1$, поэтому $\pi(\mathfrak{F}) \supset R^1(0)\, \Gamma(\operatorname{supp} \mathfrak{F}) = \Gamma\big(R^1(0) \operatorname{supp} \mathfrak{F}\big) \downarrow_0$ в силу квазиправильности и локальной выпуклости A. Таким образом а) $\Rightarrow$ б).

2. Пусть $\pi \in L(\hat{A}, A)$, $\mathfrak{F}$ компактный фильтр в A. Тогда $\hat{\mathfrak{F}} \downarrow_0 \hat{A}$ (поскольку $\|\hat{\mathfrak{F}}\|_1 \downarrow_0 R^1$, $\operatorname{supp} \hat{\mathfrak{F}} = \mathfrak{F} \cap [0] \in \mathscr{S}$). Имеем $R^1(0)\,\mathfrak{F} = \pi(\hat{\mathfrak{F}}) \downarrow_0 A$. Таким образом, б) $\Rightarrow$ а). Теорема доказана.

4.6. Теорема. *Пусть A луковичное локально выпуклое квазиправильное ЛПП. Тогда существуют луковичное пространство со сходимостью на системе фильтров Y и замкнутое подпространство K в Y такие, что $A \approx Y/K$.*

Доказательство. Положим $Y = \hat{A}$, $K = \operatorname{Ker} \pi$. В силу 4.5 и 4.3 осталось показать, что $\mathfrak{F} \downarrow_0 A \Rightarrow \exists\, \Phi \downarrow_0 \hat{A}\colon \mathfrak{F} \supset \pi(\Phi)$. Так как A луковично $\mathfrak{F} \downarrow_0 A \Rightarrow \exists\, \mathfrak{F}_1 \downarrow_0 A$, $\mathfrak{F}_1 = R^1(0)\,\mathfrak{F}_1$, $\mathfrak{F} \supset \mathfrak{F}_1$. Положим $\Phi = \hat{\mathfrak{F}}_1$. Тогда $\Phi \downarrow_0 \hat{A}$, $\mathfrak{F} \supset \mathfrak{F}_1 = R^1(0)\,\mathfrak{F}_1 = \pi(\Phi)$. Теорема доказана.

В заключение рассмотрим следующий вопрос: какие отделимые ЛПП можно получить, отправляясь от ЛВП и применяя в любом числе операции взятия индуктивного и проективного пределов (в категории отделимых ЛПП). Очевидно, что так можно получить только луковичные локально выпуклые ЛПП, но неясно, все ли они так получаются.

4.7. Определение. Замыканием класса $\mathscr{K}$ отделимых ЛПП назовем наименьший содержащий $\mathscr{K}$ класс $\overline{\mathscr{K}}$ отделимых ЛПП такой, что индуктивный предел всякого прямого и проективный предел всякого обратного спектра пространств из $\overline{\mathscr{K}}$ содержится в $\overline{\mathscr{K}}$.

4.8. Задача. *Совпадает ли замыкание $\overline{\text{ЛВП}}$ класса ЛВП всех ЛВП с классом всех отделимых луковичных локально выпуклых ЛПП?*

Из теоремы 4.6 мы выведем утверждение, более слабое, чем положительное решение задачи 4.8.

4.9. Теорема. *Класс $\overline{\text{ЛВП}}$ содержит все отделимые квазиправильные луковичные локально выпуклые пространства.*

Поскольку всякое подпространство луковичного ЛПП со сходимостью на системе фильтров является луковичным ЛПП со сходимостью на системе фильтров, достаточно доказать следующие два утверждения и применить 4.6.

4.10. Теорема. *Каждое луковичное ЛПП со сходимостью на системе фильтров принадлежит $\overline{\text{ЛВП}}$.*

4.11. Теорема. *Пусть X пригодное ЛПП, K замкнутое подпространство в X. Тогда X/K представимо в виде индуктивного предела некоторого прямого спектра подпространств пространства X.*

Доказательство 4.10. Пусть (X, Y) дуальная пара, $\mathscr{A}$ система фильтров в X удовлетворяющая условиям ($\mathscr{A}$1) и ($\mathscr{A}$2) (см. 3.1), $Y = Y_{\mathscr{A}}$ и допустим, что $Y = Y^{\#}$. Для каждого $\Phi \in \mathscr{A}$ и $A \in \Phi$ определим ЛВП

$$Y_{\Phi,A}\colon \underline{Y_{\Phi,A}} = R^1 A^0,\ \mathfrak{F} \downarrow_0 Y_{\Phi,A} \Leftrightarrow \big(\mathfrak{F} \supset R^1(0)\,[A^0],\ i_{A^0}(\mathfrak{F}) \downarrow_0 Y^0\big),$$

где $i_{A^0}\colon R^1A^0 \to Y$ тождественное вложение. Упорядочим Φ по включению: $A < B \Leftrightarrow A \supset B\ (A, B \in \Phi)$ и пусть для $A < B$ $i_{A,B}\colon Y_{\Phi,A} \to Y_{\Phi,B}$ — тождественное вложение (очевидно $\underline{Y_{\Phi,A}} \subset \underline{Y_{\Phi,B}}$ при $A < B$). Пусть $\underline{Y_\Phi} = \underline{Y}$, $\mathfrak{F} \downarrow_0 Y_\Phi \Leftrightarrow \big(\exists\, A \in \Phi\colon \mathfrak{F} \supset i_{A^0}(Y_{\Phi,A}(0))\big)$.

Легко видеть, что Y_Φ (вместе с отображениями $i_{A\circ}$) является индуктивным пределом спектра $\{Y_{\Phi,A}, i_{A,B}\}_{A,B\in\Phi}$. Упорядочивая $\mathscr{A}$ по включению $\Phi_1 > \Phi_2 \Leftrightarrow \Phi_1 \subset \Phi_2$ и полагая i_{Φ_1,Φ_2} равным тождественному отображению $Y \to Y$, получим обратный спектр $\{Y_{\Phi_1}, i_{\Phi_1,\Phi_2}\}_{\Phi_1,\Phi_2\in\mathscr{A}}$ (очевидно $i_{\Phi_1,\Phi_2} \in L(Y_{\Phi_1}, Y_{\Phi_2})$). Нетрудно проверить, что Y (вместе с тождественными отображениями) является проективным пределом этого спектра. Теорема 4.10 доказана.

Доказательство 4.11. Пусть $\mathscr{A}$—множество всех конечномерных подпространств в K, упорядоченное по включению $E_1 < E_2 \Leftrightarrow E_1 \subset E_2$, π_{E_1,E_2} — естественное отображение $X/E_1 \to X/E_2$ а π_{E_1} естественное отображение $X/E_1 \to X/K$. Легко видеть, что X/K (вместе с отображениями π_E, $E \in \mathscr{A}$) является индуктивным пределом спектра $\{X/E, \pi_{E_1,E_2}\}_{E_1,E_2\in\mathscr{A}}$. Осталось доказать следующую лемму.

4.11.1. Лемма. *Пусть X пригодное ЛПП и K конечномерное подпространство в X. Тогда K дополнимо в X.*

Доказательство. Достаточно построить проектор $P: X \to K$ (то есть отображение $P \in L(X, K)$ такое, что $P^2 = P$ и $P(X) = K$. Пусть $\dim K = n$ и $e_1, \ldots, e_n$ алгебраический базис K. По теореме Хана-Банаха (примененной к отделимому ЛВП X^0) $\forall\ 1 \leqq i \leqq n\ \exists\ \varphi_i \in L(X^0, R^1) = L(X, R^1)$: $\varphi_i(e_j) = \begin{cases} 0, & i \neq j \\ 1, & i = j \end{cases}$. Полагая $P(x) = \sum_{i=1}^{n} \varphi_i(x)\, e_i$ получаем искомый проектор. Доказательство леммы, а тем самым и теоремы 4.11 завершено.

Наряду с 4.9, имеет место и утверждение, более слабое, чем отрицательное решение задачи 4.8. Прежде чем формулировать и доказывать это утверждение, дадим два определения.

4.12. Определение. Пусть $\mathscr{K}$ — класс отделимых ЛПП. Строгим замыканием класса $\mathscr{K}$ назовем наименьший класс $\mathscr{K}_c$ отделимых ЛПП такой, что:

1. $\mathscr{K}_c \supset \mathscr{K}$;
2. Проективный предел всякого обратного спектра пространств из $\mathscr{K}_c$ содержится в $\mathscr{K}_c$;
3. Индуктивный предел всякого прямого спектра пространств из $\mathscr{K}_c$ в котором все отображения имеют нулевое ядро, содержится в $\mathscr{K}_c$.

4.13. Определение [6]. Скажем, что ЛПП X удовлетворяет условию C_q если каждый фундаментальный фильтр в X ограничен (то есть, если $\mathfrak{F} - \mathfrak{F} \downarrow_0 X \Rightarrow R^1(0)\mathfrak{F} \downarrow_0 X$).

Условие C_q было введено в [6] как необходимое и достаточное для существования изоморфизма с подпространством полного ЛПП.

4.14. Теорема. *Существует луковичное пригодное ЛПП $X \notin$ ЛВП$_c$.*

Доказательство. Пусть X — ЛПП, построенное при доказательстве теоремы 2.4 и Φ — построенный там же компактный, но не ограниченный фильтр. Легко проверить, что $\Phi - \Phi \downarrow_0 X$. Таким образом, существует луковичное пригодное ЛПП, не удовлетворяющее условию C_q. Теперь достаточно доказать следующую теорему.

4.15. Теорема. *Каждое ЛПП из* ЛВП$_c$ *удовлетворяет условию C_q.*

Это утверждение вытекает из двух простых предложений и того, что каждое ЛВП удовлетворяет условию C_q.

4.16. Предложение. *Проективный предел спектра ЛПП, удовлетворяющих условию C_q удовлетворяет условию C_q.*

Доказательство очевидно.

4.17. Предложение. *Пусть $\{Y_\alpha, i_{\alpha,\beta}\}_{\alpha,\beta\in\mathscr{A}}$ — прямой спектр ЛПП, Y, вместе с отображениями $i_\alpha \notin L(Y_\alpha, Y)$, — его индуктивный предел,* $\operatorname{Ker} i_{\alpha,\beta} = \{0\}\ \forall\ \alpha < \beta$; $\alpha, \beta \in \mathscr{A}$ *и Y_α удовлетворяет условию C_q $\forall\ \alpha \in \mathscr{A}$. Тогда Y удовлетворяет условию C_q.*

Доказательство. Пусть $\mathfrak{F}$ фундаментальный фильтр в Y. Тогда $\mathfrak{F} - \mathfrak{F} \downarrow_0 Y$ и поэтому $\exists\ \alpha_0 \in \mathscr{A}\ \exists\ \Phi_{\alpha_0} \downarrow_0 Y_{\alpha_0}: \mathfrak{F} - \mathfrak{F} \supset i_{\alpha_0}(\Phi_{\alpha_0})$. Отсюда получаем: $\exists\ F \in \mathfrak{F}$: $F - F \subset i_{\alpha_0}(Y_{\alpha_0})$, а значит, $\exists\ y_0 \in Y: y_0 - F \subset i_{\alpha_0}(Y_{\alpha_0})$. Так как $Y = \bigcup_{\alpha\in\mathscr{A}} i_\alpha(Y_\alpha)$, то $\exists\ \alpha_1 > \alpha_0: y_0 \in i_{\alpha_1}(Y_{\alpha_1}),\ F \subset i_{\alpha_1}(Y_{\alpha_1}) + i_{\alpha_0}(Y_{\alpha_0}) \subset i_{\alpha_1}(Y_{\alpha_1}) + i_{\alpha_1}\big(i_{\alpha_0,\alpha_1}(Y_{\alpha_0})\big) \subset i_{\alpha_1}(Y_{\alpha_1})$, то есть $i_{\alpha_1}(Y_{\alpha_1}) \in \mathfrak{F}$. Так как $\operatorname{Ker} i_{\alpha,\beta} = \{0\}$ при $\alpha < \beta$ то и $\operatorname{Ker} i_\alpha = \{0\}\ \ \forall\ \alpha \in \mathscr{A}$, поэтому $\mathfrak{F} = i_{\alpha_1}(\Phi_{\alpha_1})$ для некоторого фильтра в Y_{α_1}. Имеем

$$i_{\alpha_1}(\Phi_{\alpha_1} - \Phi_{\alpha_1}) = \mathfrak{F} - \mathfrak{F} \supset i_{\alpha_0}(\Phi_{\alpha_0})$$

и так как $\operatorname{Ker} i_{\alpha_1} = \{0\}$, $\operatorname{Ker} i_{\alpha_0} = \{0\}$ и $\operatorname{Ker} i_{\alpha_0,\alpha_1} = \{0\}$: $\Phi_{\alpha_1} - \Phi_{\alpha_1} \supset i_{\alpha_0,\alpha_1}(\Phi_{\alpha_0}) \downarrow_0 Y_{\alpha_1}$. Применяя C_q получаем $R^1(0)\ \Phi_{\alpha_1} \downarrow_0 Y_{\alpha_1}$ и $R^1(0)\ \mathfrak{F} = i_{\alpha_1}\big(R^1(0)\ \Phi_{\alpha_1}\big) \downarrow_0 Y$.

Предложение доказано.

Доказательство 4.15. Пусть $\mathscr{K}$ класс всех отделимых ЛПП, удовлетворяющих условию C_q. В силу 4.16 и 4.17 $\mathscr{K} = \mathscr{K}_c$ и, очевидно, $\mathscr{K} \supset$ ЛВП. Отсюда по определению строго замыкания следует, что $\mathscr{K} \supset$ ЛВП$_c$. Теорема доказана.

Литература

[1] М. П. Кац, Топологические свойства псевдотопологических линейных пространств, Math. Nachr. **75** (1976), 207—227.

[2] А. Фрёлихер и В. Бухер, Дифференциальное исчисление в векторных пространствах без нормы, Мир, Москва 1970.

[3] А. П. Робертсон и В. Дж. Робертсон, Топологические векторные пространства, Мир, Москва 1967.

[4] X. Шефер, Топологические векторные пространства, Мир, Москва 1971.

[5] М. П. Кац, Пример дуальнометрического но не (DF)-пространства. Вестн. Моск. ун-та, матем., механ. **4** (1975), 36—37.

[6] S. Gähler, W. Gähler and G. Kneis, Complection of pseudo-topological vector spaces, Math. Nachr. **75** (1976), 185—206.

Anschrift des Verfassers:

197010 Ленинград
Дибуновская 5 корп. 2 кв. 12
СССР

Some Contributions to the Theory of Pseudo-Uniform Spaces

GERT KNEIS, Berlin

1. Introduction

This paper contains some various contributions to the theory of pseudo-uniform spaces — the uniform convergence spaces in the sense of COOK and FISCHER ([2] and [3]). For missing proofs in the parts 1 and 2, the reader is referred to my paper [8]. The basic definitions concerning pseudo-topologies are assumed to be well-known; the reader not being familiar with convergence structures is referred to W. GÄHLER's book [4].

A pseudo-uniform structure on a non-empty set X is an $\wedge$-ideal $\mathfrak{u}$ of filters in $X \times X$ — that is a set of proper filters in $X \times X$ closed under finite intersections and including with any filter $\mathfrak{F}$ every filter $\mathfrak{G}$ finer than $\mathfrak{F}$, that means $\mathfrak{G} \supseteqq \mathfrak{F}$, such that the following conditions are fullfilled:

(i) For any $\mathfrak{F}, \mathfrak{G} \in \mathfrak{u}$, the filter $\mathfrak{G} \circ \mathfrak{F}$ in $X \times X$ with the base $\{G \circ F \mid F \in \mathfrak{F}, G \in \mathfrak{G}\}$ belongs to $\mathfrak{u}$ if it is proper (the sign $\circ$ denotes the relation product of subsets of $X \times X$).

(ii) For any $\mathfrak{F} \in \mathfrak{u}$, the inverse filter $\mathfrak{F}^{-1} = \{F^{-1} \mid F \in \mathfrak{F}\}$ belongs to $\mathfrak{u}$.

(iii) The principal filter $[\Delta_X]$ belongs to $\mathfrak{u}$ where Δ_X denotes the diagonal of $X \times X$.

X equipped with a pseudo-uniform structure $\mathfrak{u}$ is called a pseudo-uniform space or uniform convergence space. In the special case of $\mathfrak{u}$ being the principal $\wedge$-ideal $[\mathfrak{N}]$ generated by a fixed filter $\mathfrak{N}$ in $X \times X$ — i.e., $\mathfrak{u}$ consists of all filters finer than $\mathfrak{N}$ —, $\mathfrak{u}$ is a uniform structure. If $\mathfrak{u}$ and $\mathfrak{v}$ are pseudo-uniform structures on the same set X with $\mathfrak{u} \subseteqq \mathfrak{v}$, then $\mathfrak{u}$ is said to be finer than $\mathfrak{v}$ and $\mathfrak{v}$ is said to be coarser than $\mathfrak{u}$. A pseudo-uniform space X is called a uniform CHOQUET space if a filter on $X \times X$ belongs to the pseudo-uniform structure of X if every finer ultra filter does it.

The natural mappings between pseudo-uniform spaces are the uniformly continuous mappings. Let X and Y be spaces with the pseudo-uniform structures $\mathfrak{u}$ and $\mathfrak{v}$, respectively. A mapping $f: X \to Y$ is said to be uniformly continuous if for any $\mathfrak{F} \in \mathfrak{u}$ the filter $(f * f)(\mathfrak{F})$ belongs to $\mathfrak{v}$ where $f * f$ is defined by $(x, y) \mapsto (f(x), f(y))$; the set of all uniformly continuous mappings $f: X \to Y$ is denoted by $U(X, Y)$. Let PUN and UN be the category of all pseudo-uniform and of all uniform spaces, respectively, with the uniformly continuous mappings as morphisms. There is a natural functor μ: PUN $\to$ UN which transforms a pseudo-uniform space X with the structure $\mathfrak{u}$ into the uniform space μX consisting of the set underlying X equipped with the finest uniform structure $\mu(\mathfrak{u})$ coarser than $\mathfrak{u}$.

A pseudo-uniform structure $\mathfrak{u}$ of a space X generates a corresponding pseudo-topology $\lambda(\mathfrak{u})$ on the set underlying X: the filters $\mathfrak{F}$ on the set underlying X which are convergent to some x with respect to $\lambda(\mathfrak{u})$ are defined by $\mathfrak{F} \times [x] \in \mathfrak{u}$; the set underlying X equipped with $\lambda(\mathfrak{u})$ is denoted by λX. It is well-known that a pseudo-topology τ on a set X is a corresponding pseudo-topology of a suitable pseudo-uniform structure — i.e., τ is pseudo-uniformizable — if and only if τ has the following separation property: If there is a proper filter $\mathfrak{F}_0$ on X converging to x and y then a filter $\mathfrak{F}$ on X converges to x if and only if it converges to y. All pseudo-topological conceptions for pseudo-uniform spaces are related to the corresponding pseudo-topological space. For example, a pseudo-uniform space X with the structure $\mathfrak{u}$ is said to be separated, regular, compact, or Choquet, if X equipped with $\lambda(\mathfrak{u})$ is separated, regular, compact, or Choquet, respectively. The natural notion of regularity for pseudo-uniform spaces is the notion of uniform regularity: A space X with the pseudo-uniform structure $\mathfrak{u}$ is said to be uniformly regular if for every $\mathfrak{U} \in \mathfrak{u}$ the adherence $\overline{\mathfrak{U}}$ of $\mathfrak{U}$ with respect to the product pseudo-topology $\lambda(\mathfrak{u}) \times \lambda(\mathfrak{u})$ belongs to $\mathfrak{u}$. X is said to be uniformly t-regular if for every $\mathfrak{U} \in \mathfrak{u}$ the adherence $\overline{\mathfrak{U}}$ of $\mathfrak{U}$ with respect to the product topology $t(\lambda(\mathfrak{u})) \times t(\lambda(\mathfrak{u}))$ belongs to $\mathfrak{u}$ where $t(\lambda(\mathfrak{u}))$ denotes the finest topology on X coarser than $\lambda(\mathfrak{u})$.

Let X be a pseudo-topological (real or complex) vector space in the sense of Fischer, i.e., the linear operations are continuous. A filter $\mathfrak{U}$ in $X \times X$ belongs to the canonical pseudo-uniform structure of X in the sense of Gähler [4] if $w(\mathfrak{U})$ converges to 0 where $w: X \times X \to X$ is defined by $(x, y) \mapsto x - y$. All pseudo-uniform conceptions for pseudo-topological vector spaces are related to this canonical pseudo-uniform structure.

2. Uniform continuity of continous mappings on compact sets

Let X be a separated compact topological space. X is uniformizable in an unique way (see Bourbaki [1]) such that the notion of uniform continuity for a mapping f of X into a pseudo-uniform space Y is well defined. We have the following characterization of uniformly continuous mappings from X into an arbitrary pseudo-uniform space:

Theorem 1. *A mapping f of a compact separated topological space X into a pseudo-uniform space Y is uniformly continuous if and only if the filter $\bigcap_{x \in X} (f * f) (\mathfrak{F}_x \times [x])$ belongs to the structure of Y where $\mathfrak{F}_x$ is the neighborhood filter of X in the point $x \in X$.*

If f is continuous then $(f * f) (\mathfrak{F}_x \times [x])$ belongs to the pseudo-uniform structure, say $\mathfrak{v}$, of Y and if it is an uniform one, an arbitrary intersection of filters of $\mathfrak{v}$ belongs to $\mathfrak{v}$. Hence Theorem 1 implies the classical theorem of uniform continuity of continuous mappings on compact sets, which, however, is not valid in this form in general. Indeed, using Theorem 1, we may construct an example of a continuous mapping of the closed interval [0, 1] into a pseudo-topological vector space Y which is not uniformly continuous.

Obviously, such an example is not possible in a space Y in which the pseudo-uniform structure is a uniform one on compact sets. We have the following

Lemma. *Let Y be a uniform Choquet space with the pseudo-uniform structure $\mathfrak{u}$. If the finest uniform structure $\mu(\mathfrak{u})$ on Y coarser than $\mathfrak{u}$ is separated then the restrictions of $\mathfrak{u}$ and $\mu(\mathfrak{u})$ to an arbitrary compact set agree.*

Especially, let X be a separated compact topological space and f a continuous mapping of X into a pseudo-uniform space Y fulfilling the conditions of the Lemma. Applying the Lemma to the compact subset $f(X)$ of Y and using Theorem 1, we get

Theorem 2. *Let X be a separated compact topological space and Y be a uniform Choquet space with the pseudo-uniform structure $\mathfrak{u}$ such that $\mu(\mathfrak{u})$ is separated. Then every continuous mapping $f\colon X \to Y$ is uniformly continuous.*

As every uniform space is a uniform CHOQUET space, Theorem 2 generalizes the classical theorem of the uniform continuity of continuous mappings on compact sets.

3. Spaces of uniformly continuous mappings

Let X and Y be pseudo-uniform spaces. In contrast to uniform spaces, the set $U(X, Y)$ of all uniformly continuous mappings of X into Y is not closed in the space Y^X of all mappings of X into Y equipped with the uniform convergence. In [8] we have constructed an example of a sequence of uniformly continuous functions on the interval $[0, 1]$ with values in a pseudo-topological vector space Y, uniformly converging to a continuous, but not uniformly continuous mapping of $[0, 1]$ into Y. On the other hand, we get the result that $U(X, Y)$ is closed in Y^X by refining the pseudo-topology of uniform convergence on X.

First, we generalize the notion of uniform equicontinuity (or equiuniform continuity) to pseudo-uniform spaces:

Definition. Let X and Y be pseudo-uniform spaces with the structures $\mathfrak{u}$ and $\mathfrak{v}$, respectively. A subset H of $U(X, Y)$ is said to be uniformly equicontinuous if for every $\mathfrak{U} \in \mathfrak{u}$ the filter $[H, \mathfrak{U}]$ in $Y \times Y$ with the base $\left\{ \bigcup_{f \in H} (f * f)(U) \mid U \in \mathfrak{U} \right\}$ belongs to $\mathfrak{v}$.

In the following, two examples of uniformly equicontinuous sets are given.

a) Let X, Y, and Z be pseudo-uniform spaces. For a mapping $f\colon X \times Y \to Z$, let the partial mappings $f_x\colon Y \to Z$ $(x \in X)$ and $f_y\colon X \to Z$ $(y \in Y)$ be defined by $f_x(y) = f_y(x) = f(x, y)$. Then f is uniformly continuous if and only if the two families $\{f_x \mid x \in X\}$ and $\{f_y \mid y \in Y\}$ are uniformly equicontinuous subsets of Z^Y and Z^X, respectively.

Proof. Let $\mathfrak{u}$, $\mathfrak{v}$, and $\mathfrak{w}$ be the pseudo-uniform structures of X, Y, and Z, respectively. A base of the pseudo-uniform product structure $\mathfrak{u} * \mathfrak{v}$ of $X \times Y$ is $\{\mathfrak{U} * \mathfrak{V} \mid \mathfrak{U} \in \mathfrak{u}, \mathfrak{V} \in \mathfrak{v}\}$ where the product filter $\mathfrak{U} * \mathfrak{V}$ in $(X \times Y) \times (X \times Y)$ has the base $\{U * V \mid U \in \mathfrak{U}, \mathfrak{V} \in V\}$ with

$$U * V = \left\{\left((u_1, v_1), (u_2, v_2)\right) \mid (u_1, u_2) \in U, (v_1, v_2) \in V\right\}.$$

Now the proof is an immediate consequence of the three filter relations

$$(f * f)(\mathfrak{U} * \mathfrak{V}) \supseteq [\{f_y \mid y \in Y\}, \mathfrak{U}] \circ [\{f_x \mid x \in X\}, \mathfrak{V}],$$

$$[\{f_y \mid y \in Y\}, \mathfrak{U}] \supseteq (f * f)(\mathfrak{U} * [\Delta_Y]),$$

and

$$[\{f_x \mid x \in X\}, \mathfrak{V}] \supseteq (f * f)([\Delta_X] * \mathfrak{V}),$$

which may be proved straightforward.

b) Let Y be a pseudo-topological vector space which is quasicomplete (see [7]), locally convex, regular, and separated. Let M be a subset of Y and let $f\colon [0,1]\times M \to Y$ be bounded and uniformly continuous on every compact subset. For every uniformly continuous mapping $g\colon [0,1] \to M$, the integral $\int_0^t f(s, g(s))\, ds$ exists in the sense of Riemann. Then the integral operator $\Phi\colon U([0,1], M) \to U([0,1], Y)$ defined by $\Phi(g(t)) = \int_0^t f(s, g(s))\, ds$ maps $U([0,1], M)$ into a uniformly equicontinuous family.

There are some useful properties of uniformly equicontinuous families, which are listed in the following two theorems.

For a non empty set I and a pseudo-uniform space Y with the structure $\mathfrak{v}$, we equip an arbitrary non empty subset M of Y^I with the pseudo-uniform structure $\mathfrak{p}_I$ of the uniform convergence on I: A filter $\mathfrak{P}$ in $M \times M$ belongs to $\mathfrak{p}_I$ if the filter $\mathfrak{P}(I)$ in $Y \times Y$ with the base $\{P(I) \mid P \in \mathfrak{P}\}$ belongs to $\mathfrak{v}$ where

$$P(I) = \{(f(x), g(x)) \mid x \in I, (f, g) \in P\}.$$

On the set M, $\mathfrak{p}_I$ generates the pseudo-topology of uniform convergence on I. For another space X and a subset H of Y^X, we define $\lambda\colon X \to Y^H$ by $\lambda(x)(f) = \mathrm{ev}(f, x) = f(x)$ where $\mathrm{ev}\colon H \times X \to Y$ is the evaluation mapping.

Theorem 3. *Let X and Y be pseudo-uniform spaces. Then for a subset $H \subseteq Y^X$ there are equivalent:*

(i) *H is uniformly equicontinuous.*

(ii) *$\lambda\colon X \to Y^H$ is uniformly continuous where Y^H is equipped with the uniform convergence on H.*

(iii) *$\mathrm{ev}\colon H \times X \to Y$ is uniformly continuous where H is equipped with the uniform convergence on X.*

Theorem 4. *Let X and Y be pseudo-uniform spaces and let Y be uniformly regular. If a subset H of $U(X, Y)$ is uniformly equicontinuous then the adherence $\overline{H}^p$ of H with respect to the pointwise convergence is uniformly equicontinuous, too.*

Proof. Let $\mathfrak{u}$ and $\mathfrak{v}$ be the pseudo-uniform structures of X and Y, respectively. Because of the uniform equicontinuity of H, for every $\mathfrak{U} \in \mathfrak{u}$ there is a $\mathfrak{V} \in \mathfrak{v}$ with $[H, \mathfrak{U}] \supseteq \mathfrak{V}$. Hence for every fixed $V \in \mathfrak{V}$ there is a $U_V \in \mathfrak{U}$ such that

$$\{(h(x), h(y)) \mid (x, y) \in U_V, h \in H\} \subseteq V.$$

For any $h_0 \in \overline{H}^p$ there is a filter $\mathfrak{H}_0$ on Y^X pointwise converging to h_0 such that $H \in \mathfrak{H}_0$. For arbitrary fixed $(x, y) \in U_V$, the filter $\mathfrak{H}_0(x) \times \mathfrak{H}_0(y)$ converges to $(h_0(x), h_0(y))$ in the product pseudo-topology. Because of $H \in \mathfrak{H}_0$, the filter with the base

$$\{\{(h(x), h(y)) \mid h \in H_0\} \mid H_0 \subseteq H, H_0 \in \mathfrak{H}_0\}$$

is finer than $\mathfrak{H}_0(x) \times \mathfrak{H}_0(y)$ and hence it converges to $(h_0(x), h_0(y))$, too. On the other hand, the set V belongs to this filter, and this implies $(h_0(x), h_0(y)) \in \overline{V}$ and hence

$$\{(h_0(x), h_0(y)) \mid (x, y) \in U_V, h_0 \in \overline{H}^p\} \subseteq \overline{V}.$$

Finally, we have $[\overline{H}^p, \mathfrak{U}] \supseteq \overline{\mathfrak{V}}$. Y being uniformly regular, this proves the uniform equicontinuity of $\overline{H}^p$.

Now we are going to refine the pseudo-topology of the uniform convergence. Let X and Y be pseudo-uniform spaces.

Definition. A filter $\mathfrak{H}$ in Y^X is said to be $u_{\mathfrak{b}}$-convergent to $f \in Y^X$ if $\mathfrak{H}$ converges to f uniformly on X and there is a uniformly equicontinuous set belonging to $\mathfrak{H}$.

Theorem 5. *For pseudo-uniform spaces X and Y, $u_{\mathfrak{b}}$ is a pseudo-topology on Y^X and $U(X, Y)$ is closed in Y^X with respect to $u_{\mathfrak{b}}$.*

In the case of Y being a uniform space, $U(X, Y)$ is closed in Y^X with respect to the usual uniform convergence.

Now, let X be an arbitrary non empty set. Then the set $B(X)$ of all bounded (real or complex valued) functions on X is a closed subset of the set of all functions on X — with respect to the uniform convergence on X. Applying this and using Theorem 5, an easy calculation yields

Theorem 6. *$B_{\mathfrak{b}}(X)$, the set of all bounded uniformly continuous (real or complex valued) functions on a pseudo-uniform space X, equipped with $u_{\mathfrak{b}}$ is a locally convex pseudo-topological vector space, which is closed in the space of all functions on X.*

4. Precompactness and total boundedness

The notion of a precompact — or totally bounded — uniform space may be generalized to pseudo-uniform spaces in a natural way (see Gähler [4], for example). Let X be a pseudo-uniform space with the structure $\mathfrak{u}$. X is said to be precompact if every ultra filter on X is a Cauchy filter. X is said to be totally bounded if there is a filter $\mathfrak{U} \in \mathfrak{u}$ such that for every $U \in \mathfrak{U}$, there is a finite convering $\{K_1, \ldots, K_n\}$ of X depending on U with $K_i \times K_i \subseteqq U$ $(i = 1, \ldots, n)$ — or, equivalently, there is a finite subset E of X depending on U such that for every $x \in X$ there is an $e \in E$ with $(x, e) \in U$.

For a uniform space X we have the classical equivalences

(i) X is precompact.
(ii) X is totally bounded.
(iii) The completion of X in the category of uniform spaces is compact.

In general, instead of these equivalences, Theorem 7 is valid.

Definition. Let X be a pseudo-uniform space with the structure $\mathfrak{u}$. A separated complete uniformly regular pseudo-uniform space $\tilde{X}$ together with a uniformly continuous mapping $j: X \to \tilde{X}$ is said to be a regular completion of X if $j(X)$ is dense in $\tilde{X}$, i.e., the adherence of $j(X)$ with respect to $\lambda(\mathfrak{u})$ is equal to $\tilde{X}$ and if for any uniformly continuous mapping f of X into a separated complete uniformly regular space Y there is one and only one uniformly continuous mapping $\tilde{f} : \tilde{X} \to Y$ with $f = \tilde{f} \circ j$ (universality property). Replacing uniform regularity and density by uniform t-regularity and density with respect to $t(\lambda(\mathfrak{u}))$, respectively, a t-regular completion is defined.

Theorem 7. *Let X be a pseudo-uniform space with the structure u.*

(i) *If X is totally bounded then X is precompact.*
(ii) *Let X be totally bounded. If X has a regular or a t-regular completion $\tilde{X}$ then this completion is compact.*

Proof. (i) The proof of (i) is given by Gähler [4]. In contrast to uniform spaces, the inverse of (i) in general is not true (for an example of a compact — and hence precompact — pseudo-uniform space not being totally bounded, see Gähler [4]). (ii) Only we have to show that every ultra filter on $\tilde{X}$ is a Cauchy filter. Let $\mathfrak{H}$ be an ultra filter on $\tilde{X}$. X being totally bounded, there is a filter $\mathfrak{U} \in \mathfrak{u}$ such that for every $U \in \mathfrak{U}$ there is a finite convering $\{K_1, \ldots, K_n\}$ of X with $K_i \times K_i \subseteqq U$ $(i = 1, \ldots, n)$. Then $\{j(K_i) \mid i = 1, \ldots, n\}$ is a finite covering of $j(X)$ such that $j(K_i) \times j(K_i) \subseteqq (j * j)\,(U)\,(i = 1, \ldots, n)$. $j(X)$ being dense (t-dense) in $\tilde{X}$, we have

$$\tilde{X} = \overline{\bigcup_{i=1}^{n} j(K_i)} = \bigcup_{i=1}^{n} \overline{j(K_i)}$$

(the adherence is to be understood with respect to $\lambda(\mathfrak{u})$ or $t(\lambda(\mathfrak{u}))$, respectively). $\mathfrak{H}$ being an ultra filter, one of the sets $\overline{j(K_i)}$ $(i = 1, \ldots, n)$ belongs to $\mathfrak{H}$, say $\overline{j(K_{i_0})}$. On account of

$$\overline{j(K_{i_0})} \times \overline{j(K_{i_0})} \in \mathfrak{H} \times \mathfrak{H},$$

of

$$\overline{j(K_{i_0})} \times \overline{j(K_{i_0})} = \overline{j(K_{i_0}) \times j(K_{i_0})} \subseteqq \overline{(j * j)\,(U)},$$

and of U being arbitrary, we get

$$\mathfrak{H} \times \mathfrak{H} \supseteqq \overline{(j * j)\,(\mathfrak{U})}.$$

As $\tilde{X}$ is uniformly regular or t-regular, respectively, this proves $\mathfrak{H}$ being a Cauchy filter. (The theorem remains true without the universality property.)

Now we are going to construct the precompact reflection pr X and the totally bounded reflection tb X of an arbitrary pseudo-uniform space.

Definition. Let X be a pseudo-uniform space with the structure $\mathfrak{u}$. The set underlying X, equipped with the finest precompact pseudo-uniform structure pr $\mathfrak{u}$ coarser than $\mathfrak{u}$, is said to be the precompact modification of the space X and is denoted by pr X. The set underlying X, equipped with the finest totally bounded pseudo-uniform structure tb $\mathfrak{u}$ coarser than $\mathfrak{u}$, is said to be the totally bounded modification of the space X and is denoted by tb X.

An easy calculation shows that the structures pr $\mathfrak{u}$ and tb $\mathfrak{u}$ may be described as follows:

pr $\mathfrak{u}$ is the finest pseudo-uniform structure on the set underlying X such that all the filters of $\mathfrak{u}$ and all the filters $\mathfrak{F} \times \mathfrak{F}$ belong to pr $\mathfrak{u}$ where $\mathfrak{F}$ is an ultra filter on X. A base of the $\wedge$-ideal pr $\mathfrak{u}$ is the set of finite intersections of filters

$$\mathfrak{U}_0 \quad \text{and} \quad \mathfrak{U}_1 \circ (\mathfrak{F}_1 \times \mathfrak{F}_1) \circ \mathfrak{U}_2 \circ \cdots \circ (\mathfrak{F}_n \times \mathfrak{F}_n) \circ \mathfrak{U}_{n+1}$$

where $\mathfrak{U}_0, \ldots, \mathfrak{U}_{n+1}$ are filters of $\mathfrak{u}$ and $\mathfrak{F}_1, \ldots, \mathfrak{F}_n$ are ultra filters on X such that these relation products exist.

tb $\mathfrak{u}$ is the finest pseudo-uniform structure on the set underlying X such that all the filters of $\mathfrak{u}$ and the filter $\mathfrak{K}$ on $X \times X$ with the base

$$\left\{ \bigcup_{K \in \mathscr{K}} K \times K \mid \mathscr{K} \text{ runs over all finite coverings of } X \right\}$$

belong to tb $\mathfrak{u}$. A base of the $\wedge$-ideal tb $\mathfrak{u}$ is the set of filters on $X \times X$ consisting of all the filters of $\mathfrak{u}$ and all finite products

$$\mathfrak{U}_1 \circ \mathfrak{K} \circ \mathfrak{U}_2 \circ \ldots \circ \mathfrak{K} \circ \mathfrak{U}_{n+1}$$

where $\mathfrak{U}_1, \ldots, \mathfrak{U}_{n+1}$ are filters of $\mathfrak{u}$ such that these relation products exist.

Theorem 8. *For every uniformly continuous mapping f of a pseudo-uniform space X into a precompact (totally bounded) pseudo-uniform space Y, the mapping f of* pr X (tb X) *into Y is uniformly continuous, too.*

Proof. First, let $\mathfrak{F}$ be an ultra filter on X. In the case of Y being precompact, the ultra filter $f(\mathfrak{F})$ is a Cauchy filter on Y. Because of $(f * f)\,(\mathfrak{F} \times \mathfrak{F}) = f(\mathfrak{F}) \times f(\mathfrak{F})$ and the special form of the base of the pseudo-uniform structure of pr$X$, this proves the uniform continuity of f: pr$X \to Y$. For the second assertion, let Y be totally bounded. Then there is a filter $\mathfrak{V}$ of the pseudo-uniform structure of Y such that for every $V \in \mathfrak{V}$ there is a finite covering $L_1, \ldots, L_n$ of Y with $L_i \times L_i \subseteqq V$ $(i = 1, \ldots, n)$. The sets $K_i = f^{-1}(L_i)$ $(i = 1, \ldots, n)$ form a covering of X such that $f(K_i) = f(f^{-1}(L_i)) \subseteqq L_i$ and, consequently, $(f * f)\,(K_i \times K_i) \subseteqq V$ $(i = 1, \ldots, n)$. On account of the definition of the filter $\mathfrak{K}$ before Theorem 8 and the special form of the base of the pseudo-uniform structure of tb X, this proves the uniform continuity of f: tb $X \to Y$.

Using Theorem 8, we may describe the precompact and the totally bounded modification of a pseudo-uniform space in a functorial way. Let $\mathrm{PUN_{pr}}$ and $\mathrm{PUN_{tb}}$ denote the subcategory of PUN of all precompact and of all totally bounded pseudo-uniform spaces, respectively. Then by Theorem 8, natural functors pr: PUN $\to \mathrm{PUN_{pr}}$ and tb: PUN $\to \mathrm{PUN_{tb}}$ may be defined: If X is a pseudo-uniform space then pr$X$ and tb$X$ are the precompact and the totally bounded modification of X, and if f is a uniformly continuous mapping of X into a pseudo-uniform space Y, we take $\mathrm{pr}(f) = \mathrm{tb}(f) = f$. To avoid misunderstandings, for the following theorem let us precisely write $\mathrm{pr_{PUN}}$ instead of pr and $\mathrm{tb_{PUN}}$ instead of tb. Let us denote the well-known precompact modification of a uniform space in the category UN (see Isbell [5]) with $\mathrm{pr_{UN}}$; the subcategory of UN of all precompact uniform spaces is denoted by $\mathrm{UN_{pr}}$.

Theorem 9.

(i) $\mathrm{pr_{PUN}}$: PUN $\to \mathrm{PUN_{pr}}$ and $\mathrm{tb_{PUN}}$: PUN $\to \mathrm{PUN_{tb}}$ *are reflections.*

(ii) $\mathrm{p_{PUN}} = \mu \circ (\mathrm{pr_{PUN}}|\mathrm{_{UN}}) = \mu \circ (\mathrm{tb_{PUN}}|\mathrm{_{UN}})$.

Proof. The first part is an obvious consequence of Theorem 8. For the second part, let X be a uniform space.

$\mathrm{pr_{PUN}}$: PUN $\to \mathrm{PUN_{pr}}$ being a reflection, the diagram

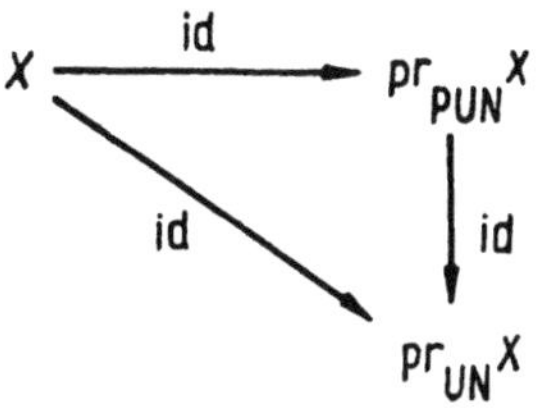

is commutative, i.e., $\mathrm{id}\colon \mathrm{pr}_{\mathrm{PUN}}X \to \mathrm{pr}_{\mathrm{UN}}X$ is uniformly continuous. $\mu\colon \mathrm{PUN} \to \mathrm{UN}$ being a reflection and $\mathrm{id}\colon \mathrm{pr}_{\mathrm{PUN}}X \to \mathrm{pr}_{\mathrm{UN}}X$ being uniformly continuous, the diagram

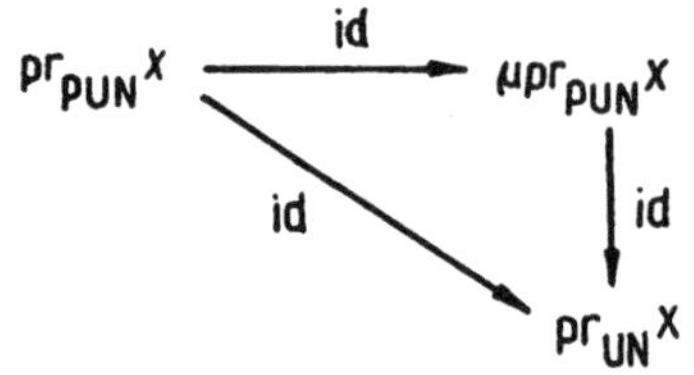

is commutative, i.e., $\mathrm{id}\colon \mathrm{pr}_{\mathrm{PUN}}X \to \mathrm{pr}_{\mathrm{UN}}X$ is uniformly continuous. $\mathrm{pr}_{\mathrm{UN}}\colon \mathrm{UN} \to \mathrm{UN}_{\mathrm{pr}}$ being a reflection, the diagram

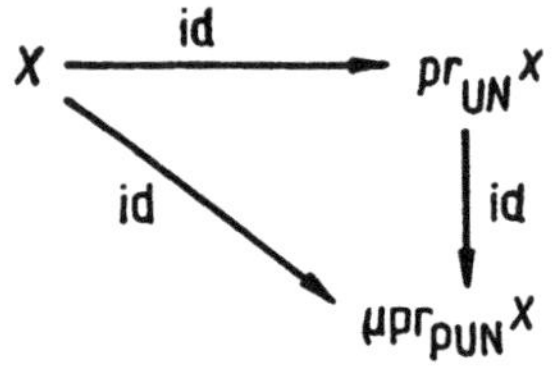

is commutative, i.e., $\mathrm{id}\colon \mathrm{pr}_{\mathrm{UN}}X \to \mathrm{pr}_{\mathrm{PUN}}X$ is uniformly continuous. For this, the following obvious fact is used: For any precompact pseudo-uniform space Y the uniform modification μY is precompact, too.

Hence the identical mapping of $\mathrm{pr}_{\mathrm{UN}}X$ into $\mu\mathrm{pr}_{\mathrm{PUN}}X$ and its inverse are uniformly continuous. This implies $\mathrm{pr}_{\mathrm{UN}} = \mu \circ (\mathrm{pr}_{\mathrm{PUN}} \mid \mathrm{UN})$. Substituting $\mathrm{tb}_{\mathrm{PUN}}$ for $\mathrm{pr}_{\mathrm{PUN}}$ in the three diagrams and using that for any totally bounded pseudo-uniform space Y the uniform modification is totally bounded, too, in an analogous way we get $\mathrm{pr}_{\mathrm{UN}} = \mu \circ (\mathrm{tb}_{\mathrm{PUN}}|_{\mathrm{UN}})$.

The problem in which we are interested is to construct a uniform compactification in the category PUN. The uniform compactification of a uniform space — the so called SAMUEL compactification — may be described as follows. Let c_{UN} be the usual completion functor in the category UN. Then the uniform compactification in UN is the composition $c_{\mathrm{UN}} \circ \mathrm{pr}_{\mathrm{UN}}$. The connection between the uniform compactification of a uniform space X and the STONE-ČECH compactification of the topological space λX corresponding to X is the following: For a uniformizable topological space Y we equip the set underlying Y with the finest uniform structure generating the topology of Y; let $\mathrm{fin}\, Y$ denote this uniform space. Then the STONE-ČECH compactification of λX is equal to

$$c_{\mathrm{UN}}\big(\mathrm{pr}_{\mathrm{UN}}(\mathrm{fin}\,(\lambda X))\big).$$

The problem is how to construct a uniform compactification in PUN by composing a suitable precompactification and a suitable completion in PUN and to compare this compactification with the compactification of KENT and RICHARDSON [6] for the associated pseudo-topological space. The application of Theorem 7 to the totally bounded modification is a step in this direction. However, the existence of a (t-)regular completion in PUN is needed. In [9], MÜLLER constructs a regular completion in the subcategory PVEK of PUN of all pseudo-topological vector spaces. In general, the problem of the existence of a (t-)regular completion for arbitrary pseudo-uniform spaces is open.

References

[1] N. Bourbaki, Éléments de mathématique, Livre III: Topologie générale, chap. I, II, Paris 1961.
[2] C. H. Cook and H. R. Fischer, On equicontinuity and continuous convergence, Math. Ann. **159** (1965), 94–104.
[3] C. H. Cook and H. R. Fischer, Uniform convergence structures, Math. Ann. **173** (1967), 290–306.
[4] W. Gähler, Grundstrukturen der Analysis. I, II, Berlin/Basel und Stuttgart 1977, 1978.
[5] J. R. Isbell, Uniform spaces, Providence 1964.
[6] D. C. Kent and G. D. Richardson, Regular compactifications of convergence spaces, Proc. Amer. Math. Soc. **31** (1972), 517–523.
[7] G. Kneis, Holomorphe Funktionen mit Werten in einem pseudotopologischen Vektorraum, Math. Nachr. **62** (1974), 147–161.
[8] G. Kneis, Contributions to the theory of pseudo-uniform spaces, Math. Nachr. **89** (1979), 149–163.
[9] B. Müller, Vervollständigungen von Limesvektorräumen, Mannheimer Manuskripte Nr. 51 (1975).

Anschrift des Verfassers:

Akademie der Wissenschaften der DDR
Zentralinstitut für Mathematik und Mechanik
108 Berlin
Mohrenstraße 39
DDR

Strict Topologies and the Continuous Convergence Structure

BERND MÜLLER, Mannheim

The concept of final topologies is very often used in Functional Analysis. One example of a final topology is the strict topology on a locally convex topological vector space E which is defined to be the finest locally convex space topology on E agreeing on every bounded subset of E with the topology of E.

In the first part of this paper we are interested in a description of the strict topology of $\mathscr{L}_{\mathfrak{s}}(E, F)$ where E is a complete barreled space, F is a BANACH space and $\mathscr{L}_{\mathfrak{s}}(E, F)$ is the set $\mathscr{L}(E, F)$ of all continuous linear mappings from E into F endowed with the topology of pointwise convergence. The well-known BANACH-STEINHAUS theorem says that the strict topology on $\mathscr{L}_{\mathfrak{s}}(E, F)$ is finer than the topology of $\mathscr{L}_{\mathfrak{co}}(E, F)$, the set $\mathscr{L}(E, F)$ carrying the topology of compact convergence. Since the continuous convergence structure on $\mathscr{L}(E, F)$ is the finest convergence structure agreeing on every bounded subset of $\mathscr{L}_{\mathfrak{s}}(E, F)$ with the topology induced by $\mathscr{L}_{\mathfrak{s}}(E, F)$, the strict topology of $\mathscr{L}_{\mathfrak{s}}(E, F)$ coincides with the finest locally convex vector space topology coarser than the continuous convergence structure. Let us denote by $\mathscr{L}_{\mathfrak{c}}(E, F)$ and $(\mathscr{L}_{\mathfrak{c}}(E, F))_{\mathfrak{r}}$ the set $\mathscr{L}(E, F)$ endowed with the continuous convergence structure and the finest locally convex vector space topology coarser than the continuous convergence structure respectively. In case E is a nuclear FRÉCHET space, a nuclear (DF)-space or a BANACH space with the bounded approximation property we will show $(\mathscr{L}_{\mathfrak{c}}(E, F))_{\mathfrak{r}} = \mathscr{L}_{\mathfrak{co}}(E, F)$ for every BANACH space F. But we will also give an example of a complete nuclear barreled space E and of a BANACH space F such that $(\mathscr{L}_{\mathfrak{c}}(E, F))_{\mathfrak{r}} \neq \mathscr{L}_{\mathfrak{co}}(E, F)$.

In chapter III we investigate the strict topology on subspaces N of $\mathscr{L}_{\mathfrak{s}}(E, \boldsymbol{R})$ where E is a BANACH space. The BANACH-DIEUDONNÉ theorem says that $\mathscr{L}_{\mathfrak{co}}(E, \boldsymbol{R})$ carries the strict topology of $\mathscr{L}_{\mathfrak{s}}(E, \boldsymbol{R})$. First we will characterize those subspaces N of $\mathscr{L}_{\mathfrak{s}}(E, \boldsymbol{R})$ for which the strict topology on N is induced by the topology of $\mathscr{L}_{\mathfrak{co}}(E, \boldsymbol{R})$. Then we are interested in BANACH spaces whose dual has the following property:

(*) For every subspace N of $\mathscr{L}_{\mathfrak{s}}(E, \boldsymbol{R})$ the strict topology on N is induced by the topology of $\mathscr{L}_{\mathfrak{co}}(E, \boldsymbol{R})$.

It turns out that the property (*) is equivalent to the property that the HAHN-BANACH theorem is true in $\mathscr{L}_{\mathfrak{c}}(E, \boldsymbol{R})$. Therefore let us say that the dual of E has the $\mathfrak{c}$-HAHN-BANACH property ($\mathfrak{c}$-HBP) iff it has the property (*). We will show that the dual of a quasi-reflexive BANACH space has the $\mathfrak{c}$-HBP and that the dual of every BANACH space containing a copy of c_0 or $\mathfrak{l}_1$ does not possess the $\mathfrak{c}$-HBP. Therefore the dual of a BANACH lattice E has the $\mathfrak{c}$-HBP iff E is reflexive. Since a BANACH space E is quasi-reflexive iff E has a finite dimensional complement in its bidual E'', one could hope that also the dual of every BANACH space F which is a complemented subspace of its bidual F'' has the c-HBP. But we will give an example of a BANACH space E with the following properties: E does not contain a copy of c_0 or $\mathfrak{l}_1$, $E'' = E \otimes \mathfrak{l}_2$ and the dual of E has not the c-HBP. All proof which are omitted in this paper can be found in [9] and [10].

All convergence spaces considered in this paper will be convergence spaces in the sense of FISCHER [4]. For a subset S of a convergence space X the adherence $a(S)$ is defined to be the set $\{y: y \in X, \exists$ a filter Φ converging to y in X with $S \in \Phi\}$. S is called dense in X if $a(S) = X$, closed if $a(S) = S$ and compact if every ultrafilter Φ with $S \in \Phi$ converges in S. A mapping f from a convergence space X into a convergence space Y is continuous if for all $x \in X$ and all filters Φ converging to x in X the filter $f(\Phi)$ generated by $\{f(L): L \in \Phi\}$ converges to $f(x)$ in Y. Let us denote by $\mathscr{C}_c(X, Y)$ the set $\mathscr{C}(X, Y)$ of all continuous mappings from X to Y endowed with the continuous convergence structure. This convergence structure is the coarsest on $\mathscr{C}(X, Y)$ for which the mapping $\omega: \mathscr{C}(X, Y) \times X \to Y$ defined by $\omega(f, x) = f(x) \ \forall\, x \in X, \forall\, f \in \mathscr{C}(X, Y)$, is continuous. This means that a filter $\mathscr{F}$ converges to f in $\mathscr{C}_c(X, Y)$ iff for all $x \in X$ and all filters Φ converging to x in X the filter $\omega(\mathscr{F}, \Phi)$ converges to $f(x)$ in Y.

All vector spaces considered in this paper will be $\boldsymbol{R}$-vector spaces. A vector space endowed with a convergence structure is called a convergence vector space (cvs) if the algebraic operations are continuous. A subspace of a cvs E will be always a vector subspace carrying the convergence structure induced by E. Let E, F and G be convergence vector spaces. We denote by $\mathscr{L}_c(E, F)$ and $\mathscr{B}_c(E, F; G)$ the subspaces of $\mathscr{C}_c(E, F)$ and $\mathscr{C}_c(E \times F, G)$ respectively consisting of the set $\mathscr{L}(E, F)$ of all continuous linear mappings from E into F and of the set of all continuous bilinear mappings from $E \times F$ into G. If E is a barreled locally convex topological vector space and F is a BANACH space it is easy to see that a filter $\mathscr{F}$ converges to T in $\mathscr{L}_c(E, F)$ iff $\mathscr{F}$ converges pointwise to T and $\mathscr{F}$ contains a pointwise bounded subset of $\mathscr{L}(E, F)$. Therefore $\mathscr{L}_c(E, F)$ carries the finest convergence structure agreeing with the topology of pointwise convergence on every bounded subset of $\mathscr{L}_s(E, F)$. We abbreviate $\mathscr{L}(E, \boldsymbol{R})$, $\mathscr{L}_c(E, \boldsymbol{R})$, $\mathscr{L}_s(E, \boldsymbol{R})$ and $\mathscr{L}_{co}(E, \boldsymbol{R})$ to $\mathscr{L}E$, $\mathscr{L}_c(E)$ or $\mathscr{L}_cE$, $\mathscr{L}_s(E)$ and $\mathscr{L}_{co}(E)$. For every cvs E there esists a natural mapping $j: E \to \mathscr{L}_c\mathscr{L}_cE$ defined by $[j(x)]\,(\zeta) = \zeta(x) \ \forall\, x \in E$, $\forall\, \zeta \in \mathscr{L}E$, which is always continuous. E is called $\mathscr{L}_c$-embedded if j is an isomorphism from E onto $j(E)$, and c-reflexive if j is an isomorphism from E onto $\mathscr{L}_c\mathscr{L}_cE$. For every $\mathscr{L}_c$-embedded cvs E we will identify E and $j(E)$.

Let X and Y be vector spaces and $\langle X, Y\rangle$ a duality (see [12]). For a subset A of X we denote by A^0 the polar of A which is defined to be the set $\{y: y \in Y, |y(a)| \leqq 1 \ \forall\, a \in A\}$ and by ΓA the convex circled hull of A. Let us mention that $\langle \mathscr{L}(E, F), E \otimes \mathscr{L}F\rangle$ is a duality if E and F are separated locally convex topological vector spaces. This duality is defined by $(x \otimes \zeta)\,(T) = \zeta\big(T(x)\big) \ \forall\, x \otimes \zeta \in E \otimes \mathscr{L}F$ and $\forall\, T \in \mathscr{L}(E, F)$.

I. A duality for $\mathscr{L}_c(E, F)$

For every cvs V the finest locally convex vector space topology on V coarser than the convergence structure of V is the topology of uniform convergence on the compact subsets of $\mathscr{L}_cV$. To describe the topology of $\big(\mathscr{L}_c(E, F)\big)_\tau$ we have to investigate the dual of $\mathscr{L}_c(E, F)$ and the compact subsets of $\mathscr{L}_c(\mathscr{L}_c(E, F))$. For this reason we are interested in convergence tensor products. For cvs E and F let us denote by $E \otimes_{\mathscr{L}} F$ the tensor product of E and F in the category of all convergence vector spaces. The existence of tensor products in this category was proved by several authors (see [1] and [6]). The convergence structure on $E \otimes_{\mathscr{L}} F$ is the finest convergence vector space structure for which the canonical mapping $\chi: E \times F \to E \otimes F$ defined by $\chi(x, y) = x \otimes y$, is continuous. As usual we write $U \otimes V$ instead of $\chi(U, V)$ for subsets U and V of E and F, respectively. Throughout this paper we assume that E and F have point-separating duals. Therefore the mapping $j: E \otimes_{\mathscr{L}} F \to \mathscr{L}_c\mathscr{L}_c(E \otimes_{\mathscr{L}} F)$ is injective. Let us denote by $E \otimes_c F$ the tensor product of E and F endowed with the convergence structure induced by the mapping j.

Theorem I.1. *The cvs $E \otimes_c F$ has the following properties:*

(i) *$E \otimes_c F$ carries the finest $\mathscr{L}_c$-embedded convergence structure for which the mapping $\chi: E \times F \to E \otimes F$ is continuous.*

(ii) *$\mathscr{L}_c(E \otimes_c F, G)$ and $\mathscr{B}_c(E, F; G)$ are isomorphic for all $\mathscr{L}_c$-embedded cvs G.*

(iii) *If E and F are locally convex topological vector spaces we have $E \otimes_c F = E \otimes_\pi F$, where $E \otimes_\pi F$ is the projective tensor product of E and F.*

Since $\mathscr{L}_c(E, \mathscr{L}_c F)$ and $\mathscr{B}_c(E, F; \mathbf{R})$ are always isomorphic (see [1] and [2]) we conclude:

Corollary. *$\mathscr{L}_c(E \otimes_c F)$ and $\mathscr{L}_c(E, \mathscr{L}_c F)$ are isomorphic for all cvs E and F. If F is $\mathfrak{c}$-reflexive $\mathscr{L}_c(E \otimes_c \mathscr{L}_c F)$ and $\mathscr{L}_c(E, F)$ are isomorphic. For convenience, let us write $\mathscr{L}_c(E \otimes_c \mathscr{L}_c F) = \mathscr{L}_c(E, F)$ if F is $\mathfrak{c}$-reflexive.*

From now on, E and F are always assumed to be complete separated locally convex topological vector spaces. Then E and F are $\mathfrak{c}$-reflexive spaces (see [3]).

Proposition I.2. *$(\mathscr{L}_c(E, F))_\tau = \mathscr{L}_{co}(E, F)$ iff, for every compact subset C of $\mathscr{L}_c(\mathscr{L}_c(E, F))$, there exist compact subsets K_1 and K_2 of E and $\mathscr{L}_c F$, respectively, such that $C \subseteq a(\Gamma(K_1 \otimes K_2))$, where the adherence is taken in $\mathscr{L}_c(\mathscr{L}_c(E, F))$.*

Theorem I.3. *If F is a Banach space then $E \otimes_c \mathscr{L}_c F$ is a dense subspace of $\mathscr{L}_c(\mathscr{L}_c(E, F))$ and $\mathscr{L}_c(\mathscr{L}_c(E, F))$ is the T_3-completion of $E \otimes_c \mathscr{L}_c F$.*

II. The strict topology of $\mathscr{L}_s(E, F)$

Throughout this section we assume that E is a complete separated locally convex topological vector space and F is a BANACH space. Let $\mathscr{L}_s(E, F)$ and $\mathscr{L}_{co}(E, F)$ be the set $\mathscr{L}(E, F)$ endowed with the topology of pointwise convergence and of compact convergence, respectively. For subsets $C \subseteq \mathscr{L}(\mathscr{L}_c(E, F))$ and $D \subseteq \mathscr{L}(E, F)$ let us denote by $\bar{C}^\sigma$ and $\bar{D}^s$ the closure of C with respect to the topology $\sigma(\mathscr{L}(\mathscr{L}_c(E, F)), \mathscr{L}(E, F))$ and of D in $\mathscr{L}_s(E, F)$ respectively. We will always consider $E \otimes \mathscr{L}F$ as a subset of $\mathscr{L}(\mathscr{L}_c(E, F))$ and $\mathscr{L}E \otimes F$ as the subspace of $\mathscr{L}(E, F)$ consisting of all operators with finite rank. For every $B \subseteq E$ let us define

$$T_B = \{T: T \in \mathscr{L}(E, F), \|T(b)\| \leq 1 \; \forall \, b \in B\}.$$

Proposition II.1. *The topology of $\mathscr{L}_{co}(E) \otimes_\pi F$ is finer than the topology induced on $\mathscr{L}E \otimes F$ by $(\mathscr{L}_c(E, F))_\tau$.*

If $\mathscr{L}_{co}(E)$ is a nuclear space, then $\mathscr{L}_{co}(E) \otimes_\pi F$ is a topological subspace of $\mathscr{L}_{co}(E, F)$. In this case $\mathscr{L}_{co}(E, F)$ and $(\mathscr{L}_c(E, F))_\tau$ induce the same topology on $\mathscr{L}E \otimes F$. If in addition E is a nuclear space $\mathscr{L}E \otimes F$ is dense in $\mathscr{L}_c(E, F)$, and in this case it is possible to show $(\mathscr{L}_c(E, F))_\tau = \mathscr{L}_{co}(E, F)$.

Theorem II.2. *Let E be a nuclear Fréchet space or a nuclear (DF)-space and let F be a Banach space. Then we have*

$$(\mathscr{L}_c(E, F))_\tau = \mathscr{L}_{co}(E, F).$$

In the following example we will show that there exist a complete nuclear barreled space E and a Banach space F such that $(\mathscr{L}_c(E, F))_\tau \neq \mathscr{L}_{co}(E, F)$. Therefore $\mathscr{L}_{co}(E, F)$ carries not always the strict topology generated by the topology of pointwise convergence on $\mathscr{L}(E, F)$.

Example 1. Let D be an uncountable set and let $\mathscr{H}$ be the family of all finite subsets of D. Define $R_i = \mathbf{R}$ for all $i \in D$. Let us denote by $\operatorname{ind}_{\mathscr{C}}$ and $\operatorname{ind}_{\mathscr{T}}$ the inductive limit taken in the category of all convergence spaces and in the category of all locally convex topological vector spaces, respectively. For every cvs V let us denote by V_τ the set V endowed with the finest locally convex vector space topology coarser than the convergence structure of V. Then we have $\mathscr{L}_c\left(\prod_{i\in D} R_i\right) = \operatorname{ind}_{\mathscr{C}, H\in\mathscr{H}} \mathscr{L}_c\left(\prod_{i\in H} R_i\right)$ and

$$\mathscr{L}_{co}\left(\prod_{i\in D} R_i\right) = \left(\mathscr{L}_c\left(\prod_{i\in D} R_i\right)\right)_\tau = \operatorname*{ind}_{\mathscr{T}, H\in\mathscr{H}} \mathscr{L}_{co}\left(\prod_{i\in H} R_i\right).$$

Since every $T \in \mathscr{L}\left(\prod_{i\in D} R_i, \mathfrak{l}_1\right)$ has finite rank and since for every $H \in \mathscr{H}$ we have

$$\mathscr{L}_c\left(\prod_{i\in H} E_i, \mathfrak{l}_1\right) = \mathscr{L}_{co}\left(\prod_{i\in H} R_i, \mathfrak{l}_1\right) = \mathscr{L}_{co}\left(\prod_{i\in H} R_i\right) \otimes_\pi \mathfrak{l}_1$$

it is easy to see that

$$\mathscr{L}_c\left(\prod_{i\in D} R_i, \mathfrak{l}_1\right) = \operatorname*{ind}_{\mathscr{C}, H\in\mathscr{H}} \mathscr{L}_c\left(\prod_{i\in H} R_i, \mathfrak{l}_1\right) = \operatorname*{ind}_{\mathscr{C}, H\in\mathscr{H}} \left(\mathscr{L}_{co}\left(\prod_{i\in H} R_i\right) \otimes_\pi \mathfrak{l}_1\right).$$

This implies

$$\left(\mathscr{L}_c\left(\prod_{i\in D} R_i, \mathfrak{l}_1\right)\right)_\tau = \operatorname*{ind}_{\mathscr{T}, H\in\mathscr{H}} \left(\mathscr{L}_{co}\left(\prod_{i\in H} R_i\right) \otimes_\pi \mathfrak{l}_1\right)$$
$$= \left(\operatorname*{ind}_{\mathscr{T}, H\in\mathscr{H}} \mathscr{L}_{co}\left(\prod_{i\in H} R_i\right)\right) \otimes_\pi \mathfrak{l}_1 = \mathscr{L}_{co}\left(\prod_{i\in D} R_i\right) \otimes_\pi \mathfrak{l}_1.$$

Since $\mathscr{L}_{co}(\mathbf{R}^D)$ is not nuclear, we have

$$(\mathscr{L}_c(\mathbf{R}^D, \mathfrak{l}_1))_\tau = \mathscr{L}_{co}(\mathbf{R}^D) \otimes_\pi \mathfrak{l}_1 \neq \mathscr{L}_{co}(\mathbf{R}^D) \otimes_\varepsilon \mathfrak{l}_1 = \mathscr{L}_{co}(\mathbf{R}^D, \mathfrak{l}_1).$$

Let us now investigate the situation when E is also a Banach space. With the same methods used in the proof of Lemma 2.4 in [7] it is possible to show:

Lemma II.3. *Let E be a Banach space, G a subspace of E with finite dimension k and let $T \in \mathscr{L}(E, E)$ be a compact operator. Assume that there exists an $\varepsilon \in \mathbf{R}$ with $0 < \varepsilon < 1$ and $k\varepsilon(1 - \varepsilon)^{-1} < 1$ such that $\|T(x) - x\| < \varepsilon \|x\|$ $\forall x \in G$. Then there exists a compact operator $S \in \mathscr{L}(E, E)$ such that*

$$S(x) = x \quad \text{for all} \quad x \in G \quad \text{and} \quad \|S\| < 2 \|T\|.$$

For every $S \in \mathscr{L}(E, E)$ let us define the mapping $S \otimes I: E \otimes \mathscr{L}F \to E \otimes \mathscr{L}F$ by

$$(S \otimes I)\left(\sum_{i=1}^{r} e_i \otimes \zeta_i\right) = \sum_{i=1}^{r} S(e_i) \otimes \zeta_i \;\forall\; \sum_{i=1}^{r} e_i \otimes \zeta_i \in E \otimes \mathscr{L}F .$$

All polars used in the next lemmas and theorems are taken with respect to the duality $\langle F, \mathscr{L}F\rangle$ or $\big\langle \mathscr{L}(E, F), \mathscr{L}\big(\mathscr{L}_c(E, F)\big)\big\rangle$.

Lemma II.4. *Let B be a finite subset of E and let V be the closed unit ball in F. Every $S \in \mathscr{L}(E, E)$ with $S(b) = b$ for all $b \in B$ has the property:*

$$(S \otimes I)(z) = z \quad \textit{for all} \quad z \in (E \otimes \mathscr{L}F) \cap \overline{\Gamma(B \otimes V^0)}^{\sigma} .$$

Proof. For every $z \in (E \otimes \mathscr{L}F) \cap \overline{\Gamma(B \otimes V^0)}^{\sigma}$ there exists a net $(z_\alpha)_{\alpha \in J}$ in $\Gamma(B \otimes V^0)$ which converges to z with respect to the topology $\sigma\big(E \otimes \mathscr{L}F, \mathscr{L}(E, F)\big)$. Assume $B = \{b_1, \ldots, b_r\}$. For every z_α choose a representation $z_\alpha = \sum_{i=1}^{r} \lambda_{\alpha,i} b_i \otimes \zeta_{\alpha,i}$ with $\zeta_{\alpha,i} \in V^0$ for $1 \leqq i \leqq r$ and $\sum_{i=1}^{r} |\lambda_{\alpha,i}| \leqq 1$. For all $T \in \mathscr{L}(E, F)$ we have: $\big(T(z_\alpha)\big)_{\alpha \in J}$ converges to $T(z)$ and from

$$T(z_\alpha) = \sum_{i=1}^{r} \lambda_{\alpha,i}\zeta_{\alpha,i}\big(T(b_i)\big) = \sum_{i=1}^{r} \lambda_{\alpha,i}\zeta_{\alpha,i}\big(T \circ S(b_i)\big) = T \circ S(z_\alpha)$$

we conclude that $\big(T(z_\alpha)\big)_{\alpha \in J}$ converges also to $T \circ S(z) = T\big((S \otimes I)(z)\big)$. This implies $T(z) = T\big((S \otimes I)(z)\big)$ for all $T \in \mathscr{L}(E, F)$). Thus we have $z = (S \otimes I)(z)$.

Lemma II.5. *Let B be a finite subset of E and let U and V be the closed unit balls in E and F, respectively. For all $0 < \varepsilon \in \mathbf{R}$ and for all $S \in \mathscr{L}(E, E)$ with $S(b) = b \;\forall\; b \in B$ we have*

$$\overline{\Gamma(\varepsilon U \otimes V^0)}^{\sigma} \cap \overline{\Gamma(B \otimes V^0)}^{\sigma} \cap (E \otimes \mathscr{L}F) \subseteqq \overline{\Gamma\big(\varepsilon S(U) \otimes V^0\big)}^{\sigma} .$$

Proof. Choose $z \in \overline{\Gamma(\varepsilon U \otimes V^0)}^{\sigma} \cap \overline{\Gamma(B \otimes V^0)}^{\sigma} \cap (E \otimes \mathscr{L}F)$. Lemma II.4 implies $(S \otimes I)(z) = z$. Assume $z \notin \overline{\Gamma\big(\varepsilon S(U) \otimes V^0\big)}^{\sigma}$. Then one can find a $T \in \mathscr{L}(E, E)$ with $T\big(\varepsilon S(U)\big) \subseteqq V$ and $T(z) > 1$. This implies $T \circ S(z) = T\big((S \otimes I)(z)\big) = T(z) > 1$. From $T \circ S(\varepsilon U) \subseteqq V$ we conclude $T \circ S \in \big(\Gamma(\varepsilon U \otimes V^0)\big)^0$ which contradicts the fact that $z \in \overline{\Gamma(\varepsilon U \otimes V^0)}^{\sigma}$.

Proposition II.6. *The set $\mathscr{K}(E, E)$ of all compact operators from E into E is dense in $\mathscr{L}_c(E, E)$ iff there exists a net $(T_\alpha)_{\alpha \in J}$ in $\mathscr{K}(E, E)$ with the following properties:*

(i) $\big(T_\alpha(x)\big)_{\alpha \in J}$ *converges to x for all $x \in E$.*

(ii) $\exists\, \mu \in \mathbf{R}$, $0 < \mu$ *with* $\|T_\alpha\| \leqq \mu \;\forall\; \alpha \in J$.

Let us remember that, for Banach spaces E and F, a filter Φ converges to 0 in $\mathscr{L}_c(E, F)$ iff it converges to 0 in $\mathscr{L}_{\mathfrak{s}}(E, F)$ and there is an $n \in \mathbf{N}$ with $nT_U \in \Phi$, where U is the closed unit ball in E and T_B is defined to be the set $\{T: T \in \mathscr{L}(E, F), \|T(b)\| \leqq 1 \;\forall\; b \in B\}$ for all subsets B of E. Therefore the continuous convergence structure is the finest convergence structure on $\mathscr{L}(E, F)$ which agrees on nT_U for every $n \in \mathbf{N}$ with the topology induced by $\mathscr{L}_{\mathfrak{s}}(E, F)$. This means that the topology of $\big(\mathscr{L}_c(E, F)\big)_r$ is the finest locally convex vector space topology agreeing on every member of $\{nT_U: n \in \mathbf{N}\}$ with the topology induced by $\mathscr{L}_{\mathfrak{s}}(E, F)$. Since $\{nT_U: n \in \mathbf{N}\}$ is countable we can apply the following result of ROELCKE [11]: The sets of the form

$$W = T_{B_0} \cap \bigcap_{n\in N} \overline{(nT_U + T_{B_n})}^{s},$$

where B_n is a finite subset of E for all $n \in N \cup \{0\}$, are a 0-neighborhood base for $(\mathscr{L}_c(E, F))_\tau$. From

$$\frac{1}{2} W = \frac{1}{2} T_{B_0} \cap \bigcap_{n\in N} \overline{\left(\frac{1}{2} nT_U + \frac{1}{2} T_{B_n}\right)}^{s} \subseteqq T_{2B_0} \cap \bigcap_{n\in N} \overline{\Gamma(nT_U \cup T_{B_n})}^{s} \subseteqq W$$

we conclude:

Lemma II.7. *The sets of the form*

$$T_{B_0} \cap \bigcap_{n\in N} \overline{\Gamma(T_{(1/n)U} \cup T_{B_n})}^{c}$$

are a 0-neighbourhood base of $(\mathscr{L}_c(E, F))_\tau$, *where* B_n *is a finite subset of* E *for all* $n \in N \cup \{0\}$.

Theorem II.8. *If the set* $\mathscr{K}(E, E)$ *of all compact operators from* E *into* E *is dense in* $\mathscr{L}_c(E, E)$ *we have for every Banach space* F:

$$(\mathscr{L}_c(E, F))_\tau = \mathscr{L}_{co}(E, F).$$

Proof. Let $W = T_{B_0} \cap \bigcap_{n\in N} \overline{\Gamma(T_{(1/n)U} \cup T_{B_n})}^{s}$ be a 0-neighbourhood in $(\mathscr{L}_c(E, F))_\tau$ where B_n is a finite subset of E for all $n \in N \cup \{0\}$. Proposition II.6 implies the existence of a net $(T_\alpha)_{\alpha\in J}$ in $\mathscr{K}(E, E)$ and of a real number $\mu > 0$, such that $\|T_\alpha\| \leqq \frac{1}{2} \mu \ \forall\ \alpha \in J$ and $(T_\alpha)_{\alpha\in J}$ converges to the identity mapping in $\mathscr{L}_{co}(E, E)$. Since $U \cap [B_n]$ is compact, where $[B_n]$ is the vector space generated by B_n in E, we conclude from Lemma II.3: For every $n \in N$ there exists an $S_n \in \mathscr{K}(E, E)$ with $\|S_n\| \leqq \mu$ and $S_n(b) = b$ for all $b \in B_n$. Lemma II.5 now implies:

$$\begin{aligned} T_{(1/n)S_n(U)} &= \left(\Gamma\left(\frac{1}{n} S_n(U) \otimes V^0\right)\right)^0 \\ &\subseteqq \left(\overline{\Gamma\left(\frac{1}{n} U \otimes V^0\right)}^{\sigma} \cap \overline{\Gamma(B_n \otimes V^0)}^{\sigma} \cap (E \otimes \mathscr{L}F)\right)^0 = \overline{\Gamma(T_{(1/n)U} \cup T_{B_n})}^{s}, \end{aligned}$$

where V is the unit ball in F. Defining $K = \bigcup_{n\in N} \frac{1}{n} \overline{S_n(U)}$ we get:

$$T_{K\cup B_0} = T_{B_0} \cap \bigcap_{n\in N} T_{(1/n)\overline{S_n(U)}} = T_{B_0} \cap \bigcap_{n\in N} T_{(1/n)\overline{S_n(U)}} \subseteqq W.$$

Since $\overline{S_n(U)}$ is a compact subset of μU, the set $K \cup B_0$ is also compact, and Theorem II.8 is proved.

Proposition II.9. *A Banach space* E *has the bounded approximation property iff the set* $\mathscr{L}E \otimes E$ *of all operators of finite rank is dense in* $\mathscr{L}_c(E, E)$.

Together with Theorem II.8 this proposition implies:

Theorem II.10. *Assume* E *is a Banach space with the bounded approximation property. Then we have* $\mathscr{L}_{co}(E, F) = (\mathscr{L}_c(E, F))_\tau$ *for every Banach space* F.

III. The strict topology on subspaces of $\mathscr{L}_{\mathfrak{s}}(E)$

In this section we assume always that E is a BANACH space. It is well-known that $\mathscr{L}_{co}(E)$ carries the strict topology of $\mathscr{L}_{\mathfrak{s}}(E)$, the set $\mathscr{L}E$ endowed with the topology of pointwise convergence which is also called the weak* topology. Now we are interested in the strict topology on subspaces N of $\mathscr{L}_{\mathfrak{s}}(E)$ and their relationship to the topology induced by $\mathscr{L}_{co}(E)$. Let us remark that the adherence $a(N)$ of a subspace N of $\mathscr{L}_{c}(E)$ is the set $\bigcup_{k\in \boldsymbol{N}} k\overline{(N \cap U^0)}^*$, where U is the closed unit ball in E, U^0 is the polar of U with respect to the duality $\langle E, \mathscr{L}E\rangle$ and $\overline{A}^*$ denotes the weak* closure of A for every subset A of $\mathscr{L}E$.

Theorem III.1. *Let E be a Banach space, N a subspace of $\mathscr{L}_cE$ and $a(N)$ the adherence of N taken in $\mathscr{L}_cE$. Equivalent are:*

(i) *The finest locally convex vector space topology on N agreeing with the weak* topology on every weak* bounded subset of N is induced by the topology of $\mathscr{L}_{co}(E)$.*
(ii) *$a(N)$ is a closed subspace of $\mathscr{L}_c(E)$.*
(iii) *$\forall\, \zeta \in \mathscr{L}(N)\ \exists\, x \in E$ such that $\zeta(f) = f(x)\ \forall\, f \in N$.*
(iv) *Every $\zeta \in \mathscr{L}(N)$ has a continuous linear extension to $\mathscr{L}_c(E)$.*

Remark. $a(N)$ is a closed subspace of $\mathscr{L}_cE$ iff $a(N) = \overline{N}^*$.

Since the strict topology on a subspace N of $\mathscr{L}_{\mathfrak{s}}(E)$ is induced by the topology of $\mathscr{L}_{co}(E)$ iff every linear mapping $\zeta: N \to \boldsymbol{R}$ which is continuous with respect to the convergence structure induced by $\mathscr{L}_c(E)$ has a continuous extension to $\mathscr{L}_c(E)$, let us define:

Definition. The dual $\mathscr{L}E$ of a BANACH space E has the *c-Hahn-Banach property* (*c-HBP*) iff every subspace N of $\mathscr{L}E$ has one of the equivalent properties of Theorem III.1.

In order to give examples of BANACH spaces E and subspaces N of $\mathscr{L}_cE$ whose adherence is not closed we need the following lemma:

Lemma III.2. *Let E be a Banach space, N a subspace of $\mathscr{L}E$ and U the closed unit ball of E. Then $a(N)$ is closed in $\mathscr{L}_c(E)$ iff there exists a $k \in \boldsymbol{N}$ such that $U^0 \cap a(N) \subseteqq k\overline{(U^0 \cap N)}^*$.*

Proof. Assume $U^0 \cap a(N) \subseteqq k\overline{(U^0 \cap N)}^*$ for some $k \in \boldsymbol{N}$. This implies $n\overline{(U^0 \cap a(N))}^* \subseteqq nk\overline{(U^0 \cap N)}^* \subseteqq a(N)\ \forall\, n \in \boldsymbol{N}$ and $a(a(N)) = \bigcup_{n\in\boldsymbol{N}} n\overline{(U^0 \cap a(N))}^* \subseteqq a(N)$. Therefore $a(N)$ is closed.

If $a(N)$ is closed we have $a(N) = \overline{N}^*$ which implies $a(N) = \mathscr{L}(E/N^0)$. Define

$$U_1 = \{y: y \in E/N^0, |f(y)| \leqq 1\ \forall\, f \in U^0 \cap a(N)\},$$
$$U_2 = \{y: y \in E/N^0, |f(y)| \leqq 1\ \forall\, f \in \overline{U^0 \cap N^0}^*\}.$$

Then U_1 is the unit ball of the quotient topology on E/N^0 and U_2 is the unit ball of a normable topology on E/N^0 coarser than the quotient topology. Since both topologies are MACKEY topologies which have the same dual, namely $a(N)$, they agree. Therefore one can find a $k \in \boldsymbol{N}$ such that $\frac{1}{k} U_2 \subseteqq U_1$. This implies $U^0 \cap a(N) \subseteqq k\overline{(U^0 \cap N)}^*$.

Example 2. Consider the BANACH space c_0 consisting of all sequences of real numbers converging to 0. Since c_0 is separable we have $a(N) = \{f : f \in \mathscr{L}(c_0), \exists$ a sequence $(f_n)_{n\in N}$ in N converging weak* to $f\}$ for all subspaces N of $\mathscr{L}(c_0)$. In [8] MAZURKIEWICZ has shown: There exists a subspace N of $\mathscr{L}(c_0)$ such that $a(N) \neq a(a(N))$. Theorem III.1 implies that $\mathscr{L}(c_0)$ does not have the c-HBP.

Example 3. Let $\mathfrak{l}_1$ be the BANACH space of all sequences $(a_n)_{n\in N}$ of real numbers such that $\sum_{n=1}^{\infty} |a_n| < \infty$. Define $\zeta_r \in \mathscr{L}(\mathfrak{l}_1)$ and $\zeta_{r,m} \in \mathscr{L}(\mathfrak{l}_1)$ by $\zeta_r((a_n)_{n\in N}) = \sum_{n=1}^{\infty} a_{rn}$ and $\zeta_{r,m}((a_n)_{n\in N}) = \sum_{n=1}^{m} a_{rn} + r \sum_{n=m+1}^{\infty} a_{rn}$. Then we have $\|\zeta_r\| = 1$ and the sequence $(\zeta_{r,m})_{m\in N}$ converges to ζ_r in $\mathscr{L}_c(\mathfrak{l}_1)$ $\forall\, r \in N$. Let N be the subspace of $\mathscr{L}(\mathfrak{l}_1)$ generated by $\{\zeta_{p,m}$: prime number, $m \in N\}$ and let $(z_\alpha)_{\alpha\in I}$ be a net in N converging weak* to ζ_p. For every $k \in N$ denote by e_k the sequence $(a_n)_{n\in N} \in \mathfrak{l}_1$ with $a_k = 1$ and $a_n = 0 \;\forall\, n \in N \setminus \{k\}$. Since $(z_\alpha(e_p))_{\alpha\in I}$ converges to $\zeta_p(e_p) = 1$ there exists a $\beta \in I$ such that $z_\alpha(e_p) > \frac{1}{2} \;\forall\, \alpha > \beta$. Choose an $\alpha > \beta$. There are real numbers $\gamma_{q,m,\alpha}$ and a finite set $M_\alpha \subseteqq N \times N$ such that z_α can be represented as $z_\alpha = \sum_{(q,m)\in M_\alpha} \gamma_{q,m,\alpha}\zeta_{q,m}$. Define $N_{p,\alpha} = \{m : (p, m) \in M_\alpha\}$ and choose a $t \in N$ with $t > \max\{m : m \in N_{p,\alpha}\}$. From $z_\alpha(e_p) = \sum_{m\in N_{p,\alpha}} \gamma_{m,q,\alpha} > \frac{1}{2}$ it follows that

$$\|z_\alpha\| \geqq |z_\alpha(e_{pt})| = \left|\sum_{m\in N_{p,\alpha}} \gamma_{p,m,\alpha} p\right| > \frac{1}{2}\, p.$$

Lemma III.2 and Theorem III.1 imply that $\mathscr{L}(\mathfrak{l}_1)$ does not have the c-HBP.

Theorem III.3. *Let E be a quasi-reflexive Banach space. Then $\mathscr{L}E$ has the c-HBP.*

Proof. Let N be a subspace of $\mathscr{L}_c(E)$. Denote by N_n the set N considered as a subspace of the normed dual E' of E. Then we have $\mathscr{L}(N) \subseteqq \mathscr{L}(N_n)$. Let V and B be the closed unit balls of N_n and its normed dual N_n', respectively. Denote by μ_N the mapping from E into $N_n' = E''/N^0$ defined by $[\mu_N(x)](f) = f(x) \;\forall\, x \in E$ and $\forall\, f \in N$. Then one can show:

$$\forall\, \zeta \in \mathscr{L}(N)\; \forall\, \varepsilon > 0\; \exists\, x_1, \ldots, x_r \in E \quad \text{such that}$$

$$\zeta\left(\frac{1}{\varepsilon} V \cap \{f : f \in N, |f(x_i)| \leqq 1 \;\forall\, 1 \leqq i \leqq r\}\right) \subseteqq [-1, 1]$$

$$\Rightarrow \forall\, \zeta \in \mathscr{L}(N)\; \forall\, \varepsilon > 0\; \exists\, x_1, \ldots, x_r \in E \quad \text{such that}$$

$$\zeta \in \varepsilon B + \Gamma\{\mu_N(x_1), \ldots, \mu_N(x_r)\}$$

$\Rightarrow \mathscr{L}(N)$ is contained in the closure of $\mu_N(E)$ in N_n'.

The space $\mu_N(E)$ is closed in N_n' iff $E + N^0$ is closed in the bidual E'' of E and $E + N^0$ is closed in E'' iff the image of N^0 is closed in E''/E. But E is quasi-reflexive which implies that E''/E is finite dimensional. Therefore $\mu_N(E)$ is closed in N_n' and one can find an $x \in E$ such that $\zeta = \mu_N(x)$ which means $\zeta(f) = f(x) \;\forall\, f \in N$.

Lemma III.4. *Let X be a closed subspace of a Banach space E and assume that $\mathscr{L}X$ fails to have the c-HBP. Then also $\mathscr{L}E$ does not have the c-HBP.*

Proof. Let π be the quotient mapping from $\mathscr{L}E$ onto $\mathscr{L}X = \mathscr{L}E/X^0$ which is continuous from $\mathscr{L}_\mathfrak{c}(E)$ onto $\mathscr{L}_\mathfrak{c}(X)$. Since $\mathscr{L}X$ does not have the $\mathfrak{c}$-HBP there exists a subspace N of $\mathscr{L}_\mathfrak{c}(X)$ and a $\zeta \in \mathscr{L}(N)$ which cannot be extended continuously to $\mathscr{L}_\mathfrak{c}(X)$. Assume that $\zeta \circ \pi$ has a continuous linear extension from $\pi^{-1}(N)$ to $\mathscr{L}_\mathfrak{c}(E)$. Then one can find an $x \in E$ such that $(\zeta \circ \pi)(f) = f(x)\ \forall\ f \in \pi^{-1}(N)$. For all $f \in \pi^{-1}(0) = X^0$ we have $0 = \zeta(\pi(f)) = f(x)$ so that $x \in X$. From $\zeta(g) = g(x)\ \forall\ g \in N$ it follows that ζ has a continuous linear extension to $\mathscr{L}_\mathfrak{c}(X)$, which is impossible.

Theorem III.5. *The dual of a Banach lattice E has the $\mathfrak{c}$-HBP iff E is reflexive.*

Proof. If E is reflexive Theorem III.3 shows that $\mathscr{L}E$ has the $\mathfrak{c}$-HBP.

If E is not reflexive it contains a copy of c_0 or $\mathfrak{l}_1$ (see [13]). From example 1 and 2 and from Lemma III.4 it follows that $\mathscr{L}E$ fails to have the $\mathfrak{c}$-HBP.

Now we will construct a Banach space E with the following properties:

(i) E does not contain a copy c_0 or $\mathfrak{l}_1$.

(ii) $E'' = E \oplus \mathfrak{l}_2$.

(iii) E does not have the $\mathfrak{c}$-HBP.

Example 4. Let J be the quasi-reflexive Banach space of order 1 constructed by R. C. James [5]. The space J consists of those sequences $y = (y_i)_{i\in N}$ of real numbers such that

$$\|y\| = \sup \left(\sum_{j=0}^{k-1} \left(\sum_{i=n_j+1}^{n_{j+1}} y_i\right)^2\right)^{1/2} < \infty,$$

where the sup is taken over all increasing finite sequences of integers $0 = n_0 < n_1 < \cdots < n_k$. For every $y = (y_i)_{i\in N} \in J$ the sum $\sum_{i=1}^{\infty} y_i$ converges and we have $\left|\sum_{i=1}^{\infty} y_i\right| \leqq \|y\|$. Define $J_n = J\ \forall\ n \in \boldsymbol{N}$ and let E be the $\mathfrak{l}_2$-sum of the spaces J_n. Every $x \in E = \left(\sum_{n\in N} J_n\right)_{\mathfrak{l}_2}$ has a representation $x = (x_{i,j})_{i,j\in N}$ such that $x_j = (x_{i,j})_{i\in N} \in J_j\ \forall\ j \in \boldsymbol{N}$ and

$$\|x\| = \left(\sum_{j=1}^{\infty} \|x_j\|^2\right)^{1/2} < \infty.$$

For every $p, m \in \boldsymbol{N}$ define $\xi_p \in \mathscr{L}E$ amd $\zeta_{p,m} \in \mathscr{L}E$ by

$$\xi_p\big((x_{i,j})_{i,j\in N}\big) = \sum_{i=1}^{\infty} x_{i,p} \quad \text{and}$$

$$\zeta_{p,m}\big((x_{i,j})_{i,j\in N}\big) = \sum_{i=1}^{m} x_{i,p} + p \sum_{i=m+1}^{\infty} x_{i,p}.$$

For all $p \in \boldsymbol{N}$ we have $\|\xi_p\| \leqq 1$ and $(\zeta_{p,m})_{m\in N}$ converges to ξ_p in $\mathscr{L}_\mathfrak{c}(E)$. Let N denote the subspace of $\mathscr{L}E$ generated by $\{\zeta_{p,m} : p, m \in \boldsymbol{N}\}$. Choose a $p \in \boldsymbol{N}$ and consider a net $(z_\alpha)_{\alpha\in I}$ in N which converges weak* to ξ_p.

Define $a = (a_{i,j})_{i,j\in N} \in E$ by

$$a_{i,j} = \begin{cases} 1 & \text{if } (i, j) = (1, p), \\ 0 & \text{otherwiese.} \end{cases}$$

Since $\big(z_\alpha(a)\big)_{\alpha\in I}$ converges to $\xi_p(a) = 1$, there exists a $\beta \in I$ such that $|z_\alpha(a) - \xi_p(a)| = |z_\alpha(a) - 1| < \frac{1}{2}\ \forall\ \alpha > \beta$ which implies $z_\alpha(a) > \frac{1}{2}\ \forall\ \alpha > \beta$. Choose an $\alpha > \beta$.

There exist real numbers $\gamma_{q,m,\alpha}$ and a finite set $M_\alpha \subseteqq N \times N$ such that z_α can be represented as $z_\alpha = \sum_{(q,m)\in M_\alpha} \gamma_{q,m,\alpha}\zeta_{q,m}$. Define $N_{p,\alpha} = \{m: (p, m) \in M_\alpha\}$. We have

$$\frac{1}{2} < z_\alpha(a) = \sum_{m\in N_{p,\alpha}} \gamma_{p,m,\alpha}\zeta_{p,m}(a) = \sum_{m\in N_{p,\alpha}} \gamma_{p,m,\alpha}.$$

Choose a $t \in N$ with $t > \max\{m: m \in N_{p,\alpha}\}$ and define $b = (b_{i,j})_{i,j\in N} \in E$ by

$$b_{i,j} = \begin{cases} 1 & \text{if } (i, j) = (t, p), \\ 0 & \text{otherwise.} \end{cases}$$

Then $z_\alpha(b) = \sum_{m\in N_{p,\alpha}} \gamma_{p,m,\alpha}\zeta_{p,m}(b) > \frac{1}{2} p$. This implies $\|z_\alpha\| > \frac{1}{2} p$. Every ξ_p belongs to $a(N)$, therefore Lemma III.2 shows that $\mathscr{L}E$ has not the c-HBP. Since $E' = \left(\sum_{n\in N} J_n'\right)_{\mathfrak{l}_2}$ and $E'' = \left(\sum_{n\in N} J_n''\right)_{\mathfrak{l}_2} = \left(\sum_{n\in N} (J_n \oplus \boldsymbol{R})\right)_{\mathfrak{l}_2} = E \oplus \mathfrak{l}_2$ are separable, E does not contain a subspace isomorphic to c_0 or $\mathfrak{l}_1$.

References

[1] P. Antoine, Etude élémentaire des catégories d'ensembles structurés, Bull. Soc. math. Belgique **18** (1966), 387–409.

[2] E. Binz und H. H. Keller, Funktionenräume in der Kategorie der Limesräume, Ann. Acad. Sci. Fenn., Ser. A I, **383** (1966).

[3] H.-P. Butzmann, Über die c-Reflexivität von $C_c(X)$, Commentarii math. Helvet. **47** (1972), 92–101.

[4] H. R. Fischer, Limesräume, Math. Ann. **137** (1959), 269–303.

[5] R. C. James, A non-reflexive Banach space isometric with its second conjugate space, Proc. Nat. Acad. Sci. U. S. A. **37** (1951), 174–177.

[6] H. Jarchow, Marinescu Räume, Commentarii math. Helvet. **44** (1969), 138–163.

[7] W. B. Johnson, H. P. Rosenthal and M. Zippin, On bases, finite dimensional decompositions and weaker structures in Banach spaces, Israel J. Math. **9** (1971), 488–506.

[8] S. Mazurkiewicz, Sur la dérivée faible d'un ensemble de fonctionelles linéaires, Studia Math. **2** (1930), 68–71.

[9] B. Müller, A Hahn-Banach theorem in conjugate Banach spaces Math. Z. **157** (1977), 285–295.

[10] B. Müller, A strict topology on $L(E, F)$ and the Banach-Steinhaus theorem (to appear).

[11] W. Roelcke, On the Finest Locally Convex Topology Agreeing with a Given Topology on a Sequence of Absolutely Convex Sets, Math. Ann. **198** (1972), 57–80.

[12] H. H. Schaefer, Topological Vector Spaces, Springer-Verlag, New York–Heidelberg–Berlin 1971.

[13] H. H. Schaefer, Banach Lattices and Positive Operators, Springer-Verlag, New York–Heidelberg–Berlin 1974.

Anschrift des Verfassers:

Universität Mannheim (WH)
Fakultät für Mathematik und Informatik
68 Mannheim
Schloß
BRD

On Derivations in Rings and Algebras Provided with Convergence Structures

WOLFGANG PREUSS, Wismar, and ANDRÁS BLEYER, Budapest

1. Introduction

Recent papers include the investigation of derivations in special fields (see [1, 4] and [5]) where the term "derivation" stands for a linear mapping of a field into itself fulfilling the product rule. In [2, 3], and [6] the derivation is defined (more general) as usual in the algebra. In the paper [2], representation theorems for sequentially continuous derivations in rings and algebras provided with convergence structures are stated. But for the application to theoretical physics, the study of closed derivations in topological algebras is more important (see [6]). In order to investigate the so-called HERMAN-POWERS-Problem concerning the closed derivations in $C[0, 1]$ (see [6]), in the paper [3] a representation theorem for closed derivations in BANACH-algebras is established. In my short lecture the papers [3] and [2] will be completed with regard to the representation of sequentially closed derivations in rings and algebras provided with convergence structures which are not necessarily topological structures.

2. Definition

A ring X will be called a "convergence ring" if X is provided with a convergence structure such that the limit is unique and the ring operations are sequentially continuous.

If X is an algebra over a scalar ring S then the term "convergence algebra" is defined in an analogic way.

An additive mapping δ in the ring X is said to be a derivation if it satisfies the following conditions:

(1) the domain $D(\delta)$ is a (convergence) subring of X;

(2) $\delta(ab) = \delta(a)\, b + a\delta(b)$ for all $a, b \in D(\delta)$.

In the case of X being a convergence algebra over a scalar ring S, δ will be called a derivation if δ is a linear mapping in X and $D(\delta)$ is a (convergence) subalgebra of X. The set of all elements $c \in D(\delta)$ with the property $\delta(c) = 0$ (0 is the zero element in X) forms a ring $K(\delta)$, the so-called "ring of constants of δ". Obviously the zero element $0 \in X$ always is in $K(\delta)$ and if $D(\delta)$ contains a unit element e, then $e \in K(\delta)$. A derivation δ in a convergence ring (or in a convergence algebra) X is defined to be a sequentially closed derivation in X if

(3) $a_n \in D(\delta)$, $a_n \to a$, and $\delta(a_n) \to b$ always imply $a \in D(\delta)$ and $b = \delta(a)$ (the symbol $\to$ always stands for the convergence in X).

Finally the symbol $P(M, a)$ stands for the set of all formal polynomials $p(a) = m_0 a^0 + m_1 a + m_2 a^2 + \cdots + m_k a^k$ with coefficients m_i in the set M.

3. Representation of sequentially closed derivations

At first we give a theorem for sequentially closed derivations in rings which is a certain generalization of the ring theorem in [2].

Theorem 1. *Suppose that X is a commutative convergence ring having the unit element e, and δ_1 is a derivation in X with $K(\delta_1) \neq \{0\}$. Under the assumptions*

(4) *there is an element $x \in D(\delta_1)$ such that either condition $\delta_1(x) = e$ or $e/\delta_1(x) \notin X$ is fulfilled;*

(5) *for every element $a \in D(\delta_1)$, there is a sequence $(p_n(x))$ in $P(K(\delta_1), x)$ such that $p_n \to a$ and $\delta_1(p_n) \to \delta_1(a)$;*

every sequentially closed derivation δ in X with $K(\delta) \supseteqq K(\delta_1)$ and $x \in D(\delta)$ has a domain $D(\delta) \supseteqq D(\delta_1)$, and the restriction of δ on $D(\delta_1)$ has the representation $\delta = f(x)\, \delta_1$ where $f(x) \in X$ depends on δ.

The principle of the proof (see [2, 3]) is very simple. By definition, the domains $D(\delta_1)$ and $D(\delta)$ are rings such that $K(\delta) \supseteqq K(\delta_1)$ *and* $x \in D(\delta_1) \cap D(\delta)$ imply $P(K(\delta_1), x) \subseteqq D(\delta_1) \cap D(\delta)$. On the other hand, $D(\delta_1)$ and $D(\delta)$ are commutative subrings of X, hence by use of (2) we obtain $\delta_1(x^n) = nx^{n-1}\delta_1(x)$ and $\delta(x^n) = nx^{n-1}\delta(x)$ for all $n = 1, 2, \ldots$ It is easy to see that $\delta(x^n) = f(x)\, \delta_1(x^n)$ where $f(x) = \delta(x)$ or $f(x) = \delta(x)/\delta_1(x)$ (corresponding to (4)) such that

(6) $\delta(p(x)) = f(x)\, \delta_1(p(x))$ for all $p(x) \in P(K(\delta_1), x)$.

Now let a be any element in $D(\delta_1)$. By assumption (5), we can find a sequence $(p_n(x)) \subset P(K(\delta_1), x)$ such that $p_n \to a$ and $\delta_1(p_n) \to \delta_1(a)$. Because of the sequential continuity of the operations in X and formula (6), we get

$$\delta(p_n) = f(x)\, \delta_1(p_n) \to f(x)\, \delta_1(a).$$

On the other hand by presupposition the derivation δ is sequentially closed such that from (3), $a \in D(\delta)$ and $f(x)\, \delta_1(a) = \delta(a)$ follow. This finishes the proof.

In Theorem 1, for the formulation of (5), $K(\delta_1) \neq \{0\}$ is to be presupposed. In the case $K(\delta_1) = \{0\}$, it is impossible to consider (5), but every other derivation δ in X fulfils the condition $K(\delta) \supseteqq K(\delta_1)$. Therefore we modify Theorem 1 for the case of X being an algebra, and we get as a generalization of [3]

Theorem 2. *Assume that X is a commutative convergence algebra over a scalar ring S having the unit element e, and δ_1 is a derivation in X fulfilling the properties*

(4) (*see Theorem* 1);

(5)* *for every element $a \in D(\delta_1)$, there is a sequence $(p_n(x))$ in $P(S, x) \cap D(\delta_1)$ such that $p_n \to a$ and $\delta_1(p_n) \to \delta_1(a)$.*

Let δ be a sequentially closed derivation in X with $\{e, x\} \cap D(\delta_1) \subset D(\delta)$. Then δ has a domain $D(\delta) \supseteqq D(\delta_1)$, and the restriction of δ to $D(\delta_1)$ has the representation $\delta = f(x)\,\delta_1$ where $f(x) \in X$ depends on δ.

The theorem can be proved analogously as Theorem 1. It is easy to see that in the case $K(\delta_1) = \{0\}$, Theorem 2 holds, too.

An interesting example for the application of such theorems can be found in [3]: Let δ be any closed derivation in $C[0, 1]$ with $1, x \in D(\delta)$. Then $D(\delta) \supseteqq C^1[0, 1]$ and $\delta = f(x)\frac{d}{dx}$ on $C^1[0, 1]$ where $f(x) = \delta(x) \in C[0, 1]$. This gives a partial answer to the so-called HERMANN-POWERS-Problem (see [6]).

It is easy to find other examples, for instance in the operational calculus.

References

[1] A. BLEYER and W. PREUSS, A note to the continuous derivation of fields, Acta Math. Acad. Sci. Hung. T. **31** (1978) (1–2), 75–79.
[2] A. BLEYER und W. PREUSS, A note to general notions of the derivation and its application, Periodica Math. Hung. **11** (1) (1980), 1–6.
[3] A. BLEYER und W. PREUSS, A remark to the characterization of closed derivations in C^*-algebras, Rivista Mat. Univ. Parma (in print).
[4] T. FÉNYES, A note on the derivations of Mikusiński's operator field, Periodica Math. Hung. **1** (1971) (3), 217–219.
[5] W. PREUSS, Remarks on operator transformations of a field of transformable operators, Acta Math. Acad. Sci. Hung. T. **31** (1978) (3–4) (1978), 269–277.
[6] S. SAKAI, Recent developments in the theory of unbounded derivations in C^*-algebras, Manuscript for the US — Japan seminar on C^*-algebras and their application to theoretical physics (April 18–22, 1977 at U. C. L. A.).

Anschrift der Verfasser:

Ingenieurhochschule Wismar
Abteilung Math./Nat.
24 Wismar
Philipp-Müller-Str.
DDR

Polytechnical University of Budapest
Department of Mathematics
1111–Budapest
Stoczek u. 2–4
Hungary

Michal-Bastiani Differentiability on Topological Groups

Klaus-Peter Rudolph, Greifswald

Abstract

The differential calculus for mappings on topological groups with values in a locally convex space is based here on two concepts: The concept of continuous differentiability of mappings between locally convex spaces, developed by Keller [4], and the concept of continuous differentiability of complex valued functions on topological groups, developed by Boseck/Czichowski [2]. It is shown that the notion of continuous differentiability in the sense of Michal/Bastiani (MB-differentiability) of mappings between locally convex spaces has a natural extension to mappings defined on topological groups. Connections between the induced notions of weakly continuous differentiability, of continuous MB-differentiability and the notion of continuous differentiability of Boseck/Czichowski are discussed for the general case as well as for special cases.

1. Continuously MB-differentiable mappings on locally convex spaces

All locally convex spaces or topological groups occuring in the sequel are supposed to be Hausdorff spaces and the vector spaces are supposed to be defined over the reals. The notion of differentiability of class $\boldsymbol{C}_c^n$ (n-times continuous differentiability of Michal/Bastiani or briefly MB-differentiability), which is given in [4], may be described in the following way:

Let $\boldsymbol{E}$ and $\boldsymbol{F}$ be locally convex spaces and $\boldsymbol{X}$ an open subset of $\boldsymbol{E}$. A continuous mapping $f\colon \boldsymbol{X} \to \boldsymbol{F}$ is said to be differentiable of class $\boldsymbol{C}_c^n$ if there exist continuous mappings

$$\boldsymbol{D}^k f\colon \boldsymbol{X} \times \boldsymbol{E}^k \to \boldsymbol{F}, \quad \boldsymbol{D}^0 f = f, \quad 0 \leqq k \leqq n,$$

such that for each $x \in \boldsymbol{X}$, $k = 0, \ldots, n-1$ and $h_1, \ldots, h_{k+1} \in \boldsymbol{E}$ in $\boldsymbol{F}$ we have

$$\lim_{t \to 0} \frac{1}{t} \left(\boldsymbol{D}^k f(x + t h_{k+1}, h_1, \ldots, h_k) - \boldsymbol{D}^k f(x, h_1, \ldots, h_k)\right) = \boldsymbol{D}^{k+1} f(x, h_1, \ldots, h_{k+1}).$$

The equivalence of the definitions, which are given here and in [4] as well follows from the universal mapping property which characterizes the convergence structure of continuous convergence and from the mean value theorem which proves the multilinearity.

2. One-parameter subgroups

To extend the notion of differentiability to mappings defined on topological groups we have to explain the meaning of direction in a topological group. By directions we shall understand the so-called one-parameter subgroups.

A one-parameter subgroup of a topological group G is a continuous homomorphism from the additive group R^+ into G. The space of all one-parameter subgroups of G is denoted by $\Lambda(G)$. The convergence structure of continuous convergence in $\Lambda(G)$ turns out to be a topology. In this topology a fundamental system of neighbourhoods of $\lambda_0 \in \Lambda(G)$ is given by all sets $U(\lambda_0) = \{\lambda; \lambda(t) \in \lambda_0(t)\, U,\ |t| \leqq 1\}$ where U belongs to a fundamental system of neighbourhoods of the unity in G. $\Lambda(G)$ equipped with this topology is denoted by $\Lambda_c(G)$.

For every one-parameter subgroup λ and $x \in G$ we may define the "conjugated one-parameter subgroup" λ^x by $\lambda^x(t) = x^1\lambda(t)\, x$. It is shown in [2] that the mapping $(x, \lambda) \to \lambda^x$ is continuous from $G \times \Lambda_c(G)$ into $\Lambda_c(G)$.

More interesting are such groups for which $\Lambda(G)$ (resp. $\Lambda_c(G)$) is a vector space (resp. a locally convex space) by

1. $(\alpha\lambda)\,(t) = \lambda(\alpha t)$

and

2. $(\lambda_1 + \lambda_2)\,(t) = \lim\limits_{n\to\infty} \left(\lambda_1\left(\frac{t}{n}\right)\lambda_2\left(\frac{t}{n}\right)\right)^n$

where the limit exists uniformly on compact subsets of R. It may be shown that this property of being a vector space (resp. a locally convex space) is preserved for closed subgroups, direct products and therefore for projective limits, as well. Since this property is valied for LIE-groups it follows that $\Lambda_c(G)$ is a locally convex space for any LP-group G (i.e. projective limit of LIE-groups). Because every locally compact group contains an open LP-subgroup, we obtain in this case $\Lambda_c(G)$ as a locally convex space, too [3]. If E^+ is the underlying ABELian group of a locally convex space E, we have a canonical topological isomorphism between E and $\Lambda_c(E^+)$ by the mapping $x \to \lambda_x$ where $\lambda_x(t) = tx$. Every nontrivial connected locally compact group has non-trivial one-parameter subgroups [3]. On the other hand there exist examples of metrizable arcwise connected and locally arcwise connected groups, ABELian or not, where $\Lambda(G)$ is trivial.

3. Continuously differentiable mappings on topological groups

Let G be a topological group, X an open subset of G and F an arbitrary locally convex space.

Definition. A continuous mapping $f\colon X \to F$ is said to be differentiable of class C_s^n (n-times weakly continuously differentiable) if there exist mappings

$$D^k f\colon X \times \Lambda^k(G) \to F, \quad D^0 f = f, \quad 0 \leqq k \leqq n,$$

such that

1. for each $x \in X$, $k = 0, \ldots, n-1$ and $\lambda_1, \ldots, \lambda_{k+1} \in \Lambda(G)$ in F we have

$$\lim_{t\to 0} \frac{1}{t}\left(D^k f(x\lambda_{k+1}(t), \lambda_1, \ldots, \lambda_k) - D^k f(x, \lambda_1, \ldots, \lambda_k)\right) = D^{k+1} f(x, \lambda_1, \ldots, \lambda_{k+1})$$

and

2. for each k, $1 \leqq k \leqq n$, $\lambda_1, \ldots, \lambda_k \in \Lambda(G)$ the mapping from X into F defined by $x \to D^k f(x, \lambda_1, \ldots, \lambda_k)$ is continuous.

Proposition. *Assume $f \in C_s^n$. Then for every fixed $\lambda \in \Lambda(G)$ the function of a real variable defined by*

$$\varphi(t) = f(x\lambda(t))$$

is of class C^n in the elementary sense in a neighbourhood of 0.

We have a Taylor's expansion

$$f(x\lambda(t)) = \sum_{k=0}^{n} \frac{t^k}{k!} D^k f(x, \lambda, \ldots, \lambda) + r_n f(x, \lambda, t)$$

where for every continuous seminorm q on F the remainder satisfies the following estimation:

$$q(r_n f(x, \lambda, t)) \leqq \frac{|t|^n}{n!} \sup_{0 \leqq \tau \leqq 1} q(D^n f(x\, \lambda(\tau t), \lambda, \ldots, \lambda) - D^n f(x, \lambda, \ldots, \lambda)).$$

Corollary. *For every continuous seminorm q on F and $\varepsilon > 0$ there exist a $\delta > 0$ and a neighbourhood U of the unity in G such that for $x' \in U$, $|t| < \delta$ we have*

$$q(r_n f(xx', \lambda, t)) \leqq \varepsilon\, |t|^n.$$

Proposition. *Assume $f \in C_s^n$. If $\Lambda(G)$ has the vector space property, then for every k, $1 \leqq k \leqq n$, and fixed $x \in X$ the mapping*

$$(\lambda_1, \ldots, \lambda_k) \to D^k f(x, \lambda_1, \ldots, \lambda_k)$$

is k-linear from $\Lambda^k(G)$ into F.

Using induction the proof is an implication of the corollary.

Definition. A continuous mapping $f: X \to F$ is said to be differentiable of class C^n (n-times differentiable in the sense of Boseck/Czichowski) if

1. f is of class C_s^n

and

2. for every k, $1 \leqq k \leqq n$, and for every choice of $\lambda_1, \ldots, \lambda_{k-1} \in \Lambda(G)$ the mapping from $X \times \Lambda_c(G)$ into F defined by $(x, \lambda_k) \to D^k f(x, \lambda_1, \ldots, \lambda_k)$ is continuous.

Remark. Obviously a continuous mapping $f: X \to F$ is differentiable of class C^n if

1. $f \in C^{n-1}$

and

2. for every fixed $\lambda_1, \ldots, \lambda_{n-1} \in \Lambda(G)$ the mapping $x \to D^{n-1} f(x, \lambda_1, \ldots, \lambda_{n-1})$ is of class C^1. (C^0 is the class of continuous mappings.)

Definition. A continuous mapping $f: X \to F$ is said to be differentiable of class C_c^n (n-times continuously MB-differentiable) if

1. f is differentiable of class C_s^n

and

2. for every k, $1 \leqq k \leqq n$, the mapping

$$D^k f: X \times \Lambda_c^k(G) \to F$$

is continuous.

Proposition. *If E^+ is the underlying Abelian group of a locally convex space E, then the n-times continuously MB-differentiable mappings on E^+ (as mappings on a group) and on E (as mappings on a locally convex space) are the same.*

This follows from the definitions and the isomorphy of $\Lambda_c(E^+)$ and E.

Proposition. *Assume $f \in C_c^n$. Then for every $s = 1, 2, \ldots$ and every fixed $x \in X$ and $\lambda_1, \ldots, \lambda_s \in \Lambda(G)$ the function defined by*

$$\varphi(t_1, \ldots, t_s) = f(x\lambda_1(t_1) \ldots \lambda_s(t_s))$$

is of class C^n in the elementary sense in a neighbourhood of 0.

In the definitions of continuous differentiability we required that $\lambda(t)$ is to be multiplied from "left" by x. In the same way we may obtain the analogous "right" notions, multiplying by x from "right". We shall write "r" for the "right" notions and "l" for the "left" notions if it is needed. Thus we shall write

$$D_l f(x, \lambda) = D_l^{\lambda} f(x)$$

and

$$D_r f(x, \lambda) = D_r^{\lambda} f(x) = Df(x, \lambda),$$

respectively, and in the general case

$$D_{\alpha_1 \ldots \alpha_s}^{\lambda_1 \ldots \lambda_s} f(x) = D_{\alpha_s}^{\lambda_s}(D_{\alpha_1 \ldots \alpha_{s-1}}^{\lambda_1 \ldots \lambda_{s-1}} f)\,(x)$$

where the α_i mean "r" or "l" if the expressions make sense.

From the definitions and the continuity of $(x, \lambda) \to \lambda^x$ immediately we have $f \in C_l^1$ if $f \in C_r^1$ and $D_l f(x, \lambda) = D_r f(x, \lambda^x)$.

For $n > 1$ the connection between "left" and "right" derivatives is the following:

Theorem. *Assume $f \in C^n$. Then there exist all derivatives of order $k \leqq n$ and for each $x \in X$ and $\lambda_1, \ldots, \lambda_k \in \Lambda(G)$ we have*

$$D_{\alpha_1 \ldots \alpha_k}^{\lambda_1 \ldots \lambda_k} f(x) = D_l^k f(x, \lambda_{i_1}, \ldots, \lambda_{i_s}, \lambda_{i_k}^x, \ldots, \lambda_{i_{s+1}}^x)$$

where $i_1 < \cdots < i_s$, $\alpha_{i_1} = \cdots = \alpha_{i_s} = l$ and $i_{s+1} < \cdots < i_k$, $\alpha_{i_{s+1}} = \cdots = \alpha_{i_k} = r$.

Corollary. *If $f \in C_c^n$, then for each k, $1 \leqq k \leqq n$, and $\alpha_1, \ldots, \alpha_k$ the mapping*

$$(x, \lambda_1, \ldots, \lambda_k) \to D_{\alpha_1 \ldots \alpha_k}^{\lambda_1 \ldots \lambda_k} f(x)$$

is continuous.

From the theorem it follows that the notions of continuous differentiability of BOSECK/CZICHOWSKI and, moreover, of continuous MB-differentiability are independent of "left", although the derivatives generally depend on "left" and "right".

4. Connections between the various concepts

As an immediate consequence of the definitions we have $C_s^n \supseteqq C^n \supseteqq C_c^n$. The following example shows that these classes are different in general.

Example. Consider the normed space consisting of all abbreviated sequences $x = (x_k)$ of real numbers with the norm $\|x\| = \max |x_k|$. Choose an arbitrary function $\varphi \in C^\infty(\mathsf{R}, \mathsf{R})$

satisfying $\varphi(0) = 0$, $\varphi^{(n)}(0) \neq 0$ for $n \geqq 1$ and $|\varphi^{(n)}(t)| \leqq C_n$ for all $t \in \mathsf{R}$ and $n = 0, 1, \ldots$ (for example $\varphi(t) = \sin t + \cos t - 1$). Choose a sequence of real numbers (α_k) with $\alpha_k \geqq 1$ and $\sum\limits_{k=0}^{\infty} \frac{1}{\alpha_k} < \infty$. For $m = 0, 1, \ldots$ we define $f_m(x) = \sum\limits_{k=0}^{\infty} \frac{\varphi(\alpha_k x_k)}{\alpha_k^{m+1}}$. Then we have

1. $f_0 \in C_s^\infty$, $f_0 \notin C^1$

and

2. $f_m \in C^\infty$, $f_m \in C_c^m$, $f_m \notin C_c^{m+1}$ for $m \geqq 1$.

On the other hand, for important classes of groups the considered notions are the same. This was first remarked for ABELian locally compact groups by RISS [5]. BAUCH [1] extended this result for the case $n = 1$ to arbitrary locally compact connected groups.

The following theorem yields a final result for arbitrary locally compact groups:

Theorem. *Let G be a metrizable group and assume $\Lambda_c(G)$ to be a Fréchet space or let G be an arbitrary locally compact group.*

Then every differentiable mapping of class C_s^n from an open subset X of G into an arbitrary locally convex space F is differentiable of class C_c^n.

Corollary *For a mapping defined on a Fréchet space, n-times weakly continuous differentiability (as mapping on the underlying Abelian group where multilinearity and continuity of derivatives are not required) implies the n-times continuous MB-differentiability (as mapping on the Fréchet space).*

Proposition. *Let G be a classical Lie-group. Then the n-times continuously differentiable mappings into an arbitrary locally convex space (as mappings on the underlying manifold) and the differentiable mappings of class $C_s^n = C_c^n$ defined above are the same.*

5. Continuously differentiable mappings between topological groups

Let G_1, G_2 be topological groups.

Definition. A continuous mapping $f: G_1 \to G_2$ is called differentiable of class C_c^n (n-times continuous MB-differentiable) if for every locally convex space F and for every mapping g from G_2 into F, differentiable of class C_c^∞, the composition $g \circ f$ is differentiable of class C_c^n.

Proposition. *Let G be an arbitrary topological group. Then the mappings*

$$(x, y) \to xy \quad \text{and} \quad x \to x^{-1}$$

from $G \times G$ and G, respectively, into G are of the class C_c^∞.

References

[1] H. F. BAUCH, Zur Fouriertransformation differenzierbarer Funktionen auf kompakten zusammenhängenden Gruppen, Diss., Univ. Greifswald 1978.

[2] H. BOSECK und G. CZICHOWSKI, Grundfunktionen und verallgemeinerte Funktionen auf topologischen Gruppen I, Math. Nachr. 58 (1973), 215—240.

[3] В. М. Глушков, Алгебры Ли локально бикомпактных групп, УМН **XII** (1957) (2), 137—143.

[4] H. H. KELLER, Differential Calculus in Locally Convex Spaces, Lecture Notes in Math. **417**, Springer-Verlag, Berlin—Heidelberg—New York 1974.

[5] J. RISS, Éléments de calcul différential et théorie des distributions sur les groupes abélians localement compacts, Acta Math. **89** (1953), 45—105.

Anschrift des Verfassers:

Ernst-Moritz-Arndt-Universität
Sektion Mathematik
22 Greifswald
Friedrich-Ludwig-Jahn-Str. 15a
DDR

Manifolds of Mappings

RUDOLF SCHMID, Zurich

0. Introduction

The basic idea for the construction of a differential structure on spaces of mappings traces back to J. EELLS [3]. He introduced a differential structure of class $\mathscr{C}^\infty$ modelled on BANACH spaces for the set of all continuous maps from a compact topological space into a finite dimensional $\mathscr{C}^\infty$ manifold. His method has been exemplary for all later attemps because it is the most natural construction of charts on mapping spaces. Especially S. SMALE [1] and R. PALAIS [9] were concerned with the corresponding problem for the space $\mathscr{C}^r(M, N)$, $r < \infty$, of all $\mathscr{C}^r$ maps from a compact manifold M into a BANACH manifold N. It turned out that, if M is a compact $\mathscr{C}^r$ manifold and N a finite dimensional $\mathscr{C}^{r+s+2}$ manifold, then $\mathscr{C}^r(M, N)$ is a BANACH manifold of class $\mathscr{C}^s$. BOURBAKI [2] states without a proof that, if M resp. N is a compact $\mathscr{C}^r$ resp. $\mathscr{C}^s$ manifold and P a finite dimensional $\mathscr{C}^t$ manifold, $1 \leqq r < s < t \leqq \infty$, the composition map

$$c: \mathscr{C}^s(N, P) \times \mathscr{C}^r(M, N) \to \mathscr{C}^r(M, P), \quad c(g, f) := g \circ f$$

is differentiable of class $\mathscr{C}^m$, $m = \min(s - r, t - s)$. For the differentiability of the composition it thus seems essential that $s > r$. This fact obviously motivates H. OMORI [8] to his definition of ILB-LIE groups, but for the proof that c is of class $\mathscr{C}^{s-r}$ (in the case were $M = N = P$ is a compact $\mathscr{C}^\infty$ manifold) OMORI refers to earlier papers and there, however, the proof of this fact is far from clear.

Let M be a compact $\mathscr{C}^\infty$ manifold and $1 \leqq r < \infty$. Then it is easy to prove that the group $\mathrm{Diff}^r(M)$ of all diffeomorphisms of class $\mathscr{C}^r$ is an open submanifold of the $\mathscr{C}^\infty$-BANACH manifold $\mathscr{C}^r(M, M)$. Composition $(g, f) \mapsto g \circ f$ and inversion $f \mapsto f^{-1}$ are continuous mappings from $\mathrm{Diff}^r(M) \times \mathrm{Diff}^r(M)$ resp. $\mathrm{Diff}^r(M)$ into $\mathrm{Diff}^r(M)$, so $\mathrm{Diff}^r(M)$ is (in its $\mathscr{C}^r$-topology) a topological group. But in general neither composition nor inversion will be differentiable, not even of class $\mathscr{C}^1$. So $\mathrm{Diff}^r(M)$ is no BANACH-LIE group. This can also be seen by a very simple heuristic argument: If $\mathrm{Diff}^r(M)$ were a BANACH-LIE group then its LIE algebra could be identified with the BANACH space $\Gamma^r(\tau_M)$ of all $\mathscr{C}^r$-vector fields on M, together with the usual LIE bracket of vector fields as LIE product. But for finite r, however, $\Gamma^r(\tau_M)$ is no LIE algebra.

Satisfying differentiability properties of the composition map between function spaces and especially a LIE group structure on the diffeomorphism group are therefore obtained only if we restrict ourselves to spaces of $\mathscr{C}^\infty$ functions.

In 1967 J. A. LESLIE [7] applied the so called $\mathscr{C}_\pi^\infty$-differentiability (cf. H. H. KELLER [6]) of mappings between FRÉCHET spaces to manifolds of mappings. He claims that the natural atlas on $\mathscr{C}^\infty(M, N)$ is of class $\mathscr{C}_\pi^\infty$, that composition

$$c: \mathscr{C}^\infty(N, P) \times \mathscr{C}^\infty(M, N) \to \mathscr{C}^\infty(M, P)$$

is $\mathscr{C}_\pi^\infty$-differentiable and that the diffeomorphism group $\mathrm{Diff}^\infty(M)$ is a $\mathscr{C}_\pi^\infty$ LIE group. Unfortunately LESLIE's paper contains many disturbing mistakes and the proofs of this main theorems are not sound. The statements mentioned above, however, have meanwhile turned out to be true and are proved by J. GUTKNECHT [5] (see section 2 and 4).

In the course of the last few years a notation of differentiability in FRÉCHET spaces, called the $\mathscr{C}_\Gamma^\infty$-differentiability, stronger than $\mathscr{C}_\pi^\infty$, has been applied to manifold of mappings and it seems that $\mathscr{C}_\Gamma^\infty$ is more adequate to the situation than $\mathscr{C}_\pi^\infty$.

1. The $\mathscr{C}_\Gamma^\infty$-differentiability in Fréchet spaces

Let E and F be FRÉCHET spaces in which an increasing sequence $(|\ |_r)_{r\in\mathbf{N}}$ of semi-norms defining the topologies of E and F is given. For every $r \in \mathbf{N}$ we put $E_r := (E, |\ |_r)$ and $F_r := (F, |\ |_r)$. Given a function $\sigma: \mathbf{N} \to \mathbf{N}$, for every $p \in \mathbf{N}$ the vector space

$$\mathscr{L}_\sigma^p(E, F) := \bigcap_{r\in\mathbf{N}} \mathscr{L}^p(E_{\sigma(r)}, F_r) \subset \mathscr{L}^p(E, F)$$

is a FRÉCHET space, its topology being defined by the sequence of semi-norms induced from the semi-normed spaces $\mathscr{L}^p(E_{\sigma(r)}, E_r)$, $r = 0, 1, \ldots$

$$\mathscr{L}_\sigma^p(E, F) = \{u \in \mathscr{L}^p(E, F) \mid |u|_r^\sigma < \infty \text{ for every } r \in \mathbf{N}\}$$

where $|u|_r^\sigma := \sup\{|u(h_1, \ldots, h_p)|_r \mid |h_i|_{\sigma(r)} \leqq 1, i = 1, \ldots, p\}$

Definition. Let U be open in E. A mapping $\Phi: U \to F$ is said to be $\mathscr{C}_\sigma^r$ if we have continuous mappings

$$D^p\Phi: U \to \mathscr{L}_\sigma^p(E, F), \quad 0 \leqq p \leqq r,$$

such that $D^0\Phi = \Phi$, and for every p, $0 \leqq p \leqq r - 1$, every $x \in U$ and every $h \in E$ the following condition is satisfied:

$$\lim_{t\to 0} \frac{1}{t} \big(D^p\Phi(x + th) - D^p\Phi(x)\big) = D^{p+1}\Phi(x)\, h.$$

The mapping Φ is said to be of class $\mathscr{C}_\Gamma^r$ if it is $\mathscr{C}_\sigma^r$ for some $\sigma: \mathbf{N} \to \mathbf{N}$. And Φ is called $\mathscr{C}_\Gamma^\infty$ if it is $\mathscr{C}_\Gamma^r$ for every $r \in \mathbf{N}$ (in this case σ may depend on r).

The notion of a differential mapping of class $\mathscr{C}_\Gamma^r$, $0 \leqq r \leqq \infty$, does not depend on the thoice of the sequences of semi-norms in the FRÉCHET spaces. For $\mathscr{C}_\Gamma^\infty$ mappings be-cween FRÉCHET spaces the usual elementary rules holds, but the "inverse mapping theorem" and the "implicite mapping theorem" in the usual form are false. Obviously every $\mathscr{C}_\Gamma^\infty$ mapping is also of class $\mathscr{C}_\pi^\infty$. The converse is not true, unless E is normable.

2. The Fréchet-manifold $\mathscr{C}^\infty(M, N)$

Let M be a compact and N a finite dimensional $\mathscr{C}^\infty$ manifold. Whatever a differential structure on $\mathscr{C}^\infty(M, N)$ may be, a chart at a point $f \in \mathscr{C}^\infty(M, N)$ would have its target in a linear space isomorphic to the tangent space $T_f\mathscr{C}^\infty(M, N)$ of $\mathscr{C}^\infty(M, N)$ at the point f. Now a simple heuristic argument shows that the elements of $T_f\mathscr{C}^\infty(M, N)$ can be identified with $\mathscr{C}^\infty$-mappings $\xi: M \to TN$ which are lifts of $f: M \to N$, i.e. $\xi \circ \tau_N = f$, where $\tau_N: TN \to N$ is the tangent bundle of N: If $\gamma: J \to \mathscr{C}^\infty(M, N)$ is a differentiable curve, defined in some neighbourhood J of 0 in $\mathbf{R}$, with $\gamma(0) = f$, then for every $x \in M$, $\gamma_x: J \to N$, defined by $\gamma_x(t) := \gamma(t)(x)$, is a curve in N and we have $\gamma_x'(0) \in T_{f(x)}N$. So we can identify the tangent vector $\gamma'(0)$ at the point f with the lift $\xi: M \to TN$ of f, defined by $\xi(x) := \gamma_x'(0)$. Still leaving $f \in \mathscr{C}^\infty(M,N)$ fixed we can identify the vector space $\mathscr{C}_f^\infty(M, TN)$ of all $\mathscr{C}^\infty$ lifts of f into TN with the space $\Gamma^\infty(f^* \tau_N)$ of all $\mathscr{C}^\infty$ sections of the vector bundle $f^* \tau_N: f^* TN \to M$, the pullback of $\tau_N: TN \to N$ with respect to $f: M \to N$ according to the following diagram

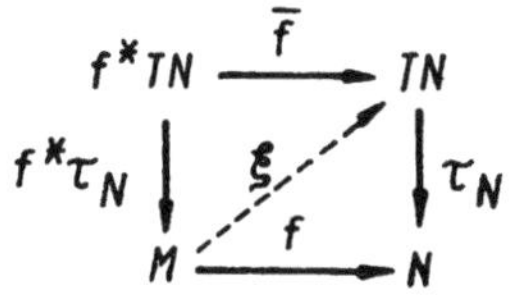

Therefore the vector space $E_f := \Gamma^\infty(f^* \tau_N) \cong T_f\mathscr{C}^\infty(M, N)$ offers himself in a natural way as a parameter space for a chart of $\mathscr{C}^\infty(M, N)$ at the point f. We endow E_f, as well as $\mathscr{C}^\infty(M, N)$, with the topology of compact $\mathscr{C}^\infty$ convergence (i.e. uniform $\mathscr{C}^\infty$ convergence since M is compact). By means of a (finite) vector bundle atlas on $f^* TN$ we get an increasing sequence $(|\ |_r)_{r \in \mathbf{N}}$ of norms on E_f which defines the topology of uniform $\mathscr{C}^\infty$ convergence and, actually, E_f is a FRÉCHET space.

Now a chart on $\mathscr{C}^\infty(M, N)$ at f is obtained if we have a spray on N with exponential map $\exp: \mathcal{O}_1 \subset TN \to N$ of class $\mathscr{C}^\infty$. From the properties of exp it follows that the mapping

$$\mathrm{Exp} := (\tau_N, \exp): \mathcal{O} \subset TN \to N \times N$$

is a $\mathscr{C}^\infty$ diffeomorphism from an open neighbourhood $\mathcal{O}$ of the zero section in TN onto the open neighbourhood Exp $(\mathcal{O})$ of the diagonal in $N \times N$. Then we have the following commutative diagram

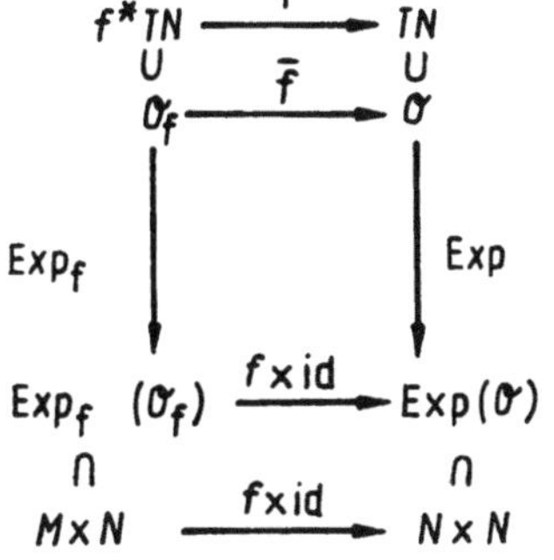

$$\mathcal{O}_f := \bar{f}^{-1}(\mathcal{O})$$

$$\mathrm{Exp}_f := (f^*\tau_N, \exp \circ \bar{f})$$

Exp_f is a $\mathscr{C}^\infty$ diffeomorphism from the open neighbourhood $\mathcal{O}_f$ of the 0-section in $f^* TN$ onto the open neighbourhood $\mathrm{Exp}_f(\mathcal{O}_f)$ of the graph of f in $M \times N$.

We define now:

$$U_f := \{h \in \mathscr{C}^\infty(M, N) \mid \mathrm{graph}\,(h) \subset \mathrm{Exp}_f(\mathcal{O}_f)\},$$

$$E_f := \Gamma^\infty(f^* \tau_N),$$

$$U_f' := \{\xi \in \Gamma^\infty(f^* \tau_N) \mid \xi(M) \subset \mathcal{O}_f\},$$

$$\Phi_f : U_f \to U_f', \quad \Phi_f(h) := \mathrm{Exp}_f^{-1} \circ (\mathrm{id}_M, h).$$

The mapping Φ_f is bijective with inverse $\Phi_f^{-1}(\xi) = pr_2 \circ \mathrm{Exp}_f \circ \xi$, for $\xi \in U_f'$ and $pr_2 : M \times N \to N$, the second projection.

Lemma. *Let (B, π, M) be an arbitrary finite dimensional vector bundle over M. Let U be an open set in B. Then the set*

$$\Gamma^\infty(U) := \{\xi \in \Gamma^\infty(\pi) \mid \xi(M) \subset U\}$$

is open in $\Gamma^\infty(\pi)$ if and only if M is compact.

If we apply this lemma to the set U_f' defined above we see that U_f' is open in $\Gamma^\infty(f^* \tau_N)$ since M is compact.

We call the triple (U_f, Φ_f, E_f) the natural chart on $\mathscr{C}^\infty(M, N)$ at the point f. It depends on the map exp (i.e. on the spray on N and on nothing else).

Theorem. *Let M be a compact $\mathscr{C}^\infty$ manifold and N a finite dimensional $\mathscr{C}^\infty$ manifold together with a $\mathscr{C}^\infty$ spray on N. Then the natural atlas $(U_f, \Phi_f, E_f)_{f \in \mathscr{C}^\infty(M,N)}$ on the set $\mathscr{C}^\infty(M, N)$ of all $\mathscr{C}^\infty$ mappings from M to N, modelled on the Fréchet spaces $E_f := \Gamma^\infty(f^* \tau_N)$ is $\mathscr{C}_\Gamma^\infty$ differentiable. The $\mathscr{C}_\Gamma^\infty$ differential structure defined thereby does not depend on the choice of the spray on N.*

One has only to verify that, given arbitrary natural charts (U_f, Φ_f, E_f), (U_g, Φ_g, E_g) on $\mathscr{C}^\infty(M, N)$, constructed by means of any two (possible different) $\mathscr{C}^\infty$ sprays on N, the transition map

$$\Gamma^\infty(f^* \tau_N) \supset \Phi_f(U_f \cap U_g) \xrightarrow{\Phi_g \circ \Phi_f^{-1}} \Phi_g(U_f \cap U_g) \subset \Gamma^\infty(g^* \tau_N)$$

is $\mathscr{C}_\Gamma^\infty$ differentiable. But this is an immediate consequence of the

Omega lemma. *Let (B, π, M) and (B', π', M) be finite dimensional $\mathscr{C}^\infty$ vector bundles over the compact base manifold M. Let $\Gamma^\infty(\pi)$, $\Gamma^\infty(\pi')$ be the corresponding Fréchet spaces of $\mathscr{C}^\infty$ sections. Let U be an open set in B and $B \supset U \xrightarrow{\Omega} B'$ a fibre preserving $\mathscr{C}^\infty$ map. Then the set $\Gamma^\infty(U)$ is open in $\Gamma^\infty(\pi)$ and the induced map*

$$\Omega_* : \Gamma^\infty(U) \to \Gamma^\infty(\pi'), \quad \Omega_* \xi := \Omega \circ \xi$$

is $\mathscr{C}_\Gamma^\infty$ differentiable and for every $p \in \mathbf{N}$ on has

$$D^p \Omega_* = (\partial^p \Omega)_*,$$

where ∂ denotes the vertical derivative.

This $\mathscr{C}^\infty$ version of the Omega lemma can be easily derived from its $\mathscr{C}^r$ version which already S. SMALE [1] stated.

Now $(\Phi_g \circ \Phi_f^{-1}) = (\mathrm{Exp}_g^{-1} \circ \mathrm{Exp}_f)_*$ holds and therefore the transition map is $\mathscr{C}_\Gamma^\infty$.

By the last theorem the set $\mathscr{C}^\infty(M, N)$ carries a natural $\mathscr{C}_\Gamma{}^\infty$ differential structure modelled on FRÉCHET spaces and it will, in the sequel, always tacitly be endowed with this structure, thus considered as a $\mathscr{C}_\Gamma{}^\infty$ FRÉCHET manifold. It is clear that $\mathscr{C}^\infty(M, N)$ is also a $\mathscr{C}_\pi{}^\infty$ FRÉCHET manifold.

The following *universal property* of the natural $\mathscr{C}_\Gamma{}^\infty$ structure on $\mathscr{C}^\infty(M, N)$ is one of the central results in the paper of J. GUTKNECHT [5].

Theorem. *Let X be a $\mathscr{C}_\Gamma{}^\infty$ manifold. A map $\Phi\colon X \to \mathscr{C}^\infty(M, N)$ is $\mathscr{C}_\Gamma{}^\infty$ differentiable if the associated map*

$$\tilde{\Phi}\colon X \times M \to N, \quad \tilde{\Phi}(x, m) := \Phi(x)\,(m),$$

is $\mathscr{C}_\Gamma{}^\infty$ differentiable.

The theorem essentially states that among all $\mathscr{C}_\Gamma{}^\infty$ structures on $\mathscr{C}^\infty(M, N)$ for which the evaluation map

$$\mathrm{ev}\colon \mathscr{C}^\infty(M, N) \times M \to N, \quad \mathrm{ev}\,(f, x) := f(x)$$

is $\mathscr{C}_\Gamma{}^\infty$ differentiable the natural $\mathscr{C}_\Gamma{}^\infty$ structure is the coarsest; for, ev is the associated map to $\mathrm{id}\colon \mathscr{C}^\infty(M, N) \to \mathscr{C}^\infty(M, N)$, $\mathrm{ev} = \widetilde{\mathrm{id}}$. The main importance of this theorem lies in the fact that it enables us to prove the differentiability of the composition map.

Theorem. *Let M, N and P be $\mathscr{C}^\infty$ manifolds, M and N compact. The composition map*

$$c\colon \mathscr{C}^\infty(N, P) \times \mathscr{C}^\infty(M, N) \to \mathscr{C}^\infty(M, P), \quad c(g, f) := g \circ f$$

is $\mathscr{C}_\Gamma{}^\infty$ differentiable.

Proof. According to the universal property of the natural $\mathscr{C}_\Gamma{}^\infty$ structure on $\mathscr{C}^\infty(M, P)$ one has to show that the map

$$\tilde{c}\colon \mathscr{C}^\infty(N, P) \times \mathscr{C}^\infty(M, N) \times M \to P$$

associated to c is $\mathscr{C}_\Gamma{}^\infty$ differentiable. But $\tilde{c}$ can be factorized as follows

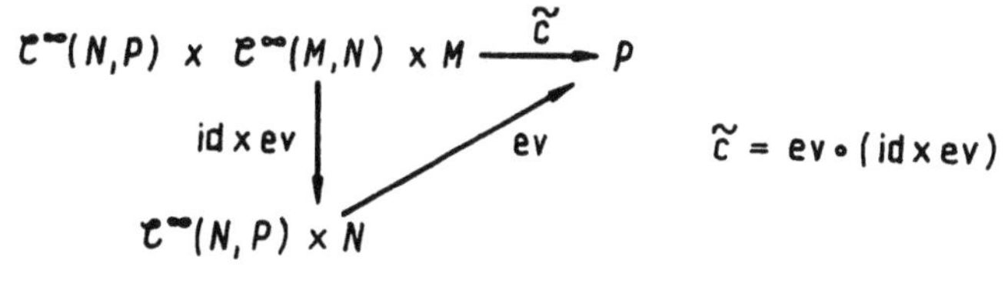

and hence $\tilde{c}$ is $\mathscr{C}_\Gamma{}^\infty$ since ev is $\mathscr{C}_\Gamma{}^\infty$.

It is clear that c is also $\mathscr{C}_\pi{}^\infty$ differentiable.

3. The tangent manifold $T\mathscr{C}^\infty(M, N)$

As in the classical theory of finite dimensional manifolds one constructs a differentiable structure on the set

$$T\mathscr{C}^\infty(M, N) := \bigcup_{f \in \mathscr{C}^\infty(M,N)} T_f\mathscr{C}^\infty(M, N)$$

and obtains the tangent manifold $T\mathscr{C}^\infty(M, N)$ of $\mathscr{C}^\infty(M, N)$ as a $\mathscr{C}_\Gamma{}^\infty$ manifold. On the other hand, as for every $f \in \mathscr{C}^\infty(M, N)$ the tangent space $T_f\mathscr{C}^\infty(M, N)$ is the vector space $\mathscr{C}_f{}^\infty(M, TN)$ of all $\mathscr{C}^\infty$ lifts of f into TN, we have the natural bijection

$$\varkappa\colon T\mathscr{C}^\infty(M, N) \to \bigcup_{f \in \mathscr{C}^\infty(M,N)} \mathscr{C}_f{}^\infty(N, TN) \subseteq \mathscr{C}^\infty(M, TN).$$

So we also can endow the set $T\mathscr{C}^\infty(M, N)$ with the natural $\mathscr{C}_\Gamma{}^\infty$ structure of the mapping space $\mathscr{C}^\infty(M, TN)$. The following theorem says that the two $\mathscr{C}_\Gamma{}^\infty$ structures on $T\mathscr{C}^\infty(M, N)$ coincide.

Theorem. *The natural bijection $\varkappa\colon T\mathscr{C}^\infty(M, N) \to \mathscr{C}^\infty(M, TN)$ is a $\mathscr{C}_\Gamma{}^\infty$ isomorphism.*

Proof. For $\eta \in T\mathscr{C}^\infty(M, N)$ let $\alpha\colon J \to \mathscr{C}^\infty(M, N)$ be a $\mathscr{C}_\Gamma{}^\infty$ differentiable curve with $\alpha(0) = f$ and $\alpha'(0) = \eta \in T_f\mathscr{C}^\infty(M, N)$. Then we have

$$\varkappa(\eta)\,(x) = \frac{\mathrm{d}}{\mathrm{d}t}\,[\alpha(t)\,(x)]_{t=0}, \quad \text{for every } x \in M,$$

To prove that $\varkappa$ is $\mathscr{C}_\Gamma{}^\infty$ differentiable we have to show (according to the universal property of the natural $\mathscr{C}_\Gamma{}^\infty$ structure on $\mathscr{C}^\infty(M, TN)$) that the map

$$\tilde{\varkappa}\colon T\mathscr{C}^\infty(M, N) \times M \to TN$$

associated to $\varkappa$, is $\mathscr{C}_\Gamma{}^\infty$ differentiable. Since the evaluation map $\mathrm{ev}\colon \mathscr{C}^\infty(M, N) \times M \to N$ is $\mathscr{C}_\Gamma{}^\infty$ differentiable its first partial derivative $T_1\,\mathrm{ev}\colon T\mathscr{C}^\infty(M, N) \times M \to TN$ is also $\mathscr{C}_\Gamma{}^\infty$ differentiable and it is easy to see that $\tilde{\varkappa} = T_1\,\mathrm{ev}$.

For the inverse $\varkappa^{-1}$ holds

$$\varkappa^{-1}(\xi) = \frac{\mathrm{d}}{\mathrm{d}t}\,(\exp_N \circ\, t\xi)|_{t=0}, \quad \text{for every } \xi \in \mathscr{C}^\infty(M, TN).$$

We first prove that the map

$$\varrho\colon \mathbf{R} \times \mathscr{C}^\infty(M, TN) \to \mathscr{C}^\infty(M, N), \quad \varrho(t, \xi) := \exp_N \circ\, t\xi,$$

is $\mathscr{C}_\Gamma{}^\infty$ differentiable. But $\tilde{\varrho}$ (associated to ϱ) can be factorized as follows

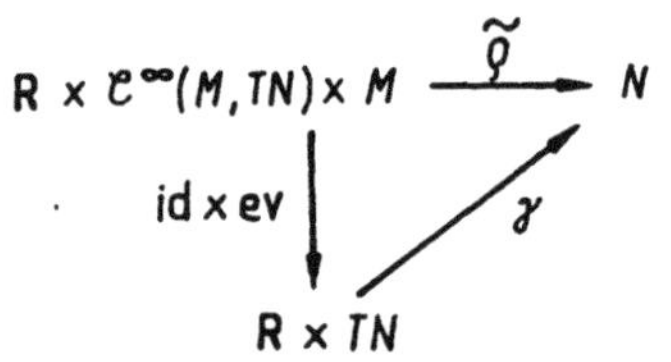

where $\gamma(t, v) := \exp_N(tv)$. Finally one shows that

$$\varkappa^{-1}(\xi) = T_1\varrho(0, \xi) \cdot 1$$

for every $\xi \in \mathscr{C}^\infty(M, TN)$. Hence $\varkappa^{-1}$ is $\mathscr{C}_\Gamma{}^\infty$ differentiable.

Since the projection $\tau_N\colon TN \to N$ is $\mathscr{C}^\infty$, the map $(\tau_N)_*\colon T\mathscr{C}^\infty(M, N) \to \mathscr{C}^\infty(M, TN) \to \mathscr{C}^\infty(M, N)$, $(\tau_N)_*(\xi) := \tau_N \circ \xi$ is $\mathscr{C}_\Gamma{}^\infty$ differentiable. But every $\xi \in \mathscr{C}^\infty(M, TN)$ is a lift of $(\tau_N)_*(\xi)$, so $(\tau_N)_*$ is the natural projection.

A $\mathscr{C}_\Gamma{}^\infty$ vector field on the $\mathscr{C}_\Gamma{}^\infty$ manifold $\mathscr{C}^\infty(M, N)$ is a $\mathscr{C}_\Gamma{}^\infty$ mapping

$$X: \mathscr{C}^\infty(M, N) \to \mathscr{C}^\infty(M, TN) = T\mathscr{C}^\infty(M, N)$$

such that $(\tau_N)_* \circ X = \mathrm{id}$, i.e. which assigns to each $f \in \mathscr{C}^\infty(M, N)$ a lift of f into TN. As in the case of finite dimensional manifolds for any two $\mathscr{C}_\Gamma{}^\infty$ vector fields X and Y on $\mathscr{C}^\infty(M, N)$ one defines the $\mathscr{C}_\Gamma{}^\infty$ vector field $[X, Y]$ by means of local charts of $\mathscr{C}^\infty(M, N)$ and $\mathscr{C}^\infty(M, TN)$.

4. Examples for submanifolds of manifolds of mappings

a) Let (B, π, M) be a $\mathscr{C}^\infty$ vector bundle over the compact base manifold M. Then the vector space $\Gamma^\infty(\pi)$ of all $\mathscr{C}^\infty$ sections of π is a closed $\mathscr{C}_\Gamma{}^\infty$ submanifold of the $\mathscr{C}_\Gamma{}^\infty$ manifold $\mathscr{C}^\infty(M, B)$, (c.f. R. Schmid [10]).

b) Let M be a compact $\mathscr{C}^\infty$ manifold and N' a (closed) submanifold of the finite dimensional manifold N. Then the $\mathscr{C}_\Gamma{}^\infty$ manifold $\mathscr{C}^\infty(M, N')$ is a (closed) $\mathscr{C}_\Gamma{}^\infty$ submanifold of $\mathscr{C}^\infty(M, N)$, (c. f. R. Schmid [10]).

c) Let M and N be compact $\mathscr{C}^\infty$ manifolds together with fixed covariant p-tensor fields $\omega \in \Gamma^\infty(\otimes^p \tau_M{}^*)$ and $\varrho \in \Gamma^\infty(\otimes^p \tau_N{}^*)$. We now consider the set $\mathscr{C}^\infty_{\omega,\varrho}(M, N)$ of those mappings from M to N which pull back the tensor field ϱ into ω,

$$\mathscr{C}^\infty_{\omega,\varrho}(M, N) := \{f \in \mathscr{C}^\infty(M, N) \mid f^*\varrho = \omega\}.$$

Theorem. *The space $\mathscr{C}^\infty_{\omega,\varrho}(M, N)$ is a closed $\mathscr{C}_\Gamma{}^\infty$ submanifold of the $\mathscr{C}_\Gamma{}^\infty$ manifold $\mathscr{C}^\infty(M, N)$.*

The essential tool in the proof of this theorem is the following lemma. Consider the covariant p-tensor fields $\Omega := p_1{}^* \varrho - p_2{}^* \varrho$ ($p_{1,2}: N \times N \to N$ the projections) on $N \times N$ and $\Omega_{\mathrm{Exp}} := \mathrm{Exp}^*\Omega$ on TN. (Exp can easily be extended to the whole of TN).

Lemma. *Given $f \in \mathscr{C}^\infty_{\omega,\varrho}(M, N)$. Let $g \in \mathscr{C}^\infty(M, N)$ so that $\xi_g := \mathrm{Exp}^{-1} \circ (f, g)$ is a lift of f into TN. Then $g^*\varrho = \omega$ if $\xi_g{}^*(\Omega_{\mathrm{Exp}}) = 0$.*

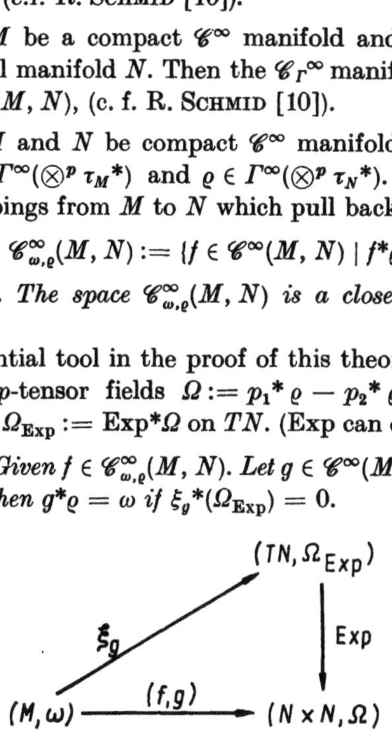

d) Let M be a compact $\mathscr{C}^\infty$ manifold. It is known that the group $\mathrm{Diff}^\infty(M)$ of all $\mathscr{C}^\infty$ diffeomorphisms of M is an open submanifold of $\mathscr{C}^\infty(M, M)$ and that the group operation c

$$c: \mathrm{Diff}^\infty(M) \times \mathrm{Diff}^\infty(M) \to \mathrm{Diff}^\infty(M), \quad c(g, f) := g \circ f,$$

is $\mathscr{C}_\Gamma{}^\infty$ differentiable. Furthermore J. Gutknecht [5] has shown that the inversion map

$$\mathrm{inv}: \mathrm{Diff}^\infty(M) \to \mathrm{Diff}^\infty(M), \quad \mathrm{inv}(f) := f^{-1},$$

is $\mathscr{C}_\Gamma{}^\infty$ differentiable. The results can be expressed in the following form:

Theorem. *Let M be a compact $\mathscr{C}^\infty$ manifold. Then the diffeomorphism group* $\mathrm{Diff}^\infty(M)$ *of M is a $\mathscr{C}_\Gamma{}^\infty$ Fréchet Lie-group.*

The tangent space $T_e \mathrm{Diff}^\infty(M) = \Gamma^\infty(\tau_M)$ of $\mathrm{Diff}^\infty(M)$ at the neutral element $e = \mathrm{id}_M$ is the vector space of all $\mathscr{C}^\infty$ vector fields on M. With respect to the usual brackets of $\mathscr{C}^\infty$ vector fields this is a Lie algebra. Since $\mathrm{Diff}^\infty(M)$ is open in $\mathscr{C}^\infty(M, M)$ and $(\tau_M)_*$: $\mathscr{C}^\infty(M, TM) \to \mathscr{C}^\infty(M, M)$ is $\mathscr{C}_\Gamma{}^\infty$, the tangent manifold $T\mathrm{Diff}^\infty(M) = (\tau_M)_*^{-1}\mathrm{Diff}^\infty(M)$ of $\mathrm{Diff}^\infty(M)$ is an open submanifold of $\mathscr{C}^\infty(M, TM)$. Just exactly in the same way as in the case of classical Lie groups to every $\xi \in T_e \mathrm{Diff}^\infty(M)$ there exists an uniquely determined right invariant (resp. left invariant) $\mathscr{C}_\Gamma{}^\infty$ vector field X_ξ (resp. X_ξ') on $\mathrm{Diff}^\infty(M)$ such that $X_\xi(e) = \xi$ (resp. $X_\xi'(e) = \xi$). Namely we define

$$X_\xi(f) := TR_f(\xi) \quad \text{and} \quad X_\xi'(f) := TL_f(\xi)$$

for every $f \in \mathrm{Diff}^\infty(M)$, where R_f: $\mathrm{Diff}^\infty(M) \to \mathrm{Diff}^\infty(M)$, $R_f(g) := g \circ f$ resp. L_f: $\mathrm{Diff}^\infty(M) \to \mathrm{Diff}^\infty(M)$, $L_f(g) := f \circ g$ is the right resp. left translation on $\mathrm{Diff}^\infty(M)$. Actually, we have

$$X_\xi(f) = \xi \circ f \quad \text{and} \quad X_\xi'(f) = Tf \circ \xi$$

for every $f \in \mathrm{Diff}^\infty(M)$.
Computing in local charts we find

$$[X_\xi, X_\eta] = X_{[\xi,\eta]} \quad \text{and} \quad [X_\xi', X_\eta'] = -X'_{[\xi,\eta]}$$

for all $\xi, \eta \in T_e \mathrm{Diff}^\infty(M) = \Gamma^\infty(\tau_M)$.

From this it follows that the Lie algebra of all right invariant (resp. left invariant) $\mathscr{C}_\Gamma{}^\infty$ vector fields on $\mathrm{Diff}^\infty(M)$ is a Lie algebra isomorphic (resp. anti-isomorphic) to the Lie algebra $\Gamma^\infty(\tau_M)$ of all $\mathscr{C}^\infty$ vector fields on M.

Remark. One also has an exponential map exp: $T_e \mathrm{Diff}^\infty(M) \to \mathrm{Diff}^\infty(M)$ but it is not a local diffeomorphism at 0, it does not map a neighbourhood of 0 in $T_e \mathrm{Diff}^\infty(M)$ onto a neighbourhood of $e = \mathrm{id}_M$ in $\mathrm{Diff}^\infty(M)$, (c.f. H. Omori [8]).

5. The Fréchet Lie group $\mathrm{Sym}^\infty(M, \omega)$ of symplectomorphisms

Let (M, ω) be a compact symplectic $\mathscr{C}^\infty$ manifold, i.e. a compact manifold M together with a closed non degenerated 2-form ω on M. A map $f \in \mathscr{C}^\infty(M, M)$ is called symplectic if $f^*\omega = \omega$. A $\mathscr{C}^\infty$ vector field ξ on (M, ω) is called locally Hamiltonian if $L_\xi\omega = 0$, where $L_\xi\omega$ is the Lie derivative of ω in direction of ξ.

We consider the group

$$\mathrm{Sym}^\infty(M, \omega) := \{f \in \mathrm{Diff}^\infty(M) \mid f^*\omega = \omega\}$$

of all symplectic $\mathscr{C}^\infty$ diffeomorphisms or $\mathscr{C}^\infty$ *symplectomorphisms* on (M, ω). According to the statements of § 4c, d the following theorem holds:

Theorem. *Let (M, ω) be a compact symplectic $\mathscr{C}^\infty$ manifold. The symplectomorphism group* $\mathrm{Sym}^\infty(M, \omega)$ *of (M, ω) is a closed $\mathscr{C}_\Gamma{}^\infty$ Fréchet Lie subgroup of* $\mathrm{Diff}^\infty(M)$. *Its Lie algebra is the Lie subalgebra of $\Gamma^\infty(\tau_M)$ consisting exactly of the locally Hamiltonian $\mathscr{C}^\infty$ vector fields on (M, ω).*

For detailed exposition of the proof the reader is referred to R. Schmid [10].

References

[1] R. ABRAHAM, Lectures of Smale on differential topology, Mimeograph. Columbia Univ. New York 1962.

[2] N. BOURBAKI, Variétés différentielles et analytiques, Fasc. rés. § 8—§ 15, Hermann, Paris 1971.

[3] J. EELLS, On the geometry of function spaces, Symposium de topologia algebraica, México, 1958, 303—307.

[4] H. R. FISCHER, Differentialrechnung in lokalkonvexen Räumen und Mannigfaltigkeiten von Abbildungen, Manuskript d. Fakultät für Math. und Informatik, Univ. Mannheim 1977.

[5] J. GUTKNECHT, Die $\mathscr{C}_{\Gamma}^{\infty}$-Struktur auf der Diffeomorphismengruppe einer kompakten Mannigfaltigkeit, Diss. ETH 5879, Zürich 1977.

[6] H. H. KELLER, Differential calculus in locally convex spaces, Lecture Notes in Math. 417, Springer-Verlag, Berlin—Heidelberg—New York 1974.

[7] J. A. LESLIE, On a differential structure for the group of diffeomorphisms, Topology 6 (1967), 263—271.

[8] H. OMORI, Infinite dimensional Lie transformation groups, Lecture Notes in Math. 427, Springer-Verlag, Berlin—Heidelberg—New York 1974.

[9] R. PALAIS, From letters of R. Palais, Unpublished manuscript, 1962.

[10] R. SCHMID, Die Symplektomorphismengruppe als Fréchet-Lie-Gruppe, Diss. Univ Zürich, 1978.

Anschrift des Verfassers:

Mathematisches Institut
Universität Zürich
8032 Zürich
Freiestraße 36
Schweiz

Bei Fragen zur Produktsicherheit wenden Sie sich bitte an:
If you have any questions regarding product safety,
please contact:

Walter de Gruyter GmbH
Genthiner Straße 13
10785 Berlin
productsafety@degruyterbrill.com